Springer Series in Optical Sciences Volume 29

Edited by Theodor Tamir

Springer Series in Optical Sciences

1 **Solid-State Laser Engineering**
By. W. Koechner

2 **Table of Laser Lines in Gases and Vapors**
3rd Edition
By R. Beck, W. Englisch, and K. Gürs

3 **Tunable Lasers and Applications**
Editors: A. Mooradian, T. Jaeger, and P. Stokseth

4 **Nonlinear Laser Spectroscopy**
By V. S. Letokhov and V. P. Chebotayev

5 **Optics and Lasers**
An Engineering Physics Approach
By M. Young

6 **Photoelectron Statistics**
With Applications to Spectroscopy and Optical Communication
By B. Saleh

7 **Laser Spectroscopy III**
Editors: J. L. Hall and J. L. Carlsten

8 **Frontiers in Visual Science**
Editors: S. J. Cool and E. J. Smith III

9 **High-Power Lasers and Applications**
2nd Printing
Editors: K.-L. Kompa and H. Walther

10 **Detection of Optical and Infrared Radiation**
2nd Printing
By R. H. Kingston

11 **Matrix Theory of Photoelasticity**
By P. S. Theocaris and E. E. Gdoutos

12 **The Monte Carlo Method in Atmospheric Optics**
By G. I. Marchuk, G. A. Mikhailov, M. A. Nazaraliev, R. A. Darbinian, B. A. Kargin, and B. S. Elepov

13 **Physiological Optics**
By Y. Le Grand and S. G. El Hage

14 **Laser Crystals** Physics and Properties
By A. A. Kaminskii

15 **X-Ray Spectroscopy**
By B. K. Agarwal

16 **Holographic Interferometry**
From the Scope of Deformation Analysis of Opaque Bodies
By W. Schumann and M. Dubas

17 **Nonlinear Optics of Free Atoms and Molecules**
By D. C. Hanna, M. A. Yuratich, D. Cotter

18 **Holography in Medicine and Biology**
Editor: G. von Bally

19 **Color Theory and Its Application in Art and Design**
By G. A. Agoston

20 **Interferometry by Holography**
By Yu. I. Ostrovsky, M. M. Butusov, G. V. Ostrovskaya

21 **Laser Spectroscopy IV**
Editors: H. Walther, K. W. Rothe

22 **Lasers in Photomedicine and Photobiology**
Editors: R. Pratesi and C. A. Sacchi

23 **Vertebrate Photoreceptor Optics**
Editors: J. M. Enoch and F. L. Tobey, Jr.

24 **Optical Fiber Systems and Their Components**
An Introduction
By A. B. Sharma, S. J. Halme, and M. M. Butusov

25 **High Peak Power Nd : Glass Laser Systems**
By D. C. Brown

26 **Lasers and Applications**
Editors: W. O. N. Guimaraes, C. T. Lin, and A. Mooradian

27 **Color Measurement** Theme and Variations
By D. L. MacAdam

28 **Modular Optical Design**
By O. N. Stavroudis

29 **Inverse Problems of Lidar Sensing of the Atmosphere** By V. E. Zuev and I. E. Naats

30 **Laser Spectroscopy V**
Editors: A. R. W. McKellar, T. Oka, and B. P. Stoicheff

31 **Optics in Biomedical Sciences**
Editors: G. von Bally and P. Greguss

32 **Fiber-Optic Rotation Sensors** and Related Technologies
Editors: S. Ezekiel and H. J. Arditty

33 **Integrated Optics: Theory and Technology**
By R.G. Hunsperger

34 **The High-Power Iodine Laser**
By G. Brederlow, E. Fill, and K. J. Witte

35 **Engineering Optics** By K. Iizuka

36 **Transmission Electron Microscopy**
By L. Reimer

37 **Opto-Acoustic Molecular Spectroscopy**
By V.S. Letokhov and V.P. Zharov

38 **Photon Correlation Techniques**
Editor: E.O. Schulz-DuBois

39 **Optical and Laser Remote Sensing**
By D.K. Killinger and A. Mooradian

V. E. Zuev I. E. Naats

Inverse Problems of Lidar Sensing of the Atmosphere

With 71 Figures

Springer-Verlag Berlin Heidelberg GmbH 1983

Professor Dr. *Vladimir E. Zuev*
Dr. *Igor E. Naats*

Institute of Atmospheric Optics, SB USSR Academy of Sciences,
Tomsk, 55, USSR

Title of the original Russian edition:
Obratnyje zadachi lasernogo zondirovanija atmosfery

ISBN 978-3-662-13539-6 ISBN 978-3-540-38802-9 (eBook)
DOI 10.1007/978-3-540-38802-9

Library of Congress Cataloging in Publication Data. Zuev, V. E. (Vladimir Evseevich). Inverse problems of lidar sensing of the atmosphere. (Springer series in optical sciences ; v. 29). Translated from Russian. Includes bibliographical references and index. 1. Atmosphere--Remote sensing. 2. Inverse problems (Differential equations) 3. Optical radar. I. Naats, I. E. (Igor Eduardovich) II. Title. III. Series.
QC871.2.Z8313 1982 551.5'028 82-10417

Originally published by Springer-Verlag Berlin Heidelberg New York in 1983.

Softcover reprint of the hardcover 1st edition 1983

Offset printing: Beltz Offsetdruck, 6944 Hemsbach/Bergstr.

2153/3130-543210

Preface

This monograph undertakes to present systematically the methods for solving inverse problems of lidar sensing of the atmosphere, with emphasis on lidar techniques that are based on the use of light scattering by aerosols. The theory of multi-frequency lidar sensing, as a new method for studying the microphysical and optical characteristics of aerosol formations, is also presented in detail. The possibilities of this theory are illustrated by the experimental results on microstructure analysis of tropospheric and low stratospheric aerosols obtained with ground-based two- and three-frequency lidars. The lidar facilities used in these experimental studies were constructed at the Institute of Atmospheric Optics SB USSR Academy of Sciences. Some aspects of remote control of dispersed air pollution using lidar systems are also considered.

A rigorous theory for inverting the data of polarization lidar measurements is discussed, along with its application to remote measurement of the complex index of refraction of aerosol substances and the microstructure parameters of background aerosols using double-ended lidar schemes. Solutions to such important problems as the separation of contributions due to Rayleigh-molecular and Mie-aerosol light scattering into the total backscatter are obtained by using this theory. Lidar polarization measurements are shown to be useful in this case. The efficiency of the methods suggested here for interpreting the lidar polarization measurements is illustrated by experimental results on the investigation of the microphysical parameters of natural aerosols and artificial smokes using polarization nephelometers.

A brief discussion is also given of the inverse problems related to the remote sensing of profiles of such atmospheric parameters as humidity, temperature, wind velocity, and characteristics of atmospheric turbulence.

We are indebted to our co-workers at the Institute of Atmospheric Optics for the fruitful collaboration that provided the results for this book, and we should like to express our particular appreciation to B.S. Kostin, E.V. Makienko, V.V. Veretennikov, and B.P. Ivanenko, all at the Laboratory of Inverse Problems of Atmospheric Optics. We are also very grateful to T.V. Kuznetsova and N.P. Malincheva for typing the manuscript, and to all those other people who helped us in preparing the graphic material.

Tomsk, March 1982 *V.E. Zuev · I.E. Naats*

Contents

1. **Introduction** 1

2. **Theory of Optical Sensing in Aerosol Polydispersed Systems** 5

2.1 Method of Integral Equations and Polydispersed Systems of Spherical Particles 6

2.2 Theory of Optical Sounding of Polydispersed Systems of Nonspherical Particles 10

2.2.1 One-Dimensional Integral Equations for Ensembles of Convex Randomly Oriented Particles 10

2.2.2 Polydispersed Systems of Ellipsoidal Particles 17

2.3 Definition of Operators of Light Scattering by Polydispersed Systems 21

2.3.1 Methods of Optical Models 21

2.3.2 Determination of the Scattering Efficiency Factors of Monodispersed Media Using the Data from Sounding of Polydispersed Systems 25

2.3.3 Determination of the Factor $\bar{T}(\bar{\ell},\lambda)$ from the Solution of the Volterra Integral Equation 30

2.4 General Characterization of the Method of Integral Equations 39

2.4.1 Effect of Smoothness of the Kernels on the Efficiency of Optical Sounding 39

2.4.2 Averaging Effects in Optical Measurements 46

2.5 Inverse Operator Method in the Theory of Optical Sounding 49

2.5.1 Optical Operators and Equations of Radiation Transfer in Spectral Intervals 49

2.5.2 Method of Smoothing Functional for Inverse Problems of Light Scattering 53

2.5.3 Determination of Discontinuous Size Spectra of Particles (Method of Histograms) 57

2.5.4 An Example of the Inversion of the Spectral Optical Depth of an Aerosol Atmosphere 63
2.5.5 Assessment of the Admissible Number of Independent Measurements for Sounding in Spectral Intervals 66
2.6 Determination of Microphysical Characteristics of Dispersed Media by the Method of Model Assessments from the Data of Optical Sounding 70
2.6.1 Parametric Families of Size-Distributions and Optical Models 70
2.6.2 Scheme of the Method of Model Assessments. An Example of Inverting the Spectral Transmission 74
2.6.3 Examples of Inversion of Lidar Measurements 76
2.6.4 Determination of Optical Constants of the Aerosol Substance by the Method of Integral Assessments 78

3. Determination of the Microphysical Parameters of Aerosol Ensembles by the Method of Multi-Frequency Laser Sensing 80
3.1 General Theory of Multi-Frequency Laser Sensing of Polydispersed Systems in the Single-Scattering Approximation 81
3.2 Single-Frequency Lidars in Studies of the Optical Properties of the Atmosphere 87
3.2.1 The Choice of the Lidar Ratio for the Accuracy of the Solution 87
3.2.2 On the Choice of the Lidar Ratio in the Case of Single-Frequency Sensing 89
3.2.3 Examples of the Interpretation of Single-Frequency Sensing Data 92
3.3 Qualitative Methods for Data Interpretation in Multi-Frequency Sensing 96
3.3.1 The Concept of an Optically Active Interval of Sensing 96
3.3.2 Some Peculiarities of the Spectral Behavior of the Backscattering Coefficient 100
3.3.3 The Logarithmic Derivative Method Applied to the Interpretation of Spectral Behaviors of Optical Characteristics 102
3.3.4 Methods for Determining Boundaries of the Size-Spectra from Spectral Optical Measurements 106
3.3.5 On the Interpretation of the Spectral Behavior of Lidar Signals 111

3.4 Determination of the Aerosol Microstructure and Index of Refraction Using Methods of Multi-Frequency Sensing 113
3.4.1 The Method of Optical Operators (Variational Approach) 113
3.4.2 The Iteration Method and Inversion of the Scattering Coefficient in the Visible 116
3.4.3 Method of Logarithmic Derivative for the Assessment of the Regularization Parameter. Sensing of Hazes in the UV Spectrum 120
3.4.4 The Parametric Form of the Iteration Method 123
3.4.5 The Method of Combined Discrepancy, an Example of Experimental Data Handling 126
3.4.6 Further Consideration of Inverse Problems in the Method of Multi-Frequency Sensing 131
3.5 Investigations of Aerosols Microstructures in the Boundary Layer Using Multi-Frequency Lidars 132
3.5.1 Methods for Determining the Local Variations of the Aerosol Microstructures 132
3.5.2 Efficiency of Multi-Frequency Sensing in Studying the "Details" of Aerosol Microstructure 138
3.6 Investigation of the Microstructure of Low Stratospheric Aerosols Using Ground Based Lidars 142
3.6.1 General Remarks 142
3.6.2 Determination of the Boundaries of Aerosol Size Spectrum 144
3.6.3 An Example of Analyzing the Data of Two-Frequency Lidar Sensing of Low Stratospheric Aerosols 149
3.7 Determination of the Microstructure of Cloud Aerosols Using Lidar Sensing 155

4. Methods for Inverting Polarization Data. The Theory of Bistatic Lidars 163
4.1 General Characterization of Integral Equations Related to the Inversion of Elements of Polydispersed Scattering Phase Matrices for the Case of Spherical Particles 164
4.2 Determination of the Aerosol Microstructures by Inverting the Polydispersed Scattering Phase Functions 167
4.2.1 Numerical Inversion of Aerosol Scattering Phase Functions 167

4.2.2 Examples of the Numerical Inversion of Scattering Phase Functions 169
4.2.3 Determination of the Lidar Ratio from Nephelometric Measurements Using the Inverse Operator Method 172
4.2.4 Examples of the Inversion of Experimentally Measured Scattering Phase Functions 173
4.3 The Use of Polarization Measurements for the Determination of Aerosol Microstructure and Index of Refraction 175
4.3.1 Methods for Determining the Optical Constants of the Aerosol Substance Based on the Use of Light Scattering Operators for Polydispersed Systems 175
4.3.2 Numerical Examples for the Determination of Optical Constants from the Polydispersed Scattering Phase Matrix 177
4.3.3 An Example of Investigation of Microphysical Parameters of Smoke Aerosols Using a Polarization Nephelometer 183
4.4 On the Inversion of Polydispersed Scattering Phase Matrices for Nonspherical Particles 186
4.4.1 General Comments 186
4.4.2 An Example Illustrating the Influence of the Particles Nonsphericity 188
4.4.3 Parametric Modification of the Integral Equations for Light Scattering by Systems of Nonspherical Particles 191
4.4.4 On the Interpretation of Depolarization of Monostatic Lidar Signals 194
4.5 Inverse Problems of the Theory of Bistatic Lidar Sensing 196
4.5.1 Lidar Equation for the Polarization Sensing and Inverse Problems of Light Scattering by Aerosols 196
4.5.2 Polarization Relations and Methods for Their Inverting 198
4.5.3 Determination of the Normalized Size-Distribution Functions from Sensing Data Using Bistatic Lidars 200
4.5.4 On the Theory of Joint Sensing Using Mono- and Bistatic Lidars. The Separation of the Contributions of Scattering Components 202
4.6 Qualitative Methods for Interpreting the Angular Characteristics of Light Scattering 206
4.6.1 Use of the Logarithmic Derivative Method to Interpret the Aureole Part of the Scattering Phase Function 206

4.6.2 The Use of Model Size-Distributions in the Method of Logarithmic Derivative ... 209
4.6.3 Inversion of Experimentally Measured Scattering Phase Functions ... 210
4.6.4 Estimation of the Boundaries of the Aerosol Size-Distributions Using the Logarithmic Derivative Method 213

5. Light Scattering by Aerosols and Lidar Sensing of the Atmosphere 215
5.1 Lidar Sensing of Water Vapor and Other Atmospheric Gases Using Differential Absorption and Scattering (DAS) ... 215
5.1.1 General Theory of the Method ... 215
5.1.2 Numerical Simulation and Estimations of the Method's Precision ... 218
5.2 Experimental Studies on Sensing Atmospheric Humidity Profiles Using Differential Absorption ... 222
5.2.1 Lidar Parameters and Measuring Technique ... 222
5.2.2 Some Experimental Results ... 223
5.3 Lidar Sensing of Atmospheric Ozone ... 226
5.4 The Method of Logarithmic Derivatives in Lidar Data Interpretation ... 229
5.4.1 Determination of Humidity Profiles of Aerosol Optical Characteristics ... 229
5.4.2 Lidar Sensing of the Atmospheric Temperature and Aerosol Optical Characteristics Using a Pure Rotational Raman Spectrum ... 231
5.4.3 Regularizing Algorithm for Processing the Sensing Data. Photon Counting Scheme ... 236
5.5 Laser Sensing of Wind Velocity and Atmospheric Turbulence Parameters ... 239
5.5.1 Lidar Sensing of Wind Velocity ... 239
5.5.2 Lidar Sensing of the Turbulent Parameters of the Atmosphere ... 243
5.5.3 Concluding Remarks ... 246

References ... 251

Subject Index ... 259

1. Introduction

The first papers devoted to atmospheric research by means of lasers appeared just after their discovery. Two circumstances were decisive in this situation. First, the development of new facilities for remotely measuring atmospheric parameters was urgently needed. Secondly, lasers were found to be very promising for use in atmospheric research.

As a matter of fact, conventional methods for measuring atmospheric parameters are as yet incapable of obtaining the bulk of specialized information (over space and time) that is required for solving the very important problem of long-range weather forecasting. In addition, the standard method being used in the world-wide meteorological network cannot provide operative information on atmospheric pollution produced by industrial facilities.

The main disadvantages of standard methods are caused by the inherently low spatial and temporal resolutions they can offer, as well as by the limited number of the parameters sounded (e.g., pressure, temperature, humidity, wind). These methods are also characterized by relatively low heights of measurements and by the low accuracy with which the humidity can be determined in the upper troposphere and stratosphere.

The principal way to overcome these difficulties is to develop basically new methods for remotely measuring the atmospheric parameters. Laser methods occupy a special place among those methods and their rapid development began as soon as the first lasers appeared [1.1].

The advantages of the laser methods for sensing the atmospheric parameters, as compared with those of other remote methods (radar, acoustic, spectroscopy of emitted radiation) are a result of the significant number of sufficiently strong interactions accompanying the propagation of light through the atmosphere [1.2]. The following effects should be mentioned among those interactions: Rayleigh scattering by molecules; light scattering by aerosols; spontaneous, stimulated, and resonance Raman effects; absorption of radiation by molecules; Doppler and collisional broadening of molecular absorption lines; Doppler frequency shifts due to scattering on moving inhomogeneities; amplitude and phase fluctuations due to turbulent effects in the atmosphere; as well as a series of nonlinear effects observed at certain powers and pulse durations of the laser radiation. Laser methods provide, in principle, the ability to study atmospheric processes in a real time scale. The information on these processes enters the receiver in a "coded" form at the speed of light. Therefore, if the methods for "decoding" it are known, i.e., one knows the solution of the corresponding inverse problems, final results can be obtained in a time interval determined by the capabilities of the digital data acquisition system used.

In general, the pulse of backscattered radiation from the atmosphere contains information on the profiles of the atmospheric parameters along the sounding paths. The spatial resolution of sounding is determined by the sounding pulse duration. Commonly used lasers have pulse durations of about tens of nanoseconds, which provide a spatial resolution of several meters if, and only if, the digital data acquisition system being used possesses a corresponding time response.

In recent years lasers have become available that deliver 10^3-10^4 pulses per second with a pulse duration of about 10^{-8} s. The use of such lasers in lidar facilities makes it possible, in principle, to obtain up to 10^4 profiles with a spatial resolution of about 1 m within one second.

Lidar methods can be expected to be widely used in the investigation and monitoring of atmospheric aerosol and gaseous pollution from industrial sources [1.2]. There is no doubt about their advantages in comparison with the methods currently in use. They are especially advantageous for investigating the dynamics of air-pollution diffusion.

Light scattering by aerosols is the phenomenon most widely used in lidar studies of the atmosphere. It allows, first of all, the study of atmospheric aerosol distribution. Here not only the distribution of aerosol mass concentration and aerosol stratification are meant, but also the spatial behavior of such microphysical parameters of atmospheric aerosols as size spectra, complex index of refraction, and shape of particles [1.3,4].

Light scattering by aerosols can also be used for sounding atmospheric humidity profiles as well as the profiles of other atmospheric molecular constituents and for measuring wind speed and characteristics of atmospheric turbulence [1.5,6]. In such measurements, aerosols serve as tracers for obtaining information on different atmospheric parameters.

The many promising prospects for using aerosol light scattering for lidar sounding of atmospheric parameters and aerosols as well as the very important role of aerosols in radiation transfer [1.7] and various physico-chemical processes in the atmosphere, including atmospheric pollution, were the main reasons that extensive investigations on the development of various lidar methods have been undertaken in the Institute of Atmospheric Optics SB USSR Academy of Sciences. This monograph summarizes the most important results of the theoretical and experimental studies that have been carried out during the last decade at this Institute under the leadership of the authors.

The book consists of five chapters. The second chapter is devoted to the general theory of optical sounding of polydispersed aerosol systems. An analysis is presented of the basic functional relationships between optical characteristics of dispersed media and their microphysical parameters. The Fredholm integral equation of the first kind is considered as a basic mathematical model. The possibilities of using this equation in problems of lidar sounding of dispersed media are studied, taking into account the morphology of aerosol particles.

The third chapter presents the theory of multifrequency lidar sounding aimed at obtaining information about aerosol microphysical parameters. In parallel with the theory, a justification of the inversion algorithm is given for numerically interpreting the data of optical measurements. Some particular applications of this method for the solution of various practical problems are also included. The capabilities of this method are illustrated with the results of three-frequency lidar sounding of microstructure parameters of tropospheric and stratospheric aerosols.

The material discussed in the third chapter clearly demonstrates the high efficiency of the multifrequency lidar technique suggested for studying aerosol microphysical parameters. There is no doubt that multifrequency lidar is the most advantageous technique for optical sounding of aerosols at the present time.

The fourth chapter presents a discussion of the inverse problem for polydispersed scattering phase matrices. The method for determining simultaneously the refractive index of an aerosol substance and the microstructure parameters of aerosol ensembles is also presented. Its possibilities are illustrated using experimental data. Some methods for solving the inverse problems of bistatic lidar sounding are presented, and the generalized mathematical scheme for inverting the data of sounding obtained with the use of combined mono- and bistatic schemes is discussed. Such a combined sounding scheme allows separating the contributions due to Rayleigh molecular and aerosol Mie scatterings and evaluating the aerosol microphysical parameters. The material presented also demonstrates the promising possibilities of using polarization characteristics of lidar returns for the remote determination of aerosol chemical composition which is of great importance in air pollution control. In addition, the polarization characteristics of lidar returns allow, in principle, the study of the shape and spatial orientation of aerosol particles.

The fifth chapter of the book presents a description of methods and results on sensing different atmospheric parameters using light scattering by aerosols. This chapter also presents the differential absorption lidar technique and its applications to the sounding of the profiles of atmospheric water vapor and ozone. The authors further describe the logarithmic derivative method and discuss its application to the determination of such atmospheric parameters as humidity, temperature, and aerosol characteristics. The questions on lidar sounding of wind speed and parameters of atmospheric turbulence are also briefly discussed.

The concluding remarks contain a summary of the material presented in the monograph and a discussion of possible future developments of the lidar method for sounding the atmospheric parameters based on the use of light scattering by aerosols.

2. Theory of Optical Sensing in Aerosol Polydispersed Systems

Optical methods for investigating aerosol formations in the atmosphere are in fact indirect methods and therefore their use in practice is, as a rule, connected with the solution of systems of functional equations. These are, first of all, the transfer equations for optical signals propagating through the scattering atmosphere, and, secondly, the relationships between optical characteristics and microphysical parameters of the aerosol systems. These relations lead, as a rule, to integral equations of the first kind in studies of microstructures of atmospheric aerosols by the methods of optical sounding. The equations are one-dimensional if sphericity of the particles is assumed.

In the general case of an arbitrary particle shape, the determination of the microstructures of such polydispersed systems requires the solution of multi-dimensional equations as well as the performance of very complicated optical experiments. It is obvious that the solution of the inverse problems of light scattering in the form of one-dimensional integration equations is the simplest approximate method for determining the microstructures of real dispersed media. This is why the first problem of optical sounding theory is to assess the possibility of investigating the microstructures of atmospheric aerosols using optical methods based on the solution of one-dimensional inverse problems of light scattering. A detailed analysis of this problem is given in this chapter by an example of the determination of the microstructure of a polydispersed system of convex particles randomly oriented in an illuminated volume.

The efficiency of the optical data inversion and, as a consequence, of the optical methods depends also on the choice of a kernel in a corresponding integral equation [2.1]. Only in the case of spherical particles can such a choice easily be made, since in this case the working apparatus for the solution of inverse problems is the Mie theory [2.2]. A detailed description of the theory and methods for determining the scattering efficiency factors of individual particles by solving corresponding inverse problems of light scattering by polydispersed systems is presented here. This makes the theory of optical sounding consistent, on the one hand, and stresses its practical applicability, on the other hand.

An integral operator for light scattering by polydispersed media that corresponds to a one-dimensional inverse problem may be one of the basic concepts of the optical sounding theory of such media. The use of this operator for describing some generalized characteristic of light scattering by a system of particles allows one to find effective approaches to the solution of the systems of transfer equations for the optical signals propagating through the dispersed media as well as to the construction of methods for the inversion of polydispersed scattering phase matrices and to the solution of various problems of applied optics.

Particular attention is paid to the methods for constructing inverse operators applicable to the sounding of the atmosphere with light pulses of short duration. Under these conditions,the size spectra of ensembles of particles have breaks, because of the smallness of the volumes illuminated, that require the integral equations to be solved on a set of broken size-distribution functions. The method of histograms can be regarded as one of the simplest methods for the solution of such inverse problems of light scattering by a polydispersed ensemble of particles which enables one to find solutions on the set of piecewise continuous functions of density. The development of this method is caused by the particular features of laser sounding of the atmosphere with light pulses, one of which is high spatial resolution. Aside from rigorous inversion methods, qualitative methods for data interpretation are also of a certain interest in the practice of optical sounding. These methods can make it possible to assess some parameters of the microstructures of the aerosol ensembles operatively during the optical experiment. The method of model assessments may be regarded as one of such methods. All the methods considered in this chapter are illustrated by appropriate examples of the inversion of optical data.

2.1 Method of Integral Equations and Polydispersed Systems of Spherical Particles

Remote determination of the aerosol microstructure is the most interesting and practically important problem of optical sounding of the atmosphere as a scattering medium, since the microstructure characterizes the properties of dispersed media most completely. In particular, a knowledge of the aerosol microstructure and especially of its spatial-temporal variations carries information about physical and chemical atmospheric processes in which the aerosol substance is involved.

Let us introduce a formal description of the microstructure of a polydispersed medium, bearing in mind, however, that the above problem becomes very complicated in a general case of an ensemble of particles having arbitrary geometrical shapes. In this connection, we shall consider, as a first step, ensembles of spherical particles.

In problems concerning propagation of radiation through the scattering atmosphere, aerosol microstructure is generally characterized by a number-density distribution of particles with respect to their sizes. In this case, the value $n(r)dr$ is the number of particles with radii in the interval $(r, r + dr)$. The function $n(r)$ is a distribution in the sense that $n(r) \geq 0$ for all r in the size interval $R = [R_1, R_2]$ and that $\int_R n(r)dr = N$ where N is any finite number even when $R_2 \to \infty$. The constant N represents a quantitative measure of a polydispersed ensemble (the total number of particles in a unit scattering volume).

The above procedure for describing the microstructure of dispersed media is usual for direct problems of light scattering dealing with the investigation of the effect of such media on the propagation of optical waves. Here the functions $n(r)$ represent any a priori models of real size spectra of particles.

Direct methods for determining microstructure parameters based on the sampling of particles from air allow only the histograms $\Delta_j(N)/\Delta_j(r)$ $(j = 1,2,\ldots,m)$ to be determined. Here $\Delta_j(N)$ is the number of particles with radii in an interval (r_j, r_{j+1}). For this reason,the functions $n(r)$ can at best be regarded only as an analytical approximation to real histograms (i.e., step-wise functions). Note further that for higher resolution of microstructure analysis, when the values $\Delta_j(r) = r_{j+1} - r_j$ become too small, the behavior of the ratios $\Delta_j(N)/\Delta_j(r)$ can be rather irregular. Therefore, it is more correct for the analysis of experimental data on the microstructures of dispersed media to assume $\Delta_j(N)$ to be a measured value provided that a proper choice of $\Delta_j(r)$ $(j = 1,2,\ldots)$ values is made. The last fact means that the choice of $\min_j \Delta_j(r)$ has been made somehow in accordance with the efficiency of particles counting.

Thus the microstructures of real dispersed media can be described most adequately by means of the integral distribution $N(r)$ which is the total number of particles in the ensemble investigated whose radii exceed the value r. The function N(r) can have discontinuities (of course, breaks of the first kind are meant). The simplest model of such a distribution is a nondecreasing step function which is just the function measured with the use of direct methods for determining the microstructures of dispersed media.

Using optical sounding of dispersed media one can infer information about size distributions N(r) by inverting the set of measured values of any characteristic of light scattering. Let us consider the values of scattering coefficient (extinction coefficient) β measured at several wavelengths $\lambda_i (i = 1,2,\ldots,n)$ from a spectral interval $\Lambda = [\lambda_{min}, \lambda_{max}]$ to be such a set. For the spectral interval Λ we shall use below the term "interval of optical sounding". It is obvious that in a general case, β is a functional of N(r), defined on a set Φ of possible distributions. Taking this into account, we shall denote the optical characteristics as $\beta(N)$. From the standpoint of physical meaning of the problem, it is natural to assume $\beta(N)$ to be a linear and continuous functional on Φ.

The analytical form of $\beta(N)$ for a polydispersed ensemble of spherical particles can be determined easily. Let us assume that a unit scattering volume contains N spherical particles whose sizes are within the interval $R = [R_1, R_2]$. The microstructure of such an ensemble can be represented by an m-dimensional vector $\mathbf{s}$ with components $s_j = \pi r_j^2 n_j (j = 1,2,\ldots,m)$, where n_j is the number of particles with radius r_j in the ensemble $(\sum_{j=1}^{m} n_j = N)$. If the ensemble of scattering particles can be regarded, on the whole, as a system of independent scatterers, then one can write

$$\beta(\mathbf{s},\lambda) = \sum_{j=1}^{m} K_j(\lambda) s_j \quad . \tag{2.1}$$

The relationship (2.1) is based on the statement that for a spherical particle the scattering cross-section can be represented as a product of its geometrical section and the scattering efficiency factor $K_j(\lambda)$ depending on the ratio of the particle size and the wavelength λ of incident radiation.

Equation (2.1) represents the functional dependence of the value of $\beta(\lambda)$ on the $\mathbf{s}$ vector. Having measured $\beta(\lambda_i)(i = 1,2,\ldots,n)$ for $n \geq m$, one can construct a set of equations for the determination of the $\mathbf{s}$ vector components. It is obvious that in this case the matrix $\{K_j(\lambda_i)\}(j = i = 1,2,\ldots,n)$ must not be degenerate. In most cases $m \gg n$, and det $\{K_j(\lambda_i)\}$ is close to zero; even if $n = m$ this results in a strong dependence of the solution vector on errors in the measurement vector β. Therefore, it is advisable to use an integral form of the functional $\beta(N)$ instead of (2.1). Such an approach has many analytical possibilities in the problems under consideration. Let us consider briefly the technique for constructing integral equations for size distributions of the N(r) type using summations like (2.1). As before, particles of the ensemble investigated are assumed to have spherical shapes and their sizes to be from the interval R. Divide R into a series of non-overlapping open subintervals (r_q, r_{q+1}) $(q = 1,2,\ldots,m)$. Then, the whole ensemble will be divided into m nonintersecting subsets each of which contains particles with radii $r_q < r < r_{q+1}$. It should be noted that such a division of the ensemble of scattering particles is unique for the given system of subintervals comprising the interval R, because the radius r describes a sphere uniquely. Obviously, in the case of particles of arbitrary shapes, the division (or the same classification) of an ensemble becomes much more difficult.

Assume the jth sphere to be from the subinterval (r_q, r_{q+1}) having a length $\Delta_q(r) = r_{q+1} - r_q$. Let the number of particles from this size interval be denoted by $\Delta_q(N)$. The division $\{\Delta_q(S)\}$, $(q = 1,2,\ldots,m)$ corresponds to the division $\{\Delta_q(N)\}$, where S is the total geometric cross-section of particles per unit volume. Introduction of the S value as a quantitative measure of a polydispersed ensemble is caused by the peculiarities of the physical problem on the diffraction of electromagnetic waves on particles of the medium. It is obvious that

$$\Delta_q(S) = \sum_j s_j \quad , \quad j \in J(\Delta_q(N)) \quad , \tag{2.2}$$

where $J(\Delta_q(N))$ is the set of indices corresponding to the spheres of the qth subset. Here the value $\Delta_q(s)$ is an element of a measure for the qth subset of particles in a given ensemble. Considering $\Delta_q(S)$ to be a continuous function of $\Delta_q(r)$ for $m \to \infty$ (N is sufficiently large) let us proceed to the limit in (2.2) denoting

$$\lim_{r_q \to r_{q+1} \to r} \Delta_q(S) = dS(r) \quad . \tag{2.3}$$

For spheres $dS(r) = \pi r^2 dN(r)$ and, hence, the full measure S of the ensemble of particles may be represented by any of the relationships

$$S = \sum_{q=1}^{m} \Delta_q(S) = \int_R dS = \int_R \pi r^2 dN(r) \quad . \tag{2.4}$$

All of these considerations can be applied to the optical characteristic β, since it is nothing but the optical measure of the ensemble of particles under consideration. Analogous summations are valid for β, namely

$$\beta = \sum_{q=1}^{m} \Delta_q(\beta) = \sum_{q}^{m} \sum_{j} K_j S_j \quad , \quad j \in J(\Delta_q(N)) \quad , \tag{2.5}$$

where $\Delta_q(\beta)$ is the optical contribution of the qth fraction of particles to the radiation scattered. Passing to the limit in (2.5) one obtains

$$\beta(\lambda) = \int_R K(r,\lambda)\pi r^2 dN(r) \quad , \quad \lambda \in \Lambda \ . \tag{2.6}$$

Thus, we have shown that the size distribution function of an ensemble of spherical particles, N(r), can be obtained using optical sounding methods as a solution of the first-kind integral equation in the form of Stieltjes integral. If we assume that the function N(r) is differentiable, then (2.6) will take the form of a Riemann integral

$$\beta(\lambda) = \int_R K(r,\lambda)\pi r^2 n(r)dr \quad , \tag{2.7}$$

as is commonly used in atmospheric optics.

The above approach, stating that the determination of microstructures of polydispersed media from the optical characteristics of light scattering can be reduced to the solution of integral equations, can be called the method of integral equations in the theory of optical sounding of dispersed media.

For monodispersed ensembles, the analytical behavior of the scattering characteristic β as a function of λ is uniquely defined by the efficiency factor $K(r,\lambda)$. The representations of β in the form of integrals can be called integral equations or simply integrals of the theory of light scattering on polydispersions. Of course, these equations do not represent the whole theory of optical sounding of dispersed media which is to be discussed below.

2.2 Theory of Optical Sounding of Polydispersed Systems of Nonspherical Particles

2.2.1 One-Dimensional Integral Equations for Ensembles of Convex Randomly Oriented Particles

The formulation of inverse problems in the theory of light scattering by a polydispersed ensemble of particles with arbitrary geometrical shapes differs qualitatively from that for spherically symmetric particles. First of all, this is caused by the fact that here the ensemble microstructure cannot be described by one-dimensional size distribution functions and, hence, the corresponding integral equations in the theory of light scattering on polydispersions should be multi-dimensional. The fact that the kernels of these equations are unknown or known only approximately, taking into account difficulties of diffraction theory, naturally makes the solution of inverse problems difficult, but it is not the main obstacle in constructing the solution algorithms.

The formulation of such inverse problems in the form of multi-dimensional integral equations is as yet premature for atmospheric investigations since not only do the mathematical difficulties become greater but also the requirements as to the accuracy and number of the measurements increase. The morphology of atmospheric sols is also little known, which makes the construction of adequate mathematical simulations of the inverse problems difficult. In this connection, we shall discuss below the question—as to how much meaning can be assigned to the solution of one-dimensional integral equation of the type of (2.6), if particles of the sounded medium have nonspherical shapes? The statement of such problems is justified by the fact that scattering characteristics measured experimentally are, as a rule, functions of only one variable (wavelength, or scattering angle) and, hence, the inversion result can be only a one-dimensional distribution function. Of course, these distributions cannot fully characterize the microstructure of dispersed media when the constituent particles are nonspherical. At the same time, as will be shown subsequently, one-dimensional size-distribution functions of aerosols obtained by inverting the optical measurements data are of obvious practical interest in the case of convex particles.

As previously (Sect.2.1), let the ensemble of scattering particles form a system of independent scatterers with the position of each particle in the volume illuminated being fixed with respect to any coordinate system. The direction of the incident flux will be defined by the unit vector $\mathbf{n}$. It is known that light scattering by an ensemble of nonspherical particles is anisotropic. That means that the scattering cross-section of the particles depends on the vector $\mathbf{n}$. Taking this into account, as well as the results of experimental studies of light scattering by nonspherical particles, one can write the summation relation analogous to (2.1)

$$\beta(\lambda,\mathbf{n}) = \sum_{j}^{N} T_j(\lambda,\mathbf{n})P_j(\mathbf{n}) \quad , \tag{2.8}$$

where $P_j(\mathbf{n})$ is the projection of the jth particle onto the plane perpendicular to $\mathbf{n}$; $T_j(\lambda,\mathbf{n})$ is the corresponding scattering efficiency factor. Integrating over all directions, one can obtain the mean value

$$\beta(\lambda) = \int_{\Omega} \beta(\lambda,\mathbf{n})(d\omega/4\pi) \quad , \tag{2.9}$$

which can be regarded as a scattering cross-section for the ensemble of particles, if they are randomly oriented within the volume illuminated. In this case light scattering is, on the average, isotropic. It is assumed therein that averaging over the orientations of the particles is equivalent to averaging over all possible directions of the vector **n**. These assumptions are, as a rule, valid under atmospheric conditions.

When passing from summation relations like (2.8) to the corresponding integrals, the choice of the vector **n** as an initial variable in our treatment is determined by the polydispersed ensembles of particles. Substituting (2.8) in (2.9) one obtains

$$\bar{\beta}(\lambda) = \sum_{j}^{N} \bar{T}_j(\lambda)\bar{P}_j \quad , \tag{2.10}$$

where $\bar{P}_j$ is the mean projection of the jth particle of the scattering ensemble onto a plane perpendicular to the vector **n** for any possible orientations of the particle with respect to that plane:

$$\bar{T}_j(\lambda) = (1/\bar{P}_j) \int_{\Omega} T_j(\lambda,\mathbf{n})P_j(\mathbf{n})(d\omega/4\pi) \quad . \tag{2.11}$$

In order to construct a one-dimensional integral representation of the β functional using (2.11) it is necessary to select, in some way, a linear parameter enabling one to divide the ensemble of particles into some separate fractions. To simplify the problem,we shall assume the particles of the ensemble to be convex and, as a consequence, their volume to be determined as $v = \bar{P}\bar{\ell}$, where $\bar{\ell}$ is the mean chord (diameter). As opposed, e.g., to spheres, where the volume is uniquely determined by the parameter r (radius), the volume of a convex body should be determined by two parameters $\bar{P}$ and $\bar{\ell}$. Therefore, the parameter $\bar{P}$ is determined by two variables v and $\bar{\ell}$ for each particle of the ensemble. The latter circumstance naturally makes the task of classifying uniquely the particles of the polydispersed system using only the linear parameter $\bar{\ell}$ very difficult and introduces uncertainty into the corresponding inverse problems. In this connection, we assume the geometry of particles to be, on the average, quite regular over the whole ensemble, meaning by this that the growth of linear size $\bar{\ell}$ is followed by the growth of mean cross-sections (or volumes), so that for any two particles of the ensemble it follows from the condition $|\bar{\ell}_j - \bar{\ell}_i| \leq \varepsilon$ that $|\bar{P}_j - \bar{P}_i| \leq \delta(\varepsilon)$.

We shall below consider this assumption to be, as a rule, valid for atmospheric aerosols.

Polydispersed systems consisting of particles with regular geometry should in their optical properties be close to the systems of spherical particles in the sense that their optical characteristics can be modeled within the framework of the Mie theory but, of course, with varying degrees of reliability.

We shall use the value $\bar{P} = \sum_j^N \bar{P}_j$ as the full measure of a polydispersed system of particles, which is analogous to the full geometrical cross-section of particles contained in a scattering volume $S = \sum_{j=1}^N s_j$. As to the choice of a size-distribution function, it is natural to use the function $\bar{P}(\bar{\ell})$ characterizing the distribution of values $\bar{P}$ with respect to the parameter $\bar{\ell}$ defined on the interval $L = [\bar{\ell}_{min}, \bar{\ell}_{max}]$.

Let us consider briefly the derivation of the integral equation for the function $\bar{P}(\bar{\ell})$ bearing in mind the problems of optical sounding of polydispersed media. For this purpose, we cover the interval L with the system of non-overlapping subintervals $\{\Delta_q(\bar{\ell})\}(q = 1,2,\dots,m)$. Denoting the number of particles with mean diameter $\bar{\ell}_q < \bar{\ell} < \bar{\ell}_{q+1}$ as $\Delta_q(N)$, one can write

$$\sum_{j=1} \bar{T}_j(\lambda)\bar{P}_j = \sum_{q=1}^{m} \sum_{j \in J(\Delta_q(N))} \bar{T}_j(\lambda)\bar{P}_j \quad .$$

Directing $\bar{\ell}_q$ to $\bar{\ell}_{q+1} = \bar{\ell}$ one can write the sum on the right-hand side of this equation in the form of the integral

$$\beta(\lambda) = \int_L d(\bar{\beta}(\bar{\ell},\lambda)) \quad , \tag{2.12}$$

where $d\ \beta(\bar{\ell},\lambda))$ is

$$\lim_{\bar{\ell}_q \to \bar{\ell}_{q+1} = \ell} \sum_{j \in J(\Delta_q(N))} \bar{T}_j(\lambda)\bar{P}_j = \sum_{j \in J(dN(\bar{\ell}))} \bar{T}_j(\lambda)\bar{P}_j \quad .$$

The sum $\sum \bar{T}_j(\lambda)\bar{P}_j$ $(j \in J(dN(\bar{\ell}))$ can be written, in turn, as a product $\bar{T}(\bar{\ell},\lambda) \sum \bar{P}_j$ assuming that $\bar{T}(\bar{\ell},\lambda) = (\sum_j \bar{P}_j)^{-1} \times \sum_j \bar{P}_j\bar{T}_j(\lambda)$. Such a representation is based on the fact that $\bar{T}_j$ and $\bar{P}_j$ are positive. Denoting further

$$\sum_{j \in J(dN(\bar{\ell}))} \bar{P}_j = d\bar{P}(\bar{\ell}) \quad , \tag{2.13}$$

we obtain the following integral equation with respect to the distribution $\bar{P}(\bar{\ell})$

$$\int_L \bar{T}(\bar{\ell},\lambda)d\bar{P}(\bar{\ell}) = \beta(\lambda) \quad , \tag{2.14}$$

where

$$\bar{T}(\bar{\ell},\lambda) = \Big(\sum_j \bar{P}_j\Big)^{-1} \sum_j T_j(\lambda,\mathbf{n})\bar{P}_j(\mathbf{n})(d\omega/4\pi) \quad . \tag{2.15}$$

Let us consider some conclusions which can be drawn from this integral equation.

First of all, as follows from (2.13) in the case of the ensemble of convex particles, the direct determination of the function $N(\bar{\ell})$ by inverting $\bar{\beta}$ is already impossible. The expression for $d\bar{P}(\bar{\ell})$ contains implicitly the differential element of the measure $dN(\bar{\ell})$. It follows from this that the estimation of the particle number density and size-distribution function with the use of optical sounding methods can be made only for ensembles of spherical particles.

As to particles with arbitrary geometry, such an estimation may be regarded as the next step in interpreting the basic result of the one-dimensional inverse problem of light scattering by polydispersed ensembles of particles obtained in the form of the distribution $\bar{P}(\bar{\ell})$. It is obvious that such an interpretation is possible provided only that some additional assumptions on the particle geometry are made. It is important to emphasize that even if the distribution $N(\bar{\ell})$ is analytically quite regular (e.g., it has derivatives) the same cannot be said about the distribution $\bar{P}(\bar{\ell})$. The latter distribution is determined not only by the distribution of the number density as a function of the mean diameter but also by variations of geometrical shape from particle to particle. This is the reason that prevents an a priori assumption on the differentiability of $\bar{P}(\bar{\ell})$ to be made. As a consequence, a Stieltjes integral describes most adequately the optical properties of scattering media composed of particles having not only different sizes but also different shapes.

Thus, if the Mie theory is used as an apparatus for the inversion of optical sounding data it is then preferable to find, first of all, the integral distribution $S(r)$. By analogy with the integral (2.6), which is the basic concept of the theory of the optical sounding of polydispersed scattering systems, the Stieltjes integral (2.14) should be considered as the reference functional relation for studying the microstructure of ensembles of convex particles. Recall that formally the size-distribution function $\bar{P}(\bar{\ell})$ is defined as $\sum_j \bar{P}_j$, where $j \in J(N(\bar{\ell}' \leq \bar{\ell})$ and $N(\bar{\ell}' \leq \bar{\ell})$ is the number of particles in the ensemble with size $\bar{\ell}'$ not exceeding $\bar{\ell}$. In general, such a distribution function is not the only characteristic of the microstructure of a scattering medium. The function $\bar{P}(\bar{\ell})$ is related, first of all, to the optical methods used for investigating the scattering media. In principle, the microstructure of any polydispersed medium can be described by the distribution of particle volume with respect to the size $\bar{\ell}$. Knowledge of the distribution $N(\bar{\ell})$ of the ensemble of spherical particles is sufficient for the construction of the distribution functions $S(r)$ and $V(r)$. As a matter of fact, in the above description we wrote the relations for $dS(r) = \pi r^2 dN(r)$ and $dV(r) = (4/3)\pi r^3 dN(r)$ which could be regarded as functional relationships between different distribution functions of the given systems.

In the case of polydispersed systems of convex particles the corresponding functional equations become more complicated. As shown above, the direct determination of the function $N(\bar{\ell})$ from the known distribution $P(\bar{\ell})$ is impossible without information on the particle morphology. But if $P(\bar{\ell})$ is known, then it is not difficult to find the distribution $V(\bar{\ell})$. In fact, substituting the expression $v_j = \bar{P}_j \bar{\ell}_j$ in (2.13), one can see that

$$d\bar{P}(\bar{\ell}) = \bar{\ell}^{-1} dV(\bar{\ell}) \quad . \tag{2.16}$$

It then follows that

$$V(\bar{\ell}) = \int_{\bar{\ell}_{min}}^{\bar{\ell}} \bar{\ell}' d\bar{P}(\bar{\ell}') \quad .$$

At the same time, (2.16) defines a one-dimensional integral equation corresponding to the distribution function $V(\bar{\ell})$

$$\int_L \bar{T}(\bar{\ell},\lambda)\bar{\ell}^{-1} dV(\bar{\ell}) = \bar{\beta}(\lambda) \quad . \tag{2.17}$$

It is interesting to note that for polydispersed systems of convex particles, one more one-dimensional distribution in addition to $\bar{P}(\bar{\ell})$ and $V(\bar{\ell})$ can be constructed, which can be found by inverting $\beta(\lambda)$. Here we mean the distribution $\theta(\bar{\ell})$, as introduced into the theory of optical sounding in [2.3]. The geometrical parameter θ is a measure of the symmetry of the particles of the dispersed medium, while the distribution $\theta(\bar{\ell})$ characterizes the ensemble average of the deviation of the particle shapes from spherical.

The measure of the symmetry of an individual particle θ_j can be based on the known isoparametric relation of the theory of convex bodies $(4\bar{P}_j)^3 - 36\pi v_j^2 \geq 0$ [2.4]. Equality is achieved only for spherical bodies. If the parameter θ_j is introduced so that $(4\bar{P}_j)^3 - 36\pi v_j^2 \theta_j = 0$, then it can be regarded as a functional defined on a set of convex bodies, with the lower limit of $\theta_j = 1$ for spheres. Since $\theta_j = (16/9\pi)\bar{P}_j\bar{\ell}_j^{-2}$, then using (2.13) one can easily show the validity of the following expression

$$d\bar{P}(\bar{\ell}) = (9\pi/16)\bar{\ell}^2 d\theta(\bar{\ell}) \quad , \tag{2.18}$$

where $d\theta(\bar{\ell}) = \sum_j \theta_j$, $j \in J(dN(\bar{\ell}))$. The distribution $\theta(\ell)$ determines an integral measure of the symmetry of particles in the ensemble with mean diameters less than $\bar{\ell}$. Expression (2.18), by analogy with (2.16), allows the function $\theta(\bar{\ell})$ to be found provided the distribution $\bar{P}(\bar{\ell})$ is known. At the same time (2.18) defines the corresponding integral equation

$$(9\pi/16) \int_L \bar{T}(\ell,\lambda)^{-2} d\theta(\bar{\ell}) = \beta(\lambda) \quad . \tag{2.19}$$

To complete this discussion, we should like to stress that the inverse solutions $\bar{P}(\bar{\ell})$ obtained here give only generalized information about real size-distribution functions of aerosol particles. The kernel $\bar{T}(\bar{\ell},\lambda)$ of the corresponding integral equation is defined as the mean value of scattering efficiency factors for particles with mean diameter in the interval $(\bar{\ell},\bar{\ell} + d\bar{\ell})$.

It is obvious that averaging over the particles' orientations used in (2.15) gives rise, simultaneously, to averaging over parameters related to the geometrical shapes of separate particles. As a consequence, the use of $\bar{T}(\bar{\ell},\lambda)$ as a kernel of the integral equation (2.14) does not require detailed information on the morphology of the dispersed medium under study. This fact is very important for optical sounding of atmospheric aerosols. Moreover, the factor $\bar{T}(\bar{\ell},\lambda)$ can be determined with the use of optical sounding methods provided the optical experiment is accompanied by aerosol sampling with subsequent microstructure analysis. Usually, the values of $\bar{P}_j$ are determined as the area of the particle projections onto the microscope table [2.5]. Using the distribution $\bar{P}(\bar{\ell})$ found for any local volume of the medium, one can formulate the corresponding inverse problem of light scattering on polydispersions, the solution of which is the function $\bar{T}(\bar{\ell},\lambda)$. Such inverse problems will be considered below.

Let us consider, in conclusion, a more exact approach to the construction of the size-distribution function for the ensemble of convex particles randomly oriented in the scattering volume of the medium. Obviously, the division of a particles ensemble into fractions according to size intervals $\{\Delta_q(\bar{\ell})\}(q = 1,2,\ldots,m)$ would be more strict and definite if it were only for particles within the given volume v. In this case, (2.12) should be written as follows

$$\bar{\beta}(\lambda,v) = \int_L d(\bar{\beta}(\bar{\ell},v,\lambda)),$$

where

$$d(\bar{\beta}(\bar{\ell},v,\lambda)) = \sum_{j \in J(dN(\bar{\ell},v))} \bar{T}_j(\lambda)\bar{P}_j$$

$$= \bar{T}(\bar{\ell},v,\lambda)v\bar{\ell}^{-1}dN(\bar{\ell},v) \quad .$$

As a result, the integral representation for the optical characteristic of light scattering by polydispersed system of particles is of the form

$$\bar{\beta}(\lambda) = \int\int_{\Omega} \bar{T}(\bar{\ell},v,\lambda)v\bar{\ell}^{-1}dN(\bar{\ell},v) \quad . \tag{2.20}$$

Here, the integration is performed over the corresponding region $\Omega(\bar{\ell},\bar{v})$. As a matter of fact, (2.20) is not defined as an integral equation with respect to the function $N(\bar{\ell},v)$. An additional definition should be made somehow in order to make the inverse problem (2.20) definite. The simplest way to do that is to assume the existence of a functional dependence of, e.g., the parameter v on the mean diameter of particles. This results, at least, in making the assumption of the constancy of geometrical shape over all particles of the ensemble investigated. In this respect, the approach discussed above and based on one-dimensional integral equations is at the moment preferable for use in atmospheric studies.

2.2.2 Polydispersed Systems of Ellipsoidal Particles

In atmospheric optics an ensemble of ellipsoidal particles can be regarded as a realistic model for scattering media. The information content of this model is much higher than that of spherical particles as applied to the interpretation of optical measurements. In particular, this model allows the extraction of more information from polarization characteristics of lidar signals when sounding the atmospheric aerosol. On the other hand, ellipsoidal particles are interesting because in this case, as for spheres, it is possible to construct an analytical solution of the diffraction problem [2.2,6,7].

As before, we shall use (2.8) as the initial relationship when constructing the integral form for the optical characteristic of a polydispersed system of ellipsoidal particles. Let the spatial position of the jth particle of the ensemble be determined with respect to a coordinate system A wherein the direction of incident radiation coincides with the x-axis. The value P_x for every particle should be found in order to make the use of (2.8) possible when proceeding to the integral. Now, this value characterizes the area of an ellipsoidal particle projection onto a plane perpendicular to the x-axis. In this case, the area is determined by the values of the semiaxes a, b and c,as well as by the particles orientation with respect to the x-axis. Since all particles of the ensemble have the same shape, then the functional dependence of P_x on the parameters a, b and c is the same for all particles and, as a consequence, there is no need to index the particle mean projections. Hence, in this case, only one function $P_x(a,b,c)$ is required, instead of the set $\{P_j(\mathbf{n})\}(j = 1,2,....)$. Here the variables a, b and c take values within the size intervals $[a_1, a_2]$, $[b_1, b_2]$ and $[c_1, c_2]$, respectively.

As is known, the equation of an ellipsoid surface in the coordinate system A' relative to its axes is written as follows

$$(x'^2/a^2) + (y'^2/b^2) + (z'^2/c^2) = 1 \quad .$$

A linear transformation of the coordinates is used when proceeding from the coordinate system A' to A, i.e.,

$$x' = \ell_1 x + m_1 y + n_1 z \quad ,$$

$$y' = \ell_2 x + m_2 y + n_2 z \quad ,$$

$$z' = \ell_3 x + m_3 y + n_3 z \quad ,$$

where (ℓ_1, ℓ_2, ℓ_3), (m_1, m_2, m_3) and (n_1, n_2, n_3) are the direction cosines for the x', y' and z' axes in the coordinate system A. Then in A, the ellipsoid surface will be described by the following equation

$$(\ell_1 x + m_1 y + n_1 z)^2/a^2 + (\ell_2 x + m_2 y + n_2 z)^2/b^2 + (\ell_3 x + m_3 y + n_3 z)^2/c^2 = 1 \quad . \tag{2.21}$$

In order to find the area of the ellipsoid projection P_x onto a plane perpendicular to x axis,it is sufficient to know the value of the ellipsoid secant along the x axis and then to use the relationship $P_x\bar{\ell}_x = V = (4/3)\pi abc$, $(a > b > c)$. In turn, the length of any ellipsoid secant $\bar{\ell}_x$ along the x axis in A coordinate system is determined as the difference $(x_2 - x_1)$, where x_1 and x_2 are the points lying on the ellipsoid surface and belonging to one and the same secant.

The equation of the ellipsoid projection boundary on the plane perpendicular to the x-axis is written as $\bar{\ell}_x = 0$. Using this relationship and (2.21) for the ellipsoid surface, one can easily find an analytical expression for the projection boundary along the x axis, for any orientation of the ellipsoid with respect to the coordinate system A. If the equation of projection P_x boundary is known, then the corresponding value of the mean secant of the ellipsoid is found as follows

$$\bar{\ell}_x = (1/P_x) \int_{P_x} \ell_x dP \quad . \tag{2.22}$$

Let us write the final formulas for calculating $\bar{\ell}_x$ and P_x, omitting intermediate steps [2.8]

$$\bar{\ell}_x = (4/3)abc(b^2c^2\ell_1^2 + a^2c^2\ell_2^2 + a^2b^2\ell_3^2)^{-1/2}$$

$$P_x = (b^2c^2\ell_1^2 + a^2c^2\ell_2^2 + a^2b^2\ell_3^2)^{1/2} \quad . \tag{2.23}$$

For convenience in the subsequent discussion it is useful to introduce the function $L(\ell_1, \ell_2, \ell_3) = b^2c^2\ell_1^2 + a^2c^2\ell_2^2 + a^2b^2\ell_3^2$ and to use the Euler angles ν, ψ and φ instead of direction cosines to describe the ellipsoid orientation (or the system A') with respect to the x axis of system A. In this case, we have

$$\begin{aligned} L(\nu,\psi,\varphi) = {} & 2c^2(a^2 - b^2) \cos\nu \cos\varphi \sin\varphi \sin\psi \cos\psi \\ & + c^2 \cos^2\nu \sin^2\psi(a^2 \cos^2\varphi + b^2 \sin^2\varphi) \\ & + c^2 \cos^2\psi(b^2 \cos^2\varphi + a^2 \sin^2\varphi) \\ & + a^2b^2 \sin^2\nu \sin^2\psi \quad . \end{aligned}$$

Now, let us proceed to the derivation of the integral expression for characteristics of light scattering. Let us consider, for simplicity, that particles of the ensemble have no preferred orientations. Using the above technique for constructing the integrals based on summations like (2.8) we obtain

$$\begin{aligned} \bar{\beta}(\lambda) = {} & (8/\pi^3) \int_0^{\pi/2} d\nu \int_0^{\pi/2} d\psi \int_0^{\pi/2} d\varphi \int_M T(a,b,c,\nu,\varphi,\psi,\lambda)\pi \\ & L^{1/2}(a,b,c,\nu,\psi,\varphi)dN(a,b,c) \quad . \end{aligned} \tag{2.24}$$

In (2.24) the three-dimensional region M is determined by the limits of possible values of the parameters a, b and c of particles in the given ensemble. The distribution N(a,b,c) characterizes the number of particles in the ensemble whose semiaxes do not exceed the values a, b and c, respectively. Equation (2.24) is not defined with respect to the function N(a,b,c). In the case of inverse problems of light scattering,(2.24) can be of interest when particles of the dispersed medium have a preferred orientation relative to the space of the volume sounded and if not only spectral but also angular characteristics of light scattering are measured experimentally. Some particular systems are also possible, for which one or more additional relationships between a, b and c can be set. All this shows the complexity of the inverse problems of light scattering by polydispersed media if the geometry of particles is to be taken into account.

Under the condition that a = b = c = r, expression (2.24) becomes identical to (2.6). Let us give, in conclusion, some expressions for the geometrical characteristics of ellipsoidal particles which could be of interest for inverse problems of the form (2.14). These characteristics are

$$\bar{\ell} = (32/3\pi^3)abc \int_0^{\pi/2} d\nu \int_0^{\pi/2} d\psi \int_0^{\pi/2} L^{-1/2}(\nu,\psi,\varphi)d\varphi \quad ,$$

$$\bar{P} = (8/\pi^2) \int_0^{\pi/2} d\nu \int_0^{\pi/2} d\psi \int_0^{\pi/2} L^{1/2}(\nu,\psi,\varphi)d\varphi \quad ,$$

$$\theta = (\pi^3/8) \int_0^{\pi/2} d\nu \int_0^{\pi/2} d\psi \int_0^{\pi/2} L^{1/2}(\nu,\psi,\varphi)d\varphi \Bigg/ \left[abc \int_0^{\pi/2} d\nu \int_0^{\pi/2} d\psi \int_0^{\pi/2} L^{-1/2}(\nu,\psi,\varphi)d\varphi\right]^2 \quad .$$

It is easily seen that the value of P for ellipsoid particles is uniquely determined by three linear parameters (the semi-axes a, b and c). Therefore, the division of the ensemble of particles into subsets (the same fractions) according to one linear parameter $\bar{\ell}$, which we have made above for constructing the distribution $\bar{P}(\bar{\ell})$, is quite relative. Keeping this in mind, we have emphasized the fact that the function $\bar{P}(\bar{\ell})$ bears only very general information about a real size-distribution of particles (or on the microstructure of a dispersed medium). This remark can also be applied to the distribution $\theta(\bar{\ell})$. Incidentally, the estimation of the measure of symmetry for ellipsoidal particles can be obtained in the following form

$$\theta \le (ab/c^2) \quad , \qquad (2.25)$$

where $a \le b \le c$ and equality is achieved in (2.25) when $a = b = c$. Estimations like (2.25) can also be obtained for other types of convex particles which are of interest for investigations of light scattering by aerosols. In particular, for the case of cubic particles we have [2.9]

$$\theta < (16/9\pi)(1 + 1/\sqrt{2}) \simeq 2.8 \quad .$$

2.3 Definition of Operators of Light Scattering by Polydispersed Systems

2.3.1 Methods of Optical Models

In the above discussion the problems of optical sounding applied to the determination of the microstructure of dispersed media were considered based on the solution of one-dimensional integral equations. The analysis of the information content of these problems when applied to atmospheric aerosol studies was emphasized. Before other problems of the theory of optical sounding are considered,the method of integral equations should be discussed from a more general point of view.

It is easily seen that (2.14) defines an integral operator $\bar{T}$ acting on the set of distributions Φ.

The integral operator $\bar{T}$ allows (2.14) to be written in operator form, i.e., $\bar{T}\bar{P} = \bar{\beta}$, $\bar{P} \in \Phi$. The integral operator $\bar{T}$ corresponding to a given kernel $\bar{T}(\bar{\ell},\lambda)$ of the initial equation (2.14) produces, in its turn, the set of optical characteristics $\bar{B}$, corresponding to the initial set of distributions Φ. If the kernel is defined in the region $[L \times \Lambda]$, then the optical characteristics $\bar{\beta}(\lambda)$ are distributions within the interval of optical sounding Λ, i.e., they satisfy, as functions, the condition $\bar{\beta}(\lambda) > 0$, $(\lambda \in \Lambda)$, and the value $\int_\Lambda \beta(\lambda)d\lambda$ is bounded for any Λ. In applied problems, sets such as $\bar{B}$ are conventionally called the sets of initial data. In our case, this is the set of optical characteristics of aerosols in the atmosphere.

The aerosol microstructure can vary due to one or more physical or chemical processes occurring in the atmosphere. Taking this into account,one can see that the formulation of problems of optical sounding based on the assumption of two sets $\bar{B}$ and Φ connected with each other by the operator $\bar{T}$ is more general and informative. The operator $\bar{T}$ defines the representation $\bar{T}: \Phi \rightarrow \bar{B}$ and represents a generalized characteristic of light scattering by polydispersed systems of particles. In the particular case of spherical particles $\bar{T} \equiv K$, where K is numerically determined by the Mie theory, provided the index of refraction of the particulate matter is known. It should be noted that the operator $\bar{T}$ and the sets $\{\Phi,B\}$ characterize a type of optical weather when used in atmospheric optics.

Introduction of the integral operator $\bar{T}$ into the theory of light scattering by polydispersed systems generalizes the method of integral equations and can be regarded as the basis for further development of the theory of optical sounding of dispersed media and its applications. Consider again (2.14). Obviously, the function $\bar{P}(\bar{\ell})$ can be found using this equation provided that the kernel $\bar{T}(\bar{\ell},\lambda)$ is known for any $\bar{\ell}$, $\lambda \in [L \times \Lambda]$. Unfortunately, in the problems of atmospheric optics dealing with light scattering by aerosols, the reliability of the a priori choice of $\bar{T}(\bar{\ell},\lambda)$ cannot be guaranteed. In this connection, the inverse problems of light scattering by aerosols are characterized by a significant initial uncertainty which should be taken into account when developing the optical sounding methods.

Two procedures are possible, in principle, for defining the operator $\bar{T}$ (the same kernel $\bar{T}(\bar{\ell},\lambda)$ which can be used for determining the dispersed medium microstructure). The first is to approximate the kernels $\bar{T}(\bar{\ell},\lambda)$ by some known functions. In particular, since the assumption of sphericity is generally used in atmospheric optics, the kernels of the Mie theory $K(r,\lambda)$ are used as $\bar{T}(\bar{\ell},\lambda)$. Of course, the possibilities of such an approach are limited. A formulation that is not based on the approximations of kernels $\bar{T}(\bar{\ell},\lambda)$ but on the possibility of constructing some new operator instead of $\bar{T}$ is more correct and informative. Such an operator should be constructed within the framework of the Mie theory, taking into account, at least in the first approach, the shape of scattering particles. A similar treatment has been suggested in [2.3,10]. The idea of the approach is to have the operator $\bar{T}$ correspond to the operator KE^{-1}, where K is the operator of light scattering by a polydispersed system of nonspherical particles, and the operator E is constructed taking into account the shape of the aerosol particles and satisfies the condition

$$\|\bar{P} - ES\| \leq \delta(\varepsilon)$$

when (2.26)

$$\|\bar{T}\bar{P} - KS\| \leq \varepsilon \quad ,$$

here S is the size-distribution function.

The existence of the operator E for a given system of particles is far from trivial and in every case requires the analysis of experimental data as well as calculation. The scattering phase matrix of a polydispersed system of nonspherical particles randomly oriented with respect to a scattering volume represents an example of such a case. This matrix consists of six different elements and, as a consequence, not every element corresponds to the matrix element of an optically equivalent system of spherical particles, which consists of only four elements. In this connection, the approximate approach to the inverse problems considered here will deal mainly with the spectral characteristics of light scattering by aerosols.

Let us consider a concrete example in order to illustrate the above. Two possibilities for approximating scattering cross-sections of nonspherical particles by those for spherical ones can be demonstrated; they are based on the results of experimental studies of light scattering by nonspherical particles and interpretation made using the Mie theory [2.5b,11]. The first one is to take the scattering cross-section of a sphere with its volume v_{0j} being equal to the volume v_j of a nonspherical particle as the cross-section of the latter. In the second case,optical equivalence is defined by the equality of mean geometrical cross-sections (or, which is the same for convex bodies, by the equality of the mean surface areas). Using these possible approximations for the scattering cross-section of an individual particle, one can easily construct the corresponding approximation analogs for polydispersed systems.

Let $\bar{\beta}_j = \beta_{0j}$ $(j = 1,2,\dots,N)$, where the scattering cross-section of a nonspherical particle is determined according to (2.8) by the expression

$$\bar{\beta}_j(\lambda) = \int T_j(\mathbf{n},\lambda)P_j(\mathbf{n})d\omega/4\pi \quad . \tag{2.27}$$

Let us assume that the partition $\{\Delta_q(\bar{\ell}_0)\}$ of some interval L_0 corresponds to the partition $\{\Delta_q(\bar{\ell})\}$ $(q - 1,2\ m)$ of the initial size interval L, so that every subinterval involves those spheres which are optically equivalent to the particles from the subinterval $\Delta_q(\bar{\ell})$. Obviously, in this case the following equality is valid

$$\sum_{j \in J(dN(\bar{\ell}))} \bar{\beta}_j = \sum_{j \in J(dN(\bar{\ell}_0))} \beta_{0j} \quad . \tag{2.28}$$

Using (2.28) and taking into account (2.27), one can write down the expression

$$\bar{T}(\bar{\ell},\lambda)d\bar{P}(\bar{\ell}) = K(\bar{\ell}_0,\lambda)dS(\bar{\ell}_0)$$

and two integral representations of the optical characteristic as well

$$\bar{\beta}(\lambda) = \int_{\bar{L}} \bar{T}(\bar{\ell},\lambda)d\bar{P}(\bar{\ell}) = \int_{L_0} K(\bar{\ell}_0,\lambda)dS(\bar{\ell}_0) \quad . \tag{2.29}$$

This result can be interpreted as follows. A polydispersed system of spherical particles with the size-distribution $S(\bar{\ell}_0)$ ($\bar{\ell}_{0min} \leq \bar{\ell}_0 \leq \bar{\ell}_{0max}$) exists whose optical characteristic $\beta_0(\lambda)$ does not differ essentially within the spectral region of sound Λ from the characteristic $\bar{\beta}(\lambda)$ of a polydispersed system of nonspherical particles. Expression (2.29) can be regarded as an equation with respect to the distribution $S(\bar{\ell}_0)$. The kernel of this equation is defined by the Mie theory and, hence, it can be solved using any numerical technique (see, e.g., Sect.2.4). For this solution to be of practical use, it is necessary to find the functional relation between $\bar{P}(\bar{\ell})$ and $S(\bar{\ell}_0)$. The condition $v_j = v_{0j}$ can be used for this purpose. Applying a summation of the type (2.13) to the particle volumes, one finds that

$$d(\bar{P}(\bar{\ell})\bar{\ell}) = d(S(\bar{\ell}_0)\bar{\ell}_0)$$

which leads to the desired expression

$$\bar{P}(\bar{\ell}) = S(\bar{\ell}_0)\bar{\ell}_0/\bar{\ell} \quad ,$$

$$\bar{\ell}_0 \in L_0 \quad , \quad \bar{\ell} \in \bar{L} \quad . \tag{2.30}$$

The boundary condition $P(\bar{\ell}_{min})\bar{\ell}_{min} = S(\bar{\ell}_{0min})\bar{\ell}_{0min}$ was used when deriving (2.30). For the second case when optical equivalence is determined by the mean geometrical cross-sections of particles, a similar relation has the form

$$\bar{P}(\bar{\ell}) = S(\bar{\ell}_0) \tag{2.31}$$

under the condition $\bar{P}(\bar{\ell}_{min}) = S(\bar{\ell}_{0min})$.

For the relationships (2.30) and (2.31) to be completely defined, a proper choice of the sizes of the optically equivalent spheres should be made. Unfortunately, assuming only convexity of the particles is not yet sufficient and some further assumptions on the geometry of the particles of the dispersed medium are necessary. If the density function $\theta(\bar{\ell})$ of the measure of symmetry θ is assumed (Sect.2.2) then one can easily find the relation between $\bar{\ell}_0$ and $\bar{\ell}$. We have

$$\bar{\ell}_0 = [\theta(\bar{\ell})]^{1/3}\bar{\ell} \quad ,$$

$$\bar{\ell}_0 = [\theta(\bar{\ell})]^{1/2}\bar{\ell} \quad , \tag{2.32}$$

for the first and second cases, respectively. Thus, we have shown that the problem on determining the function $\bar{P}(\bar{\ell})$ within the framework of the approximating approach (the same optical models method) can be reduced to the solution of integral equations with the kernels determined by the Mie theory. This means that the unknown operator $\bar{T}$ is replaced by an operator close to it but constructed using the Mie theory formulas, taking into account the morphology of the particles of the medium being investigated. Expressions (2.30-32) may be regarded as the simplest example of the operator E derived from (2.26). This operator corresponds to the transformation E: $S \rightarrow \bar{P}$. Obviously, some other considerations can be admissible when determining the optical equivalence between convex particles and spheres. But, at the same time, one has to keep in mind that in any case the particles of the medium sounded should not, on the average, deviate appreciably from spheres. It is evident that the relationships obtained above can be used not only for solving inverse problems of light scattering but also for developing optical models of the dispersed media. In particular, if the morphology of the particles is known, the function $\theta(\bar{\ell})$ can be estimated by numerical methods and the characteristics of light scattering can be determined using the relationships of this section within the framework of the Mie theory.

2.3.2 Determination of the Scattering Efficiency Factors of Monodispersed Media Using the Data from Sounding of Polydispersed Systems

In addition to the above approximate approach for determining the operator $\bar{T}$, some qualitatively new solutions of this problem of light scattering by polydispersed systems are possible. For simplicity, we shall consider the derivation of the matrix analog $\bar{T}_m$ of the integral operator $\bar{T}$ but not of the operator $\bar{T}$ itself. Here m is a finite number. The matrix operator $\bar{T}_m$ is defined on a set of vectors Φ_m($m = \dim\Phi_m$) and produces the transformation $\bar{T}_m$: $\Phi_m \rightarrow \bar{B}_m$. In such a treatment the microstructure of a dispersed medium is characterized by the vector $\mathbf{P} \in \Phi_m$. If the distribution function $\bar{P}(\bar{\ell})$ is known ($\bar{\ell} \in [\bar{\ell}_{min}, \bar{\ell}_{max}]$), it is then not difficult to construct an m-dimensional vector $\mathbf{P}$ equivalent in some sense to this distribution. A detailed study of similar questions of the problems of optical sounding of aerosols was given in [2.3].

We shall assume that some finite subset $\{\mathbf{P}_j\}$ ($j = 1,2,\ldots,n$) which consists of linearly independent vectors can always be selected from the set Φ_m. It can be shown that under these conditions the elements $t_{k\ell}$ ($k, \ell = 1,2,\ldots,m$) of the matrix $\bar{T}_m$ can be determined by the successive solution of m linear systems of the form

$$\sum_{\ell=1}^{m} t_{k\ell} P_{i\ell} = \beta_{ik} \quad , \quad i = 1,2,\ldots,n \quad ; \quad \ell,k = 1,2,\ldots,m \quad , \quad m \leq n \quad .(2.33)$$

The first index in the notations $P_{i\ell}$ and β_{ik} denotes the vector number while the second one denotes its component. If the vectors $\mathbf{P}_i$ $(i = 1,2,\ldots,n)$ are all linearly independent then it is obvious that the matrix $\{P_{i\ell}\}$ is not degenerate (i.e., $\det\{P_{i\ell}\} \neq 0$) and all the systems (2.33) for $k = 1,2,\ldots,m \leq n$ have physical meanings. The above technique for determining the operator $\bar{T}$ is based on the application of the method of linear systems to integral equations like (2.14) [2.12]. Previously (Sect.2.2) the initial information was represented by the matrix $\{t_{k\ell}\}$ and the vector $\boldsymbol{\beta}$ while now it is represented by the subsets of vectors $\{\mathbf{P}_i\}$ and $\{\boldsymbol{\beta}_i\}$ belonging to the pair $\{\Phi_n, \bar{B}_n\}$. This means that optical experiments for determining the scattering efficiency factors $\bar{T}(\bar{\ell},\lambda)$ of monodispersed systems should be accompanied by microstructure analysis.

The dimensionality of the matrix $\{t_{k\ell}\}$ obtained when solving the inverse problem (2.33) is determined by the dimensions of the vectors $\mathbf{P}_i$ which characterize the microstructure of the medium being studied. The dimensionality of this matrix determines, in turn, the number of values of the factor $\bar{T}(\bar{\ell},\lambda)$ found in the region $[L \times \Lambda]$. The dimensionality n of the corresponding vectors $\boldsymbol{\beta}_i$ agrees in experiments with the value m defined by the maximum resolution of microstructure analysis. If $n > m$, then the linear systems (2.33) can be solved with the use of the least square method which allows the effect of measurement errors on the solution $\{t_{k\ell}\}$ to be weakened somewhat.

The foregoing discussion concerned an approach which in principle can be used to solve the optical sounding problems considered here. The use of (2.33) for the numerical analysis of experimental data can meet with significant difficulties. This can be connected, first of all, with the fact that the condition of linear independence of the initial data $\{\mathbf{P}_i, \beta_i\}(i = 1,2,\ldots,n)$ may be violated in the experiment. Secondly, the modulus of $\det\{P_{i\ell}\}$ can be very small even if this condition is fulfilled, and as a consequence, the direct inversion of (2.39) is practically meaningless because of instability of the solutions obtained with respect to the errors in the initial data. These are the reasons why the numerical techniques for inverting (2.33) should use regularizing algorithms. In particular, one of them is to find the so-called normal solutions for degenerate and ill-conditioned systems. The following example is given to explain the method.

Assume that one has to solve numerically the equation $K_m \mathbf{s} = \boldsymbol{\beta}$, where $\mathbf{s} \in \Phi_m$, $\boldsymbol{\beta} \in \bar{B}_m$, and the matrix operator K_m is the analog of the integral operator $\mathbb{K}$. The normal solution of this system is the vector $\mathbf{s}^*$, from the set $\Phi_{m\beta}^{(0)}$ of solutions of the initial equation with $\boldsymbol{\beta}$ as the right-hand side, which minimizes the value $\|\mathbf{s}^{(0)} - \mathbf{s}\|$, where $\mathbf{s}^{(0)}$ is some arbitrary vector. Since in our case we have to find the operator $\bar{T}_m$,then the operator $\bar{T}^*$ should be called the normal solution. This operator belongs to the set of possible solutions of the inverse problem (2.28) - A_m^*, from one side, and has the minimum deviation from some arbitrarily fixed operator $\bar{T}_m^{(0)}$ from the other side. Of course, the choice of the operator $\bar{T}_m^{(0)}$ as well as of the vector $\mathbf{s}^{(0)}$ in the first example is conditioned by the physical content of the problem to be solved and also by the experimental possibilities. Let us discuss this question in more detail.

As is shown above, the determination of factors $\bar{T}(\bar{\ell},\lambda)$ for ensembles of particles having nonspherical shapes is valid provided the difference is not too large, on the average, over the ensemble. Otherwise, the description of scattering properties of a separate particle with the use of $\bar{T}(\bar{\ell},\lambda)$ will be too crude and the use of one-dimensional equations like (2.14), in the theory of optical sounding, will be of dubious validity. In this connection, the operator $K_m^{(0)}$ can be chosen in the first approximation as $\bar{T}_m^{(0)}$ which corresponds to a certain hypothetical ensemble of spherical particles whose optical properties do not differ strongly from those of the medium under investigation. It is appropriate to mention here that in this case the medium is not characterized only by the distribution function but it represented by the set Φ (the same Φ_m) of such functions and, as a consequence, the term "dispersed medium" implies more information content.

In order to determine the operator $K_m^{(0)}$ numerically within the framework of the Mie theory,it is necessary to know the refractive index of particulate matter and to make use of some general considerations on the analytical behavior of the size-distribution functions $\bar{P}(\bar{\ell})$ (numerical techniques to construct matrix operators of the Mie theory were given in [2.10]). Obviously, if the value $\inf\|\mathbb{K}_m^{(0)} - T_m\|$ determined by the morphology of particles is comparable with the value of the errors in the initial data $\{\mathbf{P}_i,\boldsymbol{\beta}_i\}$, then the operator $K_m^{(0)}$ can be used in calculations instead of the operator $\bar{T}_m^*$. The questions discussed here concern the development of optical modes of the dispersed media.

The determination of $\bar{T}(\bar{\ell},\lambda)$ factors was connected with the fact that particles in real scattering media could have shapes different from spherical.

It should be noted, however, that in the case of spherical particles it can also be of practical use to solve the inverse problems of light scattering by polydispersed systems like (2.33). In fact, the scattering characteristics of particles depend, to a great extent, on the optical constants of their substance in addition to their geometrical shapes.

Particles of real ensembles can have a complex chemical composition that makes an a priori assessment of the index of refraction difficult. Moreover, the refractive index can be dependent on wavelength within the interval of sounding Λ, as well as be characterized by radial inhomogeneity, and finally be dependent on the particles' radii. All these cases occur in the optics of atmospheric sols, making the inverse problems of light scattering more complicated. As a consequence,the construction of some simple optical models is required, allowing, as earlier, the use of integral equations like (2.6) and (2.14) for solving a problem of optical sounding of atmospheric aerosols. With this remark we complete the discussion of inverse problems expressed in such a general form as (2.33). In the following, when determining the kernels of integral equations like (2.14), we shall proceed from experiments which allow a more simple mathematical form for a given inverse problem of optical sounding.

Assume that in an experiment on light scattering we have a family of distribution functions $\{\bar{P}_j(\bar{\ell})\}$ which is produced by one and the same distribution law $\bar{P}(\bar{\ell})$ in such a way that $\bar{P}_j(\bar{\ell}) = \bar{P}(\bar{\ell})$ when $\bar{\ell} \in L_j = [\bar{\ell}_{min}, \bar{\ell}_j]$ and $\bar{\ell}_{min} < \bar{\ell}_1 < \bar{\ell}_2 \ldots \bar{\ell}_{max}$; $j = 1,2,\ldots$. In other words, every new size-distribution $\bar{P}_{j+1}$ does not contain the fraction of particles with the sizes from the interval $[\bar{\ell}_j, \bar{\ell}_{j+1}]$. Such an experiment is oversimplified in terms of realization in the laboratory. It is obvious that the family of size-distributions $\{\bar{P}_j(\bar{\ell})\}$ arranged in such a way forms the system of linearly independent functions and, therefore, the system (2.33) can be useful for determining $\{t_{k\ell}\}$. However, there is no particular need to do so, since the solution of this problem can be obtained with the use of simpler method. Let us consider, for simplicity, the value $\bar{\ell}_j$ to be continuously varying. Denoting $\bar{\ell}_j = \bar{y}$ one can write the following integral

$$\bar{\beta}(\lambda,\bar{y}) = \int_{\bar{\ell}_{min}}^{\bar{y}} \bar{T}(\bar{\ell},\lambda)d\bar{P}(\bar{\ell}) \quad . \tag{2.34}$$

Assuming $\bar{T}(\bar{\ell},\lambda)$ to be a differentiable function, in the region in which it is defined, one can obtain from (2.34)

$$d\bar{\beta}(\bar{y}/\lambda)/d\bar{P}(\bar{y}) = \bar{T}(\bar{y}/\lambda) \quad . \tag{2.35}$$

Equation (2.35) is the formal solution of (2.34) on the set of data $\{\Phi, \bar{B}\}$ with respect to the function $\bar{T}(\bar{\ell},\lambda)$, where Φ is the family of size-distribution functions, produced by the function $\bar{P}(\bar{\ell})$ and by the system of intervals $(\bar{\ell}_{min}, \bar{y})(\bar{y} \leq \bar{\ell}_{max})$; $\bar{B}$ is the corresponding set of optical characteristics $\bar{\beta}(\lambda/\bar{y})$. If the distribution function $\bar{P}(\bar{\ell})$ is differentiable and, hence, the density function can be introduced, then (2.35) can be written in the simple form

$$d\bar{\beta}(y/\lambda)/dy = \bar{p}(\bar{y})\bar{T}(\bar{y}/\lambda) \quad . \tag{2.36}$$

Equation (2.34) is not only of purely theoretical interest. For example, in [2.13] a description of the experimental set-up and the technique for determining the microstructure of artificial fog during its dissipation in the fog chamber was given. In such experiments the optical characteristics of the scattering medium are functions of the time t, and formally they can be expressed in the form of an integral

$$\beta(t,\lambda) = \int_{R_1(t)}^{R_2(t)} K(r,\lambda)s(r)dr \quad .$$

Since the fog particles are spherical, the factors $K(r,\lambda)$ can be determined from the Mie theory, and as a result, this expression can be regarded as the equation with respect to the density $s(r)$. Measuring $\beta(t)$ experimentally during the time interval $[t_1, t_2]$, the choice of which is conditioned by the sedimentation rate of the fog droplets, one can find the initial size-distribution function $s(r)$. Formally the solution of such an inverse problem of light scattering is given by an expression similar to (2.36).

The example considered above shows the variety of the applications of the integral equations method in optical sounding. The experiment considered was performed so that it appeared to be possible to represent the initial integral equation for $s(r)$ in a form which allowed a simple analytical solution. As a matter of fact, the equations of the (2.34) type are Volterra equations of the first kind with respect to $\bar{P}(\bar{\ell})$ [also $\bar{T}(\bar{\ell},\lambda)$], and are better conditioned relative to the function $s(r)$, than are the Fredholm equations (2.14).

2.3.3 Determination of the Factor $\bar{T}(\bar{\ell},\lambda)$ *from the Solution of the Volterra Integral Equation*

The following description discusses the determination of the factors $\bar{T}(\bar{\ell},\lambda)$ from sounding data on polydispersed media within spectral intervals.

As is shown in [2.14-16], if the inverse problems of light scattering by polydispersed ensembles of particles are formulated for spectral characteristics, then the corresponding integral equations have some interesting analytical peculiarities. Specifically, for ensembles of spherical particles,these peculiarities are caused by the fact that the kernels $K(r,\lambda)$ of the corresponding equations as functions of two variables are homogeneous to zero power, i.e., they do not depend on r and λ separately but on their ratio. Note that in the Mie theory this ratio is written in the form of the diffraction Mie parameter $x = 2\pi r\lambda^{-1}$. One can assume, based on the experiments on light scattering by particles of nonspherical shapes (see e.g., [2.17]), that this property of the efficiency factors $K(r,\lambda)$ holds also for the factors $\bar{T}(\bar{\ell},\lambda)$. Just as in the Mie theory, the diffraction parameter $\bar{x} = 2\pi\bar{\ell}\lambda^{-1}$ may be introduced and one may state that the properties of the function of two variables $\bar{T}(\bar{\ell},\lambda)$ are identical to those of the function of one variable $\bar{T}(\bar{x})$. Proceeding from that we shall prove below the following statement: The bulk of the initial information $\{\mathbf{P}_j\boldsymbol{\beta}_j\}$ from spectral optical measurements required for determining the efficiency factor $\bar{T}(\bar{\ell},\lambda)$ when solving the inverse problem (2.33) can be limited to a single realization. In other words, one spectral characteristic measured experimentally and one initial ensemble of particles [the same single size-distribution $\bar{P}(\bar{\ell})$] are sufficient for determining $\bar{T}(\bar{\ell},\lambda)$ by inverting the optical characteristic $\bar{\beta}$ of a polydispersed medium. There is no doubt about the importance of this statement for optical studies in the atmosphere.

Thus we shall assume that values of the factor $\bar{T}(\bar{\ell},\lambda)$ from (2.14) can be calculated using $\bar{T}(\bar{x})$ for all $\bar{\ell} \in [\bar{\ell}_{min}, \bar{\ell}_{max}]$ and $\lambda \in \Lambda = [\lambda_{min}, \lambda_{max}]$, here $\bar{x} \in [2\pi\bar{\ell}_{min}\lambda^{-1}_{max}, 2\pi\bar{\ell}_{max}\lambda^{-1}_{min}]$. We shall denote the interval where the variable $\bar{x}$ is defined as $\bar{X}$. Let $\lambda_i \in [\lambda_{min}, \lambda_{max}]$ and $i = 1,2,\dots,n$. Then one can write

$$\bar{\beta}_i = \int_L \bar{T}(2\pi\bar{\ell}\lambda_i^{-1})d\bar{P}(\bar{\ell}) \qquad i = 1,2,\dots,n \ . \tag{2.37}$$

Let us introduce the values $\xi = \bar{\ell}_{min}\bar{\ell}^{-1}_{max}$, $y_i = \lambda_1\lambda_i^{-1}$,

$$f_i = \lambda_1\bar{\beta}_i/\lambda_i \ , \quad \bar{\ell}y_i/\bar{\ell}_{max} = \bar{z} \ , \qquad (\xi y_i \le \bar{z} \le y_i) \ .$$

The validity of the following expression can easily be proved

$$f_i = \int_{\xi y_i}^{y_i} \bar{T}(\alpha\bar{z})d\Phi(\bar{z},y_i) \quad i = 1,2,\ldots,n \quad ,$$

where $\Phi(\bar{z},y_i) = \bar{\ell}_{max}\bar{P}(\bar{\ell}_{max},\bar{z}y_i^{-1})$ and $\alpha = 2\pi\bar{\ell}_{min}\lambda_1^{-1}$. Assuming that $\bar{\ell}_{min} \ll \bar{\ell}_{max}$ (i.e., $\xi \cong 0$) and that λ is continuous in Λ, one can write the following integral equation

$$\int_0^y \bar{T}(\bar{z})d\Phi(\bar{z},y) = f(y) \quad (0 \le y \le 1) \quad . \tag{2.38}$$

Thus we have shown that the factor $\bar{T}(\bar{x})$ can be found from the solution of the integral equation (2.38) provided the size-distribution is known. Equation (2.38) is a Volterra integral equation of the first kind.

The integral equation obtained is of major importance for the theory of inverse problems in the optics of dispersed media considered here. It shows first of all that the mathematical problem of determining the scattering efficiency factors of particles by the methods of optical sounding of polydispersed ensembles within the spectral intervals may be reduced to the solution of the Volterra integral equation. Secondly, the bulk of information required may be limited by one representation of the scattering ensemble, which describes the morphology of particles of the dispersed medium investigated.

Obviously, (2.38) as a functional equation with respect to $\bar{T}(\bar{z})$ is better conditioned than the corresponding equation (2.14) with respect to $\bar{P}(\bar{\ell})$, although both of them are produced by the same polydispersed integral $\bar{\beta}$. The latter is a Fredholm integral equation of the first kind. As to (2.38), it can be reduced under certain conditions to a Volterra integral equation of the second kind. In the process, one constructs, at the same time, a regularizing operator. Besides that, the solution of (2.38) by reducing it to a linear system is simpler since the matrix to be inverted is triangular. The influence of the initial errors on the accuracy of the inverse solutions can easily be taken into account when inverting such matrices. Since a vast literature is devoted to the methods of solving Volterra equations, we shall not discuss these questions further, and give only some particular cases which are interesting from the point of view of optical sounding of aerosols.

It is necessary to mention, above all, that since the microstructure of a polydispersed ensemble under investigation is determined experimentally (counters, impactors, and so on) then, naturally, the analytic form for representing the size-distribution function is a histogram (the same piecewise constant functions). Proceeding from that fact, one can construct reasonably simple and effective methods for solving (2.38).

For the sake of clarity, we shall assume that the microstructure of the ensemble is sufficiently regular and that it can be described by the density function $\bar{P}(\bar{\ell})$. In other words, we admit that $\bar{P}(\bar{\ell})$ is a continuous function and $d\bar{P}(\bar{\ell}) = \bar{p}(\bar{\ell})d\bar{\ell}$. In this case (2.38) becomes a Riemann integral. Let the distribution function $\bar{p}(\bar{\ell})$ be defined experimentally and represented by a piecewise constant function. The values $\bar{\ell}_i$ $(i = 1,2,\ldots,n)$ denote the discontinuities of $\bar{p}(\bar{\ell})$ at which it takes the values $\bar{p}_i$ (more correctly, $\bar{p}_i = \bar{p}(\bar{\ell}_i + 0)$). Obviously, the simplest case of such distributions is the constant continuous distribution $\bar{p}(\bar{\ell}) = \bar{p}_0 = \text{const.}$ over the whole interval $L = [\bar{\ell}_{min}, \bar{\ell}_{max}]$. This is the case from which we shall start.

In this model the constant $\bar{p}_0$ (the value of the distribution density) is determined by the integral $m^{-1}(L) \int_L \bar{p}(\bar{\ell})d\bar{\ell}$, where $m(L)$ denotes the interval length. Using then (2.14) one can write

$$\bar{\beta}(\lambda) = \bar{p}_0 \int_L \bar{T}(\bar{\ell},\lambda)d\bar{\ell} \quad . \tag{2.39}$$

Using the homogeneity of the function $\bar{T}(\bar{\ell},\lambda)$, one can easily show that

$$\bar{\beta}(\lambda) - \lambda\bar{\beta}'(\lambda) = \bar{p}_0[\bar{\ell}_{max}\bar{T}(2\pi\bar{\ell}_{max}\lambda^{-1}) - \bar{\ell}_{min}\bar{T}(2\pi\bar{\ell}_{min}\lambda^{-1})] \quad .$$

In the further discussion,we shall use this expression in the following form

$$\bar{T}(\bar{x}) - \xi\bar{T}(\xi\bar{x}) = \psi(\bar{x}) \quad , \tag{2.40}$$

where $\bar{x} = 2\pi\bar{\ell}_{max}\lambda^{-1}$, $\psi(\bar{x}) = (\bar{p}_0\bar{\ell}_{max})^{-1}(\bar{\beta} - \lambda\bar{\beta}')_{\lambda=2\pi\bar{\ell}_{max}\bar{x}^{-1}}$.

If λ successively takes all values from the interval $[\lambda_{min},\lambda_{max}]$ then (2.40) may be regarded as a functional equation with respect to the function $\bar{T}(\bar{x})$ on the interval $[2\pi\bar{\ell}_{min}\lambda_{max}^{-1}, 2\bar{\ell}_{max}\lambda_{min}^{-1}]$. The following recursive scheme can be used for solving (2.40) with respect to $\bar{T}(\bar{x})$:

$$\bar{T}_1(\bar{x}) = \psi(\bar{x})$$

$$\bar{T}_2(\bar{x}) = \bar{T}_1(\bar{x}) + \xi\bar{T}_1(\xi\bar{x}) \quad ,$$

$$\bar{T}_3(\bar{x}) = \bar{T}_2(\bar{x}) + \xi\bar{T}_2(\xi\bar{x}) \ \ldots \ \text{and so on} \quad .$$

This scheme is equivalent to the representation of $\bar{T}(\bar{x})$ in the form of a power series

$$\bar{T}(\bar{x}) = \sum_{k=0} \xi^k \psi(\xi^k \bar{x}) \quad , \tag{2.41}$$

which converges when $\xi < 1$ and $\psi(\bar{x}) \to 0$ for $\bar{x} \to 0$. Expression (2.41) is of particular interest in all cases for which the size-distribution function $\bar{p}(\bar{\ell})$ is close to uniform. It is not difficult to generalize the above recursive scheme for the case of histograms. Let the distribution function take the values $\bar{p}_1, \bar{p}_2 \ldots \bar{p}_m$ at the discontinuities. As in the above we have

$$\begin{aligned} &\bar{T}_1(\bar{x}) = \psi(\bar{x}) \\ &\bar{T}_2(\bar{x}) = \psi(\bar{x}) + \sum_{j=1}^{m} b_j \xi_j \bar{T}_1(\xi_j \bar{x}) \\ &\ldots\ldots\ldots\ldots\ldots\ldots\ldots \\ &\bar{T}_q(\bar{x}) = \psi(x) + \sum_{j=1}^{m} b_j \xi_j \bar{T}_{q-1}(\xi_j \bar{x}) \quad . \quad \text{Where} \\ &b_j = (\bar{p}_j - \bar{p}_{j+1}) \bar{p}_m^{-1} \quad , \quad \xi_j = \bar{\ell}_j (\bar{\ell}_{max})^{-1} \quad , \\ &\psi(\bar{x}) = (\bar{p}_m \bar{\ell}_{max})^{-1} (\bar{\beta} - \lambda \bar{\beta}')_{\lambda = 2\pi \bar{\ell}_{max} \bar{x}^{-1}} \quad . \end{aligned} \tag{2.42}$$

The recursive scheme (2.42) converges when $\max_j |b_j| \xi_{m-1} < 1$. The series analogous to (2.41) can also be written for (2.42), but we shall confine ourselves to (2.42) because this series is cumbersome, and also due to the fact that this scheme is of more practical use in calculations.

The method of approximate solution of (2.38) is based on the assumption that the density function is a piece-wise-constant function. Taking this into account, the method can be conditionally called the histogram method for solving inverse problems like (2.38). Later we shall give some numerical examples determining scattering factors of particles, which were calculated by using analytical relations of this nature. Particles of scattering ensembles in these examples will be considered spherical. In this case the "reference information" for such an inverse problem can easily be obtained numerically. We have to note, on the other hand, that the experimental determination of the efficiency factors K(x) of the Mie theory can be of practical use even if the index of refraction of particulate matter is unknown. Moreover , in some real situations, the index of refraction may depend on the radius of the particles, with an unknown form of this dependence. If this is so, then as before the inverse problem must be stated for the determination of the mean factor K(x) from the spectral measurements accompanied by the microstructure analysis which can be used for solving problems of optical sounding of such

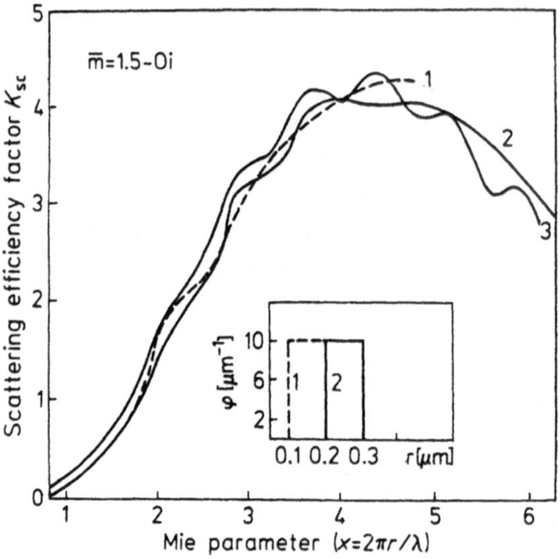

Fig.2.1. The scattering efficiency factors $K_{sc}(x)$ (curves 1,2) reconstructed from $\bar{K}(\lambda)$ according to (2.43), curve 3 is the reference efficiency factor

polydispersed aerosol ensembles. Some other examples from atmospheric optics can be presented here which show that experimental determination of scattering efficiency factors is preferable in some cases.

Let us consider the solution of the following equation as the first numerical example:

$$\int_{R_1}^{R_2} K_{sc}(r,\lambda)\varphi(r)dr = \bar{K}_{sc}(\lambda) \quad , \tag{2.43}$$

where $\varphi(r)$ is a normalized size-distribution function of spherical particles equal to $S^{-1}\pi r^2 n(r)$. Figure 2.1 illustrates attempts to reconstruct the monodispersed efficiency factors $K_{sc}(x)$ using polydispersed factors $\bar{K}_{sc}(\lambda)$. Two functions $\varphi(r)$ = const. = $10\mu m^{-1}$ defined on the size intervals [0, 0.2μm] and [0.2μm, 0.3μm], respectively, were taken as the reference models.

The reconstruction of the factor $K_{sc}(x)$ was made using the values of $\bar{K}_{sc}(\lambda)$ calculated for the spectral interval Λ = [0.27μm, 2.25μm]. The wavelength increment $\Delta\lambda$ used in calculations was 0.05μm for λ up to 0.6μm, then it was increased to 0.1μm and beginning from λ = 1μm to $\Delta\lambda$ = 0.3μm. The overall number of $\bar{K}_{sc}(\lambda)$ values calculated for this spectral region was equal to 17.

It should be noted that this class of inverse problems of the optics of dispersed media requires that a large number of $\bar{K}(\lambda)$ values be calculated for the spectral interval of optical sounding as compared with those problems where the size-distribution functions are to be found. This peculiarity is caused by the fact that the functions K(x) (e.g., in the Mie theory) have a more complicated analytical form than that of the functions s(r). Of course, much more information is required from experiments in order to reveal the fine structure of the factor K(x) with respect to the diffraction parameter x. Curve 1 in Fig.2.1 corresponds to the solution of the functional equation (2.40) in the region to the left of the point $x_2 = 2\pi R_2 \lambda_{min}^{-1} \cong 4.7$. For the second model (Curve 2) this equation is already defined for x up to 7. The exact values of the function $K_{sc}(x)$ are also given in Fig.2.1 (Curve 3). Only two terms of the series (2.41) were taken for the calculations made here, and this appeared to be sufficient because of its rapid convergence. A characteristic feature of the solutions obtained is the weaker oscillating components than those in the initial factor $K_{sc}(x)$ (Curve 3). To achieve better correspondence,one must have higher resolution within the sounding interval Λ especially in the IR region. In other words, all the peculiarities of the spectral behavior of $K_{sc}(\lambda)$ within the Λ interval should be identified in order to reveal the fine diffraction structure of $K_{sc}(x)$. This last remark is of particular importance when reconstructing highly oscillatory factor, such as the backscattering efficiency factor of spherical particles $K_\pi(x)$ with the real part of index of refraction from 1.5 to 1.6 and the imaginary one $\varkappa \lesssim 0.005$. Such an example is presented in Fig.2.2, where the broken line represents the solution while the initial factor $K_\pi(x)$ is shown by the dashed line. The model size distribution used was $\varphi(r) = \text{const.}$ within the interval [0.1μm, 0.3μm] ($\xi \simeq 0.33$). The solution in this example is represented by the first approximation $K^{(1)}$ [the second term of (2.41)]. It is interesting to note that the approximation $K_\pi^{(1)}$ can be regarded as a simplified model of the initial factor $K_\pi(x)$ when calculating the optical characteristics and as an inverse solution of problems of light scattering when sounding aerosols with the use of lasers. This is caused by the fact that in this example the absolute value of $[\bar{K}_\pi(\lambda) - \bar{K}_\pi^{(1)}(\lambda)]$ does not exceed 2 or 3 percent of the value of $\bar{K}_\pi(\lambda)$ over the entire spectral region [0.27μm, 2.25μm]. In other words, the optical characteristics

$$\beta_\pi(\lambda) = \int_R K_\pi(r,\lambda)s(r)dr \qquad \text{and}$$

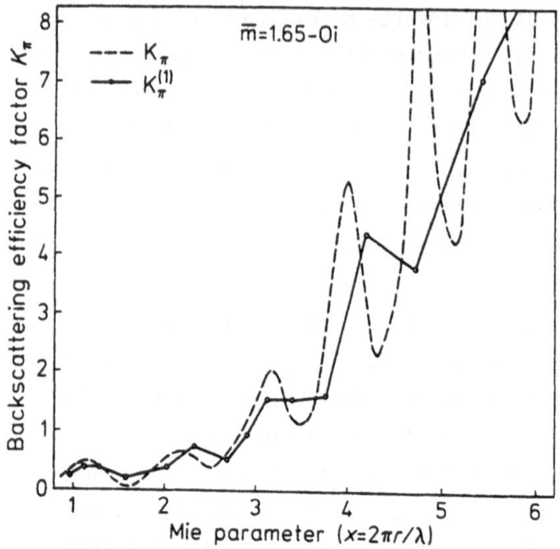

Fig.2.2. An example for restoring the backscattering efficiency factor using (2.43). The solid line is the first-order approximation $K_\pi^{(1)}(x)$, the dashed line represents the reference efficiency factor K_π

$$\beta_\pi^{(1)}(\lambda) = \int_R K_\pi^{(1)}(r,\lambda)s(r)dr$$

do not differ essentially from each other. There is no need to emphasize the usefulness of such simplified models of the kernel $K_\pi(x)$ for systems of operational processing of lidar returns and for determining aerosol microstructure. It is particularly important to stress the fact that the models discussed in this book allow the construction of such models using experimental material, i.e., taking into account properties of real scattering media.

The above examples show that it is not very difficult to obtain information about K(x) by inverting (2.43) provided that the initial size-distribution function $\varphi(r)$ is uniform. In this connection the question arises: how can the errors in the approximation of real distributions by uniform ones affect the accuracy of the reconstruction of K(x) factors? Proceeding from the functional relation (2.43) one can assume that the form of the weighting function $\varphi(r)$ should not strongly affect the spectral behavior of $\bar{K}(\lambda)$. Evidently, the length of the averaging interval $[R_1, R_2]$ makes the main effect in this case. The wider this interval, the stronger is the smoothing effect of the integral (2.43) on the monodispersed factor $K(r,\lambda)$ and the lower are the possibilities to obtain information on the peculiar features of $K(r,\lambda)$ when inverting the polydispersed factor $\bar{K}(\lambda)$. In particular, the role played by the interval

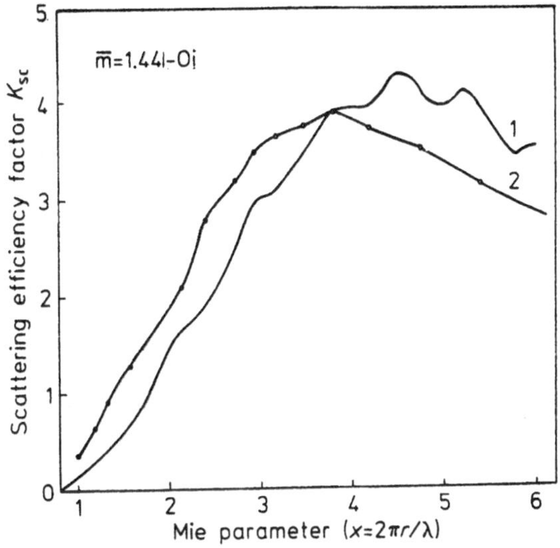

Fig.2.3. Numerical solution of (2.43) (Curve 2) in the case when the model size-distribution $\varphi(r)$ does not fit the reference curve (Curve 1 - reference factor)

length R in the calculation schemes (2.41) and (2.42) can be seen from the value of the parameter $\xi = R_1 \cdot R_2^{-1}$. The smaller the value of ξ, the fewer terms in (2.41) have to be used in calculations for a given accuracy of $\psi(x)$ and vice-versa. On the other hand, the fewer the number of terms in the series (2.41) taken in the solution, the simpler are its analytic properties. Let us discuss the data presented in Fig.2.3 when taking these remarks into account. This figure illustrates the reconstruction of $K_{sc}(x)$ when the uniform size-distribution is used as a model one. At the same time the initial scattering efficiency factor $K_{sc}(\lambda)$ of a polydispersed ensemble corresponds to a gamma distribution (haze H [2.18]). This unimodal size distribution is located mostly within the interval [0.1μm, 0.5μm]. (95% of the total geometrical cross-section $S = \int_R s(r)dr$). An exact value of the factor $K_{sc}(x)$ is shown by Curve 1 in Fig.2.3, and the solution of (2.40) is given by Curve 2. The spectral interval used in the calculations was the same as before. Additional calculations have shown that, in principle, one can make the reconstruction of the scattering efficiency factor of a monodispersed ensemble better if a smaller interval R is chosen. For example, the interval [0.1μm, 0.3μm] can be chosen. It is localized in the vicinity of the distribution mode r_s = 0.2μm. In general, the inverse problems dealing with the reconstruction of the efficiency factors of monodispersed media from optical characteristics of polydispersed ones are better conditioned for narrow distributions than for broad ones. Thus, the uniform size-distribution can be recommended as a model for estimating the efficiency factors K(x), in a first approxi-

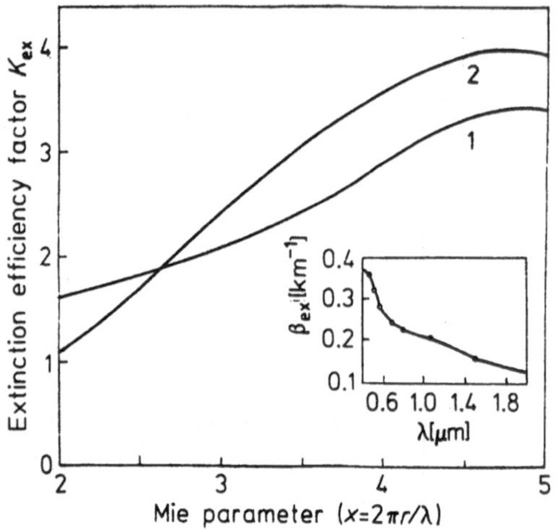

Fig.2.4. Results of the inversion of $\beta_{ex}(\lambda)$ with respect to $K_{ext}(x)$ (Curve 1); Curve 2 shows $K_{sc}(x)$ for m = 1.4 - 0i

mation, from spectral optical measurements made in polydispersed media as follows from the calculational analysis. A corresponding example is given in Fig.2.4. Measurements of the extinction coefficient $\beta_{ext}(\lambda)$ were made in the region [0.4μm; 2.24μm] along a horizontal path in the ground layer of the atmosphere [2.19]. Since the measurements of microstructure were not available, nothing certain can be said about the size-distribution s(r) and its boundaries. It was known that optical sounding of the atmosphere was carried out at a visual range about $S_M > 10$ km and the atmospheric optical situation could be classified as the haze. Let us assume that the effective size interval for particles of such haze is from 0.1μm to 0.3μm. Estimation of the total geometrical cross-section of the aerosol particles sounded will be made according to the relationship $\bar{K}_m \cdot S \geq \max \beta_{ext}(\lambda)$, where $\bar{K}_m$ is the value of the absolute maximum of the factor $\bar{K}_{ext}(\lambda)$ for a polydispersed medium. As a numerical estimation has shown, the value $\bar{K}_m$ is generally in the interval from 2.6 to 3.3, and the narrower the aerosol size-distribution is the larger the corresponding $\bar{K}_m$ value. Assuming $\bar{K}_m \simeq 3$, we find S to be equal approximately to 0.12 km^{-1} (as to the "H-haze" model recommended for such optical situations $S \simeq 0.096$ km^{-1}). Assuming further the size-distribution of the aerosol sounded to be uniform over the interval [0.1μm, 0.3μm] with the total measure S = 0.12 km^{-1}, one can estimate the efficiency factor $K_{sc}(x)$ of a monodispersed system, that corresponds to this experiment. Curve 1 in Fig.2.4 represents the solution of (2.40) for the spectral characteristic $\beta_{ext}(\lambda)$ measured (Fig.2.4). For comparison, the efficiency factor $K_{sc}(x)$ of spherical particles with the index of refraction

$\bar{m} = 1.4 - 0i$ is also given. If only small deviations of the particle shapes from spherical occur then $\bar{m} = 1.4 - 0i$ may be considered as an approximate estimation of the index of refraction of the aerosol substance. Terminating our discussion of such examples, we should like to note that qualitative methods for interpreting the inverse problems of the class considered here may use not only uniform size-distribution functions but analytically more complicated models as well. The uniform size-distribution is convenient when it is to be set a priori, because in this case the most general characteristic of microstructure is required, namely, the measure of distribution and its location on the size scale. The choice of other models requires more definiteness.

2.4 General Characterization of the Method of Integral Equations

2.4.1 Effect of Smoothness of the Kernels on the Efficiency of Optical Sounding

The methods of direct microstructure analysis based on direct sampling of particles from the air does not give any information on sols in their natural "undisturbed" state. Optical methods have certain advantages in this respect because in this case the information obtained experimentally characterizes the medium in its undisturbed state. However, it is not trivial to realize such possibilities. In the description to follow,a brief analysis of the possibilities of the above sounding systems as a probe of dispersed media is given.

The resolution of any given method, i.e., its capability to reveal "details" of the local structure of the size-distribution function is of great importance when investigating the aerosol microstructure. When planning optical experiments, one has, first of all, to know what defines this resolution and how the parameters of the optical facilities should be chosen. To solve these problems one should proceed from the analysis of functional relations connecting experimentally measured optical characteristics and the size-distributions sought. The problem stated will be discussed below only qualitatively based on the assumption that particles of the media under investigation differ insignificantly from spheres, and hence the integral equation $Ks = \beta$ with the kernels defined by the Mie theory is applicable for determining an ensemble microstructure. In order to clarify the peculiar features of this functional equation and how they affect the information capabilities of the optical methods, the study of the transformation K: $\Phi \rightarrow B$ properties should be undertaken for a given operator K and the initial set of distributions Φ.

In problems dealing with determining the microstructure of a scattering ensemble, the set Φ is comprised of positive bounded functions, with zero boundary values. In some cases the functions s(r) can be very "simple" analytically. In particular, some model size-distributions (such as Junge, log-normal and so on) which are widely used in atmospheric optics are special examples from this class of distributions. In the general case, the microstructures of real scattering media are described by functions of more general form such as the multimodes size distribution functions. Let the size distribution s(r) be defined by the sum of two functions $\bar{s}(r)$ and g(r), defined in their turn on the size interval $R = [R_1, R_2]$. The first one will be called regular while the second will be called an irregular component of the size-distribution function s(r). The concept of "regularity" is too relative in meaning and should be defined separately for each particular case. In our treatment, we deal mainly with positive functions s(r), and hence, in the first approach we shall require the regular component $\bar{s}(r)$ to be positive. The condition of its positiveness is not necessary for the irregular component g(r); this function has only to be limited in absolute value. Let us take some multi-modes function in R as the initial size-distribution function s(r) in order to illustrate the above. Let the regular component be a uniform size distribution and $\bar{s}(r) = \text{const} = (1/m(R) \int_R s(r)dr$, where m(R) is the length of the interval R. Then the irregular component can be written as $g(r) = s(r) - \bar{s}(r)$. Obviously, the function g(r) changes its sign in the interval R and, in the first approximation, has the form of an oscillation with respect to the value $\bar{s}(r) = \text{const}$. A more complicated behavior of the irregular component of a size distribution corresponds to a more complex microstructure of the medium investigated. Within the framework of this approach, one may consider the information on local peculiarities of the microstructures of the dispersed media to be contained, first of all, in the irregular component g(r) while the microstructure as a whole is characterized by the component $\bar{s}(r)$. Let us assign more definite meaning to the concept "irregularity" and introduce some quantitative estimates of it.

For simplicity, let the distribution function s(r) be differentiable. Then the components $\bar{s}(r)$ and g(r) can be compared with each other by the degree of their smoothness within R. It is common to characterize the smoothness of functions with the use of functionals from their derivatives which we denote as $F_R(s')$ [$F_R(s') \geq 0$ for all $s \in \Phi$]. The relation

$$F_R(\bar{s}') \ll F_R(g') \qquad (2.44)$$

should also be valid for the components $\bar{s}(r)$ and $g(r)$. This relation characterizes the functions $\bar{s}(r)$ and $g(r)$ in more detail. The norms of the derivatives in the spaces C_1 or L_2 are usually taken as the functionals $F_R(s')$.

Consider now the representation $K: \Phi \to B$, taking into account the fact that the size-distributions $s \in \Phi$ have the components $\bar{s}$ and g. It is obvious that in this case the characteristic β may be represented by two components β_s and β_g corresponding to the $\bar{s}$ and g components of a size distribution. Since the component $g(r)$ is oscillating, the absolute value of $\int_R K(r,\lambda)$ $g(r)dr$ can be very small in some cases, in particular when the size-distribution function $s(r)$ has a complicated analytical form. For example, the value of $\int f(t) \sin(\omega t)dt$ can be made arbitrarily small for any continuous function $f(t)$ by making a suitable choice of the parameter ω within the limited interval. A rigorous proof can be given that the integral vanishes when $\omega \to \infty$ [2.20].

Thus, β_g may appear to be very small while the component $g(r)$ itself plays an important role in the size distribution $s(r)$ sought. As a consequence, the component $\beta_g(\lambda)$ disappears in the experiments, being hidden under the measurement errors $\sigma(\lambda)$, and any oscillating components of the solution of the equation $Ks = \beta$ which are found have nothing to do with the real size distribution of the particles. In this connection the limitations of the smoothness of solutions of equation $Ks = \beta$ obtained with the use of experimental data are obviously very important. In the theory of operator equations of the same type, this statement is known as the regularization principle. We have just demonstrated this principle in connection with the integral equations method. Now the assessment of the information possibilities of any given method for sounding the dispersed media microstructure can be reduced to the estimation of maximum admissible smoothness for the solution of a corresponding integral equation. In the general case, it is difficult to make such estimates. In our further discussion we shall give estimates based on some special assumptions, the introduction of which allows less general but more certain conclusions to be drawn.

All the properties of the equation $Ks = \beta$ discussed above can be represented in a more formalized manner. As a matter of fact, the following statement has been proved above. Two size distributions can be selected from the class Φ, such that the relation $\|Ks_1 - Ks_2\|_B \leq \delta(\sigma)$ does not imply in any way that the value $\|s_1 - s_2\|$ is small. This means that the problem of solving equation $Ks = \beta$ in the class Φ is an "ill-posed" problem when there are errors in the measured value β [2.21]. For the inverse problem to be "correct", the set of possible solutions Φ should be restricted to be compact, which is practically done when solving the equation $Ks = \beta$ numerically with the use of regularization methods considered in detail in the next section. Let us consider again the investigation of the resolution of optical sounding methods for determining the microstructures of the dispersed media.

One can note, first of all, that if the optical characteristic $\beta(\lambda)$ has a highly smooth behavior in the sounding interval Λ, i.e., it can be characterized by small value of $F_\Lambda(\beta')$, then a good resolution of the microstructure can hardly be expected from the solution of the equation $Ks = \beta$. As a rule, in real situations it appears to be difficult to judge the smoothness of $\beta(\lambda)$ because of the presence of a random component $\sigma(\lambda)$ in measurements which is usually not smooth. As is easily seen from (2.7), the analytical behavior of $\beta(\lambda)$ should be defined to a great extent by the behavior of the kernel $K(r,\lambda)$ with respect to λ, and, hence, the smoothness of $\beta(\lambda)$ cannot be less than that of $K(r,\lambda)$ in any case. In this connection, one can state that the smoothness of the solution of the equation $Ks = \beta$ should not be less than that of its kernel [2.22]. This results in the fact that the information content of one or another optical method for investigating microstructure is defined, to a great extent, by the smoothness of the kernel of a corresponding integral equation.

If the kernel $K(r,\lambda)$ is very smooth and the family of functions $K_\lambda(r)$ consists of functions differing insignificantly from each other for λ varying in the interval Λ, then the corresponding optical experiment has low information content with respect to the microstructure of a dispersed medium. The smoother the kernel $K(r,\lambda)$, the stronger is the averaging effect of the integral operator K in the set Φ and, as a consequence, the lower is the extent to which the peculiarities of the behavior of $s(R)$ within R are revealed in those of the spectral behavior of $\beta(\lambda)$ in Λ. The methods for estimating the averaging effect of the operator K in the transformation $K: s \to \beta$ will be given below for optical sounding within spectral intervals Λ, while now we give the characteristic of analytical properties of the efficiency factors $K(r,\lambda)$ defined by the Mie theory.

As is known, the Mie theory is widely used as an apparatus for interpreting the aerosol optical characteristics and, hence, it is interesting to assess the possibilities of the method of integral equations when kernels

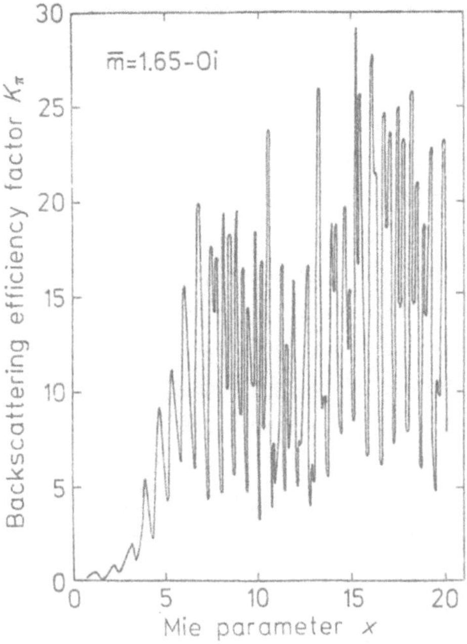

Fig.2.5. An example of the backscattering efficiency factor $K_\pi(x)$ illustrating the oscillatory behavior of the kernel

for the case of spherical particles are used. One observes that the factors $K(r,\lambda)$ are distinctly oscillating when the real part of the index of refraction of the aerosol substance $\bar{m} \geq 1.4$, and the imaginary one $\varkappa \leq 0.01$. Such an example is presented in Fig.2.5, where the backscattering efficiency factor $K_\pi(x)(x = 2\pi r\lambda^{-1})$ for $\bar{m} = 1.65$ and $\varkappa = 0$ is shown. It is obvious that the oscillations dominate in the behavior of this factor, and smoothness of such a kernel is very low. A particular result of this is the fact that multifrequency lidars are too accurate systems for the remote determination of the aerosol microstructure [2.10]. The above is valid for the case of polydispersed systems of spherical particles. However if the shapes of the particles deviate significantly from spherical, then the factors $\bar{T}(\bar{\ell},\lambda)$ in equations of the type (2.14) will be much smoother than those from the Mie theory. This is caused, firstly, by the necessary averaging over orientations, see (2.15), and, secondly, by the fact that light scattering by spheres evidently has a more complicated diffraction picture. Such a conclusion was drawn by PINNICK et al. [2.17] who carried out the experiments on light scattering by a monodispersed ensemble of nonspherical randomly oriented particles.

Particles of real media are naturally not always spherical, and therefore the information capabilities of optical sounding systems for investigations of microstructure in these cases are lower than in the case of systems of spherical particles. This conclusion is a result of the limited possibilities of the use of one-dimensional integral equations in the theory of optical sounding of dispersed systems of particles with arbitrary shapes. Thus, analyzing equations of the theory of optical sounding with the kernels defined by the Mie theory, one can assess possible characteristics of particular optical systems. Besides, it should be noted that in principle some smoother functions than $K(r,\lambda)$ can be used as the kernels for the inversion of optical measurements data based on the use of integral equation $Ks = \beta$. It is very important to take this aspect into account when preparing the computer systems for operational processing of the optical measurement data. The problem is that the Mie series converge very slowly, especially when $\bar{m} > 1.5$ and $\varkappa$ vanishes. For this reason, calculations of the kernel $K_\pi(r,\lambda)$ as well as the numerical solution of equation $K_\pi s = \beta_\pi$ requires much computer time. It is obvious that constructing simpler (smoother) analogs $\bar{K}(r,\lambda)$ of the kernels of equation $K_\pi s = \beta_\pi$ would allow more efficient processing of the information measured. In this connection, some remarks on the approaches to this problem should be made.

As is known, the oscillating component of the efficiency factor $K(r,\lambda)$ is caused by the diffraction maxima and is interpreted from a physical standpoint as a result of surface waves [2.6]. In the case of nonspherical particles, the diffraction picture is much simpler because of the averaging over their orientations with respect to the space of a scattering volume. Proceeding from this fact, a method for constructing the smoothed analogy $\bar{K}(\bar{\ell},\lambda)$ of the kernels $K(r,\lambda)$ has been suggested in [2.10] based on the use of (2.15) for the factors $\bar{T}(\bar{\ell},\lambda)$ and taking into account the particle geometry. In particular, for more or less oval particles (e.g., ellipsoids) the following formula may be used

$$\bar{K}(\bar{\ell},\lambda) = \int_{\Omega'} K(p\bar{\ell},\lambda) q \, d\omega/4\pi \ , \tag{2.45}$$

where p and q are some weighting functions defined in the space of the Euler angles, i.e., $\int_{\Omega'} p \, d\omega/4\pi = \int_{\Omega'} q \, d\omega/4\pi = 1$. The expressions for the weigthing functions p and q for ellipsoids were given in [2.10], i.e.,

$$p(\nu,\psi,\varphi) = L^{-1/2}(\nu,\psi,\varphi)\left[(8/\pi^3)\int_0^{\pi/2} d\varkappa \int_0^{\pi/2} d\psi \int_0^{\pi/2} L^{-1/2}(\nu,\psi,\varphi) d\varphi\right]^{-1}$$

$$q(\nu,\psi,\varphi) = L^{1/2}(\nu,\psi,\varphi)\left[(8/\pi^3)\int_0^{\pi/2} d\nu \int_0^{\pi/2} d\psi \int_0^{\pi/2} L^{1/2}(\nu,\psi,\varphi)d\varphi\right]^{-1} .$$

In the case of spherical particles $p = q = 1$ and hence $\bar{K}(\bar{\ell},\lambda) = K(\bar{\ell}_0,\lambda)$. It then follows that the averaging effect of (2.45) on the efficiency factors of the Mie theory is weaker for particles having shapes close to spherical.

It should also be noted that some alternative methods for simplifying the Mie theory can be found in papers dealing with the particular features of the angular distribution of light scattered by the ensemble of nonspherical randomly oriented particles. Thus, in [2.23] the method of simply neglecting the oscillating terms in the amplitude Mie functions $a_n(x)$ and $b_n(x)$ was suggested, based on the fact that such phenomena as the glory effect [2.24] do exist for these ensembles. Since the oscillating components of these functions are caused by the surface waves, giving rise to a glory phenomenon also, such an approach is from a physical point of view meaningful and probably justified. But in the case of the inverse problems of light scattering by polydispersed systems, this ignorance of the oscillating terms is not admissible at all. In fact, if the functions $a_n(x)$ and $b_n(x)$ are replaced by constants, then the corresponding scattering efficiency factors become independent of particle sizes and, as a consequence, integral equations of the (2.7) type lose any practical sense. In this respect, the method based on the local smoothing slowly converging series is more justified in spite of being purely formal [2.25]. As applied to this problem, it can be expressed as follows

$$\bar{K}(r,\lambda) = (1/\Delta(r)) \int_{-\Delta(r)/2}^{\Delta(r)/2} K(r + t,\lambda)dt \quad , \tag{2.46}$$

where $\Delta(r)$ is the length of the smoothing interval. This method was used in the theory of light scattering by polydispersed media for the development of qualitative methods and also for the interpretation of optical measurements.

Considering the construction of smoothed analogs $\bar{K}(r,\lambda)$ for kernels $K(r,\lambda)$ of the initial integral equations of the optical sounding theory, one can see that in this case the replacement of equation $Ks = \beta$ by $\bar{K}s = \beta$ is meant. This transition is not always followed by a loss of optical resolution, when investigating the microstructure of a medium. In fact, if the size-distribution $s_0(r)$ sought is a priori known to be smooth enough,then it is quite reasonable to restrict oneself to the solution of an integral equation with the smooth kernel $\bar{K}(r,\lambda)$. However, if the medium under investigations has a more complicated structure or if it is important to study the transformation of one fraction of particles into another, then, naturally, an attempt should be undertaken to realize fully the possibilities of a given optical method. Of course, in this case such simplifications are not acceptable.

Formally the replacement of the equation $Ks = \beta$ by the equation $\bar{K}s = \beta$ is equivalent to the approximation of one operator by another. In the case of any particular problem, e.g., when the class Φ is known and the errors of optical measurements Φ are set, such an approximation can be considered acceptable provided the condition

$$\inf_{s \in \Phi} (\|Ks - \bar{K}s)\| / \|s\|) \leq \delta(\sigma) \tag{2.47}$$

is fulfilled. Note that (2.47) can be valid even if the value $\|\bar{K}(r,\lambda) - K(r,\lambda)\|_{C[R \times \Lambda]}$ is not small. In other words, in all the above cases nothing was said about the direct approximation of one type of kernels by another. This emphasizes once more the creative character of the theory being developed here and, above all, of the concept of an integral operator for light scattering by polydispersed systems.

In summary, we should state that the lack of spatial symmetry of particle shapes in the polydispersed ensemble gives rise to a decrease in the optical resolution when studying the microstructure of the ensemble. This is true for the case when the particles are randomly oriented with respect to the illuminated volume. The effect of averaging over particle orientations, being absent in the case of ensembles of spherical particles, gives rise to a decrease of the information potentials of the method of integral equations in the case of nonspherical particles.

2.4.2 Averaging Effects in Optical Measurements

Unfortunately, the nonsphericity of the particle shapes is far from being the only reason requiring us to simplify the integral equations. Let us consider, for example, an optical experiment wherein the aerosol optical depth $\tau^{(a)}(\lambda)$ is measured using a spectrophotometer along a path of fixed length. Since the filters used have finite spectral width then the measured values $\tau^{(a)}(\lambda)$ are related to the true $\tilde{\tau}^{(a)}(\lambda)$ according to the following integral relation

$$\tilde{\tau}^{(a)}(\lambda) = (1/\Delta\lambda) \int_{-\Delta\lambda/2}^{\Delta\lambda/2} \tau^{(a)}(\lambda - \lambda')f(\lambda')d\lambda' \tag{2.48}$$

where $f(\lambda)$ is the spectral transmission function of a filter centered at λ. This means that the aerosol microstructure can be determined from the integral equation

$$\tilde{\tau}^{(a)}(\lambda) = c \int_R \bar{K}_{ext}(r,\lambda)s(r)dr \quad , \tag{2.49}$$

assuming sphericity of the aerosol particles. Here c is a constant, and $\bar{K}_{ext}(r,\lambda)$ is a smoothed extinction efficiency factor averaged according to (2.48). Since the kernel $\bar{K}_{ext}(r,\lambda)$ is smoother than the initial $K_{ext}(r,\lambda)$, the efficiency of its use in aerosol microstructure analysis is, of course, lower. In this respect,one can discuss the loss of information inherent in this optical sounding method caused by the finite spectral resolution of a measuring device. In this connection, it is very important to emphasize that, in general, the spectral behavior of the aerosol optical depth $\tau^{(a)}(\lambda)$ should be measured with a high spectral resolution [the same can be said about the directed scattering coefficient $D_{11}(\lambda)$].

As the experimental investigations show, the behavior of $\tau^{(a)}(\lambda)$ may be quite irregular and,at the same time,irregularities in optical characteristics can be interpreted in terms of a discrete microstructure of the aerosol. The stronger the irregularities in optical characteristics, the larger the information content of optical measurements data on the local details of the microstructure that can be obtained. In this connection, the problem on determining the value of $\max\Delta(\lambda)$ for a given optical method, taking into account the expected properties of the dispersed medium, is extremely important.

It should be stressed that the finiteness of the photometer spectral resolution is not the only possible source of information losses in the case of the microstructure investigations considered above. Thus, for example, when the chemical composition of the aerosol substance varies along the sounding path (the situation often occurs in nature), the index of refraction of the aerosol substance is a function of the distance between two points z_1 and z_2 and, as a consequence, the measured values $\tilde{\tau}^{(a)}(\lambda)$ should be given in the integral form $\int_{z_1}^{z_2} \beta_{ext}(z',\lambda)dz'$.

Let us try to formulate mathematically the problem of measuring $\tilde{\tau}^{(a)}(\lambda)$ taking into account inhomogeneities of the refractive index of the aerosol substance along the sounding path. For this purpose, we determine the mean value of the volume extinction coefficient as follows

$$\bar{\beta}_{ext}(\lambda) = \int_{z_1}^{z_2} \beta_{ext}(z',\lambda)dz'/(z_2 - z_1) = \tilde{\tau}^{(a)}(\lambda)/(Z_2 - z_1) \quad .$$

Then, assuming the spectral behavior $\tilde{\beta}_{ext}(\lambda)$ known in the interval Λ, one can find a corresponding size-distribution $\bar{s}(r)$ that characterizes the microstructure of aerosols, as the average over the path between the points z_1 and z_2, from equation $\bar{K}_{ext}\bar{s} = \bar{\beta}_{ext}$ whose kernel is

$$\bar{K}_{ext}(r,\lambda) = \int_{z_1}^{z_2} K_{ext}(\bar{m}(z),r,\lambda)dz/(z_2 - z_1) \quad . \tag{2.50}$$

The smoothing of the initial kernel resulting from (2.50) again leads to a loss of information about the microstructures of the media sounded. Such examples illustrate the difficulties of practical realization of the possibilities of a given optical sounding method. Therefore, it is obvious that in any case experiments based on the use of the integral equation method should be arranged in order to avoid accompanying averaging effects whenever possible. As is shown above, any averaging (smoothing) of measured characteristics of light scattering results in a smoothing of the kernels of the corresponding integral equations. It is also evident that for the investigations of microstructures of polydispersed media, those experiments are preferable for which the corresponding kernels are the least smooth, all other conditions being equal. When it is not possible to make such a choice, it is necessary to choose the parameters of a measuring system in a proper way, taking into account that the efficiency factors $K(r,\lambda)$ are determined by the geometrical arrangement of the experiment. When we talk about sounding within spectral intervals we must, above all, select the wavelengths of operation, or, equivalently, select the most informative interval for sounding the given dispersed medium [2.26]. If, for example, the interval R of particles sizes is known and sounding is possible in any subinterval Λ_j; then the most informative interval is that for which the functional $F(K_r')$ is maximum in value. In the theory of optical sounding, this problem is known as a problem of the choice of optically active sounding intervals [2.15,27].

It is not difficult to show, based on the studies of the functional $F(K_r')$, that,in the case of spherical particles,the most informative region of λ values is such that the values of r and λ are comparable. In the theory of diffraction of electromagnetic waves this is known as the Mie region. If the condition $\lambda \gg r$ is fulfilled, i.e., in the case of Rayleigh-Hans scattering, the kernel $K(r,\lambda) \to ax^4$, when $\lambda \to \infty$ and, as a consequence, polydispersed integrals of the type (2.6) take the form $\beta(\lambda) = aS(2\pi\lambda^{-1})^4\overline{r^4}$, where a is a constant depending on the index of refraction, and $\overline{r^4}$ is the fourth moment of the size distribution $\varphi = S^{-1}s(r)$. These relationships determine entirely the information on the microstructure of a polydispersed ensemble, which can be obtained by using the optical sounding methods at large values of λ. On the other hand, for $\lambda \ll r$, i.e., in the case of geometrical optics, $K(r,\lambda) \to K_\infty$ = const. and, hence, $\beta(\lambda) \to K_\infty S$ = const. As a consequence, sounding at small values of λ carries almost no information on the microstructure of the medium sounded.

It should be noted, however, that the above statements concern the investigations of microstructure parameters of the dispersed media with the use of optical sounding methods. The use of the above λ regions in determining the optical constants of particulate matter is, of course, helpful [2.28].

2.5 Inverse Operator Method in the Theory of Optical Sounding

2.5.1 Optical Operators and Equations of Radiation Transfer in Spectral Intervals

In the foregoing discussion of the method of integral equations, the optical characteristics of light scattering by the media sounded were assumed to be directly measurable in experiments. At the same time, in the experiments on light scattering, the light fluxes are, as a rule, measured (the same optical signals) and then the information on the optical characteristics of scattering media is inferred from these signals using corresponding equations. This is so, for example, with the lidar equation of the theory of laser sounding of the atmosphere.

The lidar equation relates the amplitude of a backscattered signal to the atmospheric backscattering β_π and the extinction coefficients β_{ext}. It is obvious that this equation is not defined with respect to the optical characteristics of light scattering and its use in practice is impossible unless additional information is available. The relationships set a priori between the above optical characteristics are sometimes used as such an additional information, which results, in these cases, in significant errors in data interpretation. The method of integral equations developed above allows the principal solution of this problem to be given within the framework of the theory of optical sounding, i.e., it enables one to point the way for avoiding this indefiniteness in the corresponding functional equations. The method developed in [2.21] was used as a basis for rigorous theory of multiwavelengths laser sounding of dispersed media. Its main idea can be demonstrated with the following analytic considerations.

Let the optical sounding of any volume of a scattering medium be carried out at wavelengths λ_i $(i = 1,2,\ldots,n)$ from the spectral interval Λ. Then the power of a signal backscattered by the volume of the atmosphere located at a distance z from the receiver is determined in the single scattering approximation as follows

$$P(z,\lambda) = P_0(\lambda)B(\lambda)z^{-2}\left[\beta_\pi^{(R)}(z,\lambda) + \beta_\pi^{(a)}(z,\lambda)\right]$$

$$\exp\left\{-2\int_{z_0}^{z}\left[\beta_{sc}^{(R)}(z') + \beta_{sc}^{(a)}(z')\right]dz'\right\} . \tag{2.51}$$

This equation involves four optical characteristics, such as Rayleigh $\beta_{sc}^{(R)}$ and aerosol $\beta_{sc}^{(a)}$ volume coefficients of total scattering and the corresponding backscattering volume coefficients $\beta_\pi^{(R)}$ and $\beta_\pi^{(a)}$. It is obvious that this functional equation is not defined with respect to optical characteristics of the atmosphere and, hence, that information about them cannot be inferred from the measurements of $P(z,\lambda)$ directly. For (2.51) to be defined, a proper supplementary definition must be given in the whole region of z and λ, i.e., in the two-dimensional region $\left[[z_0, z_m] \times [\lambda_{min}\lambda_{max}]\right]$. The supplementary definition may be performed by introducing supplementary functional relations between the optical characteristics sought. As to the molecular component of a scattering signal, the relation between $\beta_{sc}^{(R)}(z,\lambda)$ and $\beta_\pi^{(R)}(z,\lambda)$ can readily be determined according to the Rayleigh theory as follows

$$\beta_\pi^{(R)}(z,\lambda) = k\beta_{sc}^{(R)}(z,\lambda) \tag{2.52}$$

where k is a constant independent of z and $\lambda(k \cong 3/2)$. Now let us consider the aerosol component of scattered signal and see whether the analogous relationship between $\beta_{sc}^{(a)}(z,\lambda)$ and $\beta_\pi^{(a)}(z,\lambda)$ can be constructed or not. Obviously, if one had to construct such a relationship for a monodispersed ensemble of particles, it could readily be written down within the framework of the Mie theory. In fact, as follows from the Mie theory, the efficiency factors $K_\pi(r,\lambda)$ and $K_{sc}(r,\lambda)$ for monodispersed scattering medium are analytically expressed in terms of the same functions $a_n(x)$ and $b_n(x)$, viz.

$$(1/x^2)\left|\sum_{n=1}^{\infty} (-1)^n(2n+1)[a_n(x) - b_n(x)]\right|^2 \quad \text{and}$$

$$(2/x^2)\sum_{n=1}^{\infty}(2n+1)[|a_n(x)|^2+|b_n(x)|^2] \quad ,$$

respectively, and, hence, both of them are equally determined. Of course, it is hardly possible to find the relation between the functions $K_\pi(r,\lambda)$ and $K_{sc}(r,\lambda)$ in an explicit analytical form like (2.51) due to analytical difficulties but this makes no difference. The main point here is the fact that one and the same reference data are required for seeking numerical values of $K_\pi(r,\lambda)$ and $K_{sc}(r,\lambda)$ for all values of r and λ, i.e., the knowledge of the refractive index of the particle substance is necessary.

Now let us consider polydispersed ensembles of particles. Using the operators for light scattering by polydispersed systems introduced above, one may represent the optical characteristics $\beta_{sc}^{(a)}(\lambda)$ and $\beta_\pi^{(a)}(\lambda)$ in the form $(K_{sc}s)(\lambda)$ and $(K_\pi s)(\lambda)$, respectively. Let the reference set of distributions Φ, where the operators K_{sc} and K_π are defined, allow the existence of inverse operators K_{sc}^{-1} and K_π^{-1}. The inverse operators enable one to represent the size-distribution function as follows: $s = (K^{-1}\beta)(r)$, $r \in R$. In this case,the following expression is valid $(K_{sc}s)(\lambda) = (K_{sc}K_\pi^{-1}\beta_\pi)(\lambda)$, which obviously leads to the desired functional relation between $\beta_\pi^{(a)}(\lambda)$ and $\beta_{sc}^{(a)}(\lambda)$

$$\beta_{sc}^{(a)}(\lambda) = (W\beta_\pi^{(a)})(\lambda) \quad , \quad \lambda \in \Lambda \tag{2.53}$$

where W is the product $K_{sc}K_\pi^{-1}$. The operator W, constructed in such a way, sets up a mutually unique relation between the spectral behaviors of two optical characteristics of a given polydispersed ensemble of particles. Equations (2.52,53) make an additional definition of the initial functional equation of multiwavelengths sounding (2.51). Introduction of these relationships is necessary for the correct formulation of the inverse problem of multiwavelengths sounding.

For the operator W (matrix analog) to be constructed numerically, assumptions on the particles' sphericity and the knowledge of the refractive index of the aerosol substance are necessary. The further definition of (2.51) can be achieved by setting some supplementary relations between optical characteristics of aerosol and molecular components of light scattering, provided any reasonable assumptions can be made or, otherwise, by means of some particular assumptions. In the following discussion we shall frequently be concerned with such questions, while we now let the contribution of molecular light scattering be negligible as compared with that of the aerosol component. In this case,the equation resulting from (2.51) is completely defined with respect to aerosol optical characteristics. Using

(2.53) and an implicit form of functional equations, one can easily obtain the following equation:

$$F(P,\beta_\pi^{(a)}, (W\beta_\pi^{(a)})_\lambda, z,\lambda) = 0 , \quad \lambda \in \Lambda , \; z \in Z \tag{2.54}$$

which is completely defined with respect to $\beta_\pi^{(a)}(z,\lambda)$. Using the function $\beta_{sc}^{(a)}(\lambda)$ found from (2.54), one can obtain the spectral behavior of the scattering coefficient $\beta_{sc}^{(a)}(\lambda)$ according to (2.53). The method described above allowed aerosol optical characteristics β_π and β_{sc} to be determined using the data of multiwavelengths sounding. Since the method is based on constructing the inverse operator K_π^{-1} then it may be called, on the whole, the inverse operator method. This method is a further development of the integral operator method. The inverse operator method is widely used for solving various problems of atmospheric optics. An example can be given by considering the inverse problem of radiation transfer within the spectral interval Λ. In the simplest case of a plane parallel aerosol atmosphere, the radiation transfer equation can be written as follows [2.29]

$$F(I,\beta_{ext},\beta_{sc}, D_{11},\xi,\eta,\varphi,\lambda) = 0 , \quad \lambda \in \Lambda , \tag{2.55}$$

where D_{11} is a vector whose components are, in fact, the scattering coefficients $D_{11}(\theta_j)$ for scattering angles $\theta_j (j = 1,2,\dots,m)$. Since for a polydispersed medium $D_{11} = Qs$, where Q is the corresponding matrix operator, one can construct the relationships $\beta_{ext}(\lambda) = (K_{ext}Q^{-1}D_{11})(\lambda)$, $\beta_{sc} = (K_{sc}Q^{-1}D_{11})(\lambda)$ using this or another method for inversion. These expressions are identical to the scalar form of (2.53). Using these relationships and (2.55), one can construct a well-defined system of equations for determining the vector D_{11} (the same polydispersed scattering phase function $\mu(\theta)$) based on measurements of the angular distribution $I(\xi,\eta,\varphi)$ of the fluxes scattered by the medium under investigation. Leaving aside the questions of applications of the inverse operator method to the problem of radiation transfer through the aerosol atmosphere, note one more application of this method which is of great importance for optics of dispersed media.

We should like to emphasize that the introduction of inverse operators for solving functional equations, depending on several optical characteristics of light scattering, gives rise to the construction of new integral operators of the theory of light scattering by polydispersed ensembles, e.g., the operator W in our first example. A new class of transformations corresponds to this operator, viz. $W: B_\pi \rightarrow B_{ext}$, where B_π and B_{ext} are the sets of possible values of $\beta_\pi(\lambda)$ and $\beta_{ext}(\lambda)$, respectively. The operator W, if known, provides a calculational method for seeking one characteristic based on the measurements of another. Thus, in the second instance, the operators W (i.e., $K_{ext}Q^{-1}$ and $K_{sc}Q^{-1}$) allow the spectral characteristics of light scattering to be determined from the data of nephelometric measurements of aerosol scattering phase functions.

The study of operators like W and the construction of methods for their numerical determination is, in fact, one of many problems of the theory of optical sounding of dispersed media. Note that operators like W can be determined from the data of corresponding optical experiments. If, e.g., the pair $\{B_1, B_2\}$ is measured, then one can easily find the operator W_{12} as an inverse solution of a problem like (2.33). In practice, this circumstance is of particular concern for optical research since the measurements of optical characteristics of dispersed media are not difficult per se as opposed, e.g., to the microstructure analysis of these media.

The above operators can also be determined, if the algorithm for calculating the efficiency factors of a polydispersed ensemble of particles is known, as in the case of spherical ones.

2.5.2 Method of Smoothing Functional for Inverse Problems of Light Scattering

It is important to observe that the choice of method for numerically constructing the operator K^{-1} depends significantly on properties of the integral operator K in the set Φ of possible solutions of the equation $Ks = \beta$. In particular, as we have shown above, one can select two distributions s_1 and s_2 from the set $\Phi \in C$, such that $\|Ks_1 - Ks_2\| \leq \varepsilon$, while the norm $\|s_1 - s_2\|$ may be arbitrarily large. This well-known property of the operator K means that K^{-1} is not bounded on C and the problem of solving equation $Ks = \beta$ on $\Phi \in C$ is an "ill-posed" problem. For its solution, the regularizing algorithms are used. The principles for constructing them were described in [2.30,31].

One of the most analytically verified principles is the variational principle which, in particular, is used in the method of smoothing functional (Tikhonov's method). For the equation $Ks = \beta$, this functional is written as $T_\alpha(s) = \rho^2(Ks,\beta) + \alpha\Omega^2(s)$, where $\rho(Ks,\beta)$ is the discrepancy and $\Omega^2(s) = p_0\|s\|^2_{L_2} + p_1\|s'\|^2_{L_2}$ is the stabilizing functional ($\alpha > 0$). The value $\Omega(s)$ is the measure of the function s smoothness. One can easily show that all functions s which obey the condition $\Omega(s) \leq \Omega_0$ = const., form a compact set of distributions.

Note that the restriction imposed on the norm of the functional derivative of $\Omega(s)$ is very important for inverting the characteristics of light scattering by aerosols. Only in this case can the uniform convergence of the regularized solutions $s_\alpha(r)$ to the actual one $s_0(r)$ be ensured for diminishing errors of measurements. The limitation imposed only on the norm $\|s\|_{L_2}$ provides convergence only on the average, which may be sufficient for calculating certain functionals of the solution (or certain microstructure parameters). However the uniform convergence of approximations is of particular concern in the inverse problems where the shape of the distribution sought, as well as its local details (the so-called fine structure of the size spectrum),are to be determined. For this reason we shall not consider, in the further discussions, the inversion schemes which are based only on the limitation of the solution norm, in spite of the fact that such schemes are relatively simple and widely used. A detailed analysis of the dependence of the convergence of regularized solutions on the initial restrictions and errors of measurements can be found elsewhere (see, e.g., [2.31]).

According to the variational principle, the function s , which minimizes the functional $T_\alpha(s)$ on Φ is taken as a regularized solution of the initial equation $Ks = \beta$. Let s_0 be the exact solution of equation $Ks = \beta_0$, when the right-hand side β_0 is fixed. Denote, as before, the optical measurements' errors by σ, assuming that $\|\beta_0 - \beta\| \leq \sigma$. Since the function s_α is taken as a regularized solution then, naturally, the question arises: what is the smoothness of s_α as compared with that of the exact size-distribution s_0 for the given σ and α? To answer this question we write down the inequality

$$\|Ks_\alpha - \beta\|^2 + \alpha\Omega^2(s_\alpha) \leq \|Ks_0 - \beta\|^2 + \alpha\Omega^2(s_0) \quad , \tag{2.56}$$

which follows from the fact that s_α minimizes the functional $T_\alpha(s)$ on Φ. From (2.56) follows the inequality

$$\|Ks_\alpha - \beta\|^2 + \alpha\Omega^2(s_\alpha) \leq \sigma^2 + \alpha\Omega(s_0) \quad ,$$

which, in its turn, leads to estimations

$$\Omega^2(s_\alpha) \leq \alpha^{-1}\sigma^2 + \Omega^2(s_0) \quad , \tag{2.57}$$

$$\|Ks_\alpha - \beta\|^2 \leq \sigma^2 + \alpha\Omega^2(s_0) \quad . \tag{2.58}$$

Inequality (2.57) answers the above question and, to a great extent, predetermines the choice of a regularizing parameter α when using the smoothing functional method. If α is too small as compared with σ^2 then the term

$\alpha^{-1}\sigma^2$ can significantly exceed $\Omega^2(s_0)$ and, as a consequence, the smoothness of the regularized solution $\Omega^2(s_\alpha)$ will be determined mainly by noise and not by the smoothness of a real size distribution $\Omega^2(s_0)$. A reasonable upper limit of the value of α is determined by the obvious condition that the discrepancy $\rho^2(Ks_\alpha,\beta) = \|Ks_\alpha - \beta\|^2$ should not exceed σ^2 significantly. It is difficult to say anything definite about the properties of the regularized solution. Nothing remains but to impose some reasonable restriction on the smoothness of the "exact" solution s_0 assuming that $\Omega(s_0) \leq \Omega_0$. The constant Ω_0 should be considered known a priori. If the smoothness of a real distribution is limited by the value of Ω_0, then it is natural to demand that the smoothness of s_α does not exceed the value Ω_0. If is sufficient to assume for this purpose that

$$\alpha = \sigma^2\Omega_0^{-2} \quad . \tag{2.59}$$

Condition (2.59) leads, on the other hand, to inequalities which could be considered quite acceptable, viz.

$$\begin{aligned} &\|Ks_\alpha - \beta\| \leq \sqrt{2}\sigma \\ &\|Ks_\alpha - \beta_0\| \leq (1 + \sqrt{2})\sigma \quad , \end{aligned} \tag{2.60}$$

as follows from (2.58) and (2.59). Equation (2.58) means that α decreases no faster than σ^2.

The above analysis demonstrates quite clearly the idea of regularizing methods based on the variational principle as well as the difficulties with their use in practice. The choice of a "proper" value of the regularizing parameter α is difficult and should be accompanied in every case by an analysis of properties of the regularized solutions, taking into account whatever a priori information on the real distribution s_0 may be available. It is obvious that in any concrete case the optimal value of the regularizing parameter α exists as determined from the condition $\min\|s_\alpha - s_0\|$, but this value remains unknown as does the exact solution s_0 itself. In this connection,the development of methods for the approximate evaluation of the regularizing parameter becomes of particular concern. These questions will be considered below (Chap.3) in less general but more informative form, while we now discuss some peculiarities of calculational schemes for regularizing algorithms.

Recall above all that the functional $T_\alpha(s)$ is defined formally on a class of functions which is more restricted than the set of size-distribution functions Φ. Indeed, for $T_\alpha(s)$ to be bounded in value, the corresponding size-distribution function $s \in \Phi$ should possess derivatives such that an integral of their second powers exists. But as has been shown above, the microstructures of dispersed media may be represented in general by more complicated functions, including discontinuous ones. Therefore, the use of the functional $T_\alpha(s)$ and an assumption on the existence of the integral $\int_R [s']^2 dr$ (in the Riemann sense) limits naturally the possibilities of the method of integral equations. Unfortunately, this restriction appears to be insufficient for the construction of simple schemes for the inversion of optical measurements data. And only if one assumes the first derivative s' to be differentiable in addition to the differentiability of the size-distribution function itself, can one manage to solve numerically the variational problem for the functional $T_\alpha(s)$ simply. As a matter of fact, the function s_α minimizing the functional $T_\alpha(s)$ in the class C_2 can then be found from the Euler equation

$$(K^*K + \alpha D_2)s = K^*\beta \quad , \qquad (2.61)$$

where K^* is the operator conjugate to K, and D_2 is the Sturm-Liouville operator. The calculational scheme corresponding to this equation is the following

$$s_\alpha = (K^*K + \alpha D_2)^{-1}K^*\beta = K_\alpha^{-1}\beta \quad , \qquad (2.62)$$

where K_α^{-1} is the regularized operator inverse to that in equation $Ks = \beta$. It is obvious that (2.62) is determined in the class of, restricted in value, doubly differentiable functions. On the other hand, the corresponding subset C_2^+ with the elements being positive functions should be singled out from the class C_2 since the size-distribution s(r) sought must be positive within the limits of a given size interval R.

It is interesting to note that if the value of α is too large (as compared, e.g., with $\sigma^2\Omega_0^2$), then corresponding solutions s_α [in the scheme (2.62)] are, as a rule, positive over the whole region R. But the decrease of α leads, finally, to strengthening of the oscillating components of the solution and to breakdown of the condition of positiveness. In this connection, the choice of the smallest value of α can be related when inverting $\beta(\lambda)$ with the requirement for the regularized solutions from the family ϕ_α to be positive for successive values $\alpha_1 > \alpha_2, \ldots, \alpha_{min}$. Recall that the maximum value of α is limited by the condition $\rho^2(Ks,\beta) \le \delta(\sigma^2)$, where $\delta(\sigma^2)$ is of the same order of smallness as σ^2 is. In the simplest case, one may assume $\delta(\sigma^2) = c\sigma^2$, where $c \ge 1$. Note, however, that the solution of equation $Ks = \beta$ in the class of too smooth size-distribution functions such as the class C_2^+ are not always admissible when studying the microstructures of polydispersed media.

2.5.3 Determination of Discontinuous Size Spectra of Particles (Method of Histograms)

As one can show by numerical analysis, the scheme (2.62) is quite good for the inversion of optical characteristics of light scattering by aerosols if $\beta(\lambda)$ is measured at several points (usually 7 to 12 points are sufficient) with an accuracy not worse than 5%, and if the initial size-distribution function possesses not more than 2 or 3 local maxima in the size interval R, which should not be very narrow (e.g., 1μm or 2μm). It should be noted in this connection that if one has more accurate measurements and/or the behavior of the measured characteristics $\beta(\lambda)$ is too complicated, more sophisticated methods for inversion should be used. The method of histograms can be regarded as one such method [2.3,10].

The main idea of this method is to construct the set of piecewise continuous distributions $\tilde{\Phi}$ whose elements $\tilde{s}(r)$ have a finite number of breaks of the first kind over the entire size interval R. If one permits the density functions s(r) and especially n(r) to be not differentiable at every point of the interval R, one makes the information potentialities of the method richer. As the analysis of experimental material shows, the histograms for aerosol ensembles $[\Delta(N)/\Delta(r)]$ are rarely regular with respect to r, and therefore, they can hardly be approximated by smooth functions. In particular, the reader is referred to [2.32], where examples of such size spectra are found for a stratospheric aerosol.

The questions being discussed here are of particular concern for the inverse problems of laser sounding of aerosols. Pulsed lidar systems are capable of providing high spatial resolution, i.e., small illuminated volumes of the atmosphere. Obviously, when the concentration of particles in the volume sounded is low,it is hardly possible to characterize the microstructure of the ensemble by a density function and, as a consequence, the inverse problems should be stated in a more generalized form. If the methods of integral equations are being used, then the Stieltjes integral should be used [2.3]. In our further discussion we shall consider relatively simple variants of the inverse problems, when the microstructure may be described with density functions which are not as smooth as those usually occurring in direct problems of aerosol light scattering [2.18].

The density functions s(r) used in the following discussion will be chosen with such analytical properties that $\tilde{s}''(r)$ represents, up to a constant factor, the histogram $[\Delta N(r)/\Delta(r)]$ characterizing the distribution of particles number density as a function of their sizes. This is achieved by the use of a piecewise-quadratic approximation for real size distributions. The smoothness of $\tilde{s}(r)$ in this method is characterized by the functional $\Omega^2(\tilde{s}) = p_0\ (R)\int_R \tilde{s}^2\ dr + p_1\ (L)\int_R (\tilde{s}')^2 dr$, where the first integral is meant to be of Reimann kind, while the second is of Lebesgue type.

The use of the Lebesgue integral allows the dimensionality of the space L_2 to be unchanged when developing algorithms for solving problems. This, in turn, leads to the most simple calculational schemes. As a matter of fact, the solutions obtained by such an approach bear information on the microstructure of the dispersed medium sounded at almost every point of the interval of possible sizes R, i.e., excluding some finite number of points. It should be emphasized here that this result is in good agreement with the characteristic features of problems in the theory of optical sounding of polydispersed systems. Indeed, as is shown above, the most adequate representation of β is in terms of a Stieltjes integral which enables one to take into account the discontinuities in microstructures of real dispersed media as well as the morphology of the particles.

The density function s(r) as a characteristics of microstructure was introduced into the corresponding integral equations quite recently. It is sufficient to recall that the Stieltjes integral $\int_R K(r,\lambda)dS(r)$ may, in principle, be represented as a sum of the integral $\int_R K(r,\lambda)s(r)dr$ and some function $\beta^*(\lambda)$, which is defined by jumps in S(r) at the points of its breaks and can be regarded as an irregular component of the optical characteristics of light scattering [2.3,10].

The function s(r) may be considered as the derivative of S(r) over the whole interval R, excluding the above mentioned discontinuities. In this respect the use of equations $KS = \beta$ instead of $Ks = \beta$ is, in principle, equivalent to excluding the points at which the behavior of S(r) is irregular and to the use of smoother functions for the description of microstructures as well. Naturally, this leads to a decrease in the information content of optical data inversion. It should be noted, however, that some uncertainties are inherent also in the methods for direct microstructure analysis. In fact, the points of breaks in histograms may surely by interpreted as a result of absence of information on particles of corresponding sizes. All the above considerations were used as a basis for the method of histogram, whose basic relations are given below.

Let m inner points $\{r_\ell\}$ $(\ell = 1,2,\ldots,m)$ of the interval $R = [R_1, R_2]$ be chosen and m nonintersecting intervals $\Delta_\ell = (r''_\ell - r'_\ell)$ be constructed, where $r'_\ell = (r_{\ell-1} + r_\ell)/2$, $r''_\ell = (r_\ell + r_{\ell+1})/2$, $r'_1 = (R_1 + r_1)/2$, $r''_m = (r_m + R_2)/2$.

Let the system of numbers $\{s_\ell\}$ ($\ell = 1,2,\ldots,m$) be related to the points $\{r_\ell\}$ and let this system be considered as components of the solution vector $\mathbf{s}$ of the inverse problem $K\mathbf{s} = \beta$ for the finite dimensional case. The piece-wise-quadratic distribution $\tilde{s}(r)$ coincides in each of the intervals $(r'_\ell\ r''_\ell)$ with the parabola $\tilde{s}_\ell(r) = s_{\ell-1}\tilde{a}_\ell(r) + s_\ell\tilde{b}_\ell(r) + s_{\ell+1}\tilde{c}_\ell(r)$ where $\tilde{a}_\ell(r)$, $\tilde{b}_\ell(r)$, $\tilde{c}_\ell(r)$, $c_\ell(r)$ are interpolating Lagrange polynomials of the second power for the points $r_{\ell-1}$, r_ℓ and $r_{\ell+1}$, respectively.

It can easily be seen that the size-distribution $\tilde{s}(r)$ constructed in such a way possesses the following analytic properties [2.3]: 1) $\tilde{s}(r)$ is continuous on the interval R with the exception of the points r'_ℓ and r''_ℓ ($\ell = 1,2,\ldots,m$) where it has breaks of the first kind; 2) $\tilde{s}(r)$ is differentiable on R excluding the same points and $\tilde{s}'(r'_\ell - 0) = \tilde{s}'(r'_\ell + 0)$; $\tilde{s}'(r''_\ell - 0) = \tilde{s}'(r''_\ell + 0)$; 3) the second derivative $\tilde{s}''(r)$ is almost everywhere a piecewise continuous function and is equal, up to a constant factor, to the distribution of the particle number density with respect to sizes within the interval R.

As an example of the last property, the case of spherical particles is mentioned for which the distribution functions s(r) and n(r) are related by a simple analytical expression, viz. $s(r) = \pi r^2 n(r)$. It can readily be shown that

$$\Delta_\ell(N)/\Delta_\ell(r) = (1/2\pi h^2)(s_{\ell-1} - 2s_\ell + s_{\ell+1}) \quad , \tag{2.63}$$

where $h = r_{\ell+1} - r_\ell$ for equidistant points $\{r_\ell\}$ ($\ell = 1,2,\ldots,m$). By analogy, one has for the distribution of s within the interval R

$$\Delta_\ell(S)/\Delta_\ell(r) = (1/24)(s_{\ell-1} + 22s_\ell + s_{\ell+1}) \quad . \tag{2.64}$$

The above piecewise functions are the final results in solving the inverse problem of light scattering, and therefore the method considered can be called the method of histogram. The size-distributions $\tilde{s}(r)$ used in the above discussion play a purely subsidiary role and serve as a necessary mean to proceed to the matrix analog $K_m\mathbf{s} = \boldsymbol{\beta}$ ($\mathbf{s} \in \Phi_m$) of the integral equation $K\tilde{s} = \beta(\tilde{s} \in \Phi)$. The variational problem for the quadratic form $T_{\alpha m}(s)$ on ψ_m corresponds uniquely to that for the functional $T_\alpha(\tilde{s})$ on the set of distributions $\tilde{\Phi}$. The corresponding vector s_α minimizing this form is determined according to the following scheme

$$s_\alpha = (K_m^* K_m + \alpha H)^{-1} K_m^* \boldsymbol{\beta} \quad , \tag{2.65}$$

where K_m is the matrix analog of corresponding operator K, while H is the smoothing matrix in the method of histograms defined as

$$H = p_0 h T_0 + p_1 h^{-1} T_1 \quad . \tag{2.66}$$

Here p_0 and p_1 are the scaling factors equal to $(R_2 - R_1)$ and $(\lambda_{max} - \lambda_{min}) \times (R_2 - R_1)^2$, respectively. The matrices T_0 and T_1 have the forms

$$T_0 = \begin{bmatrix} \frac{103}{4} & \frac{37}{3} & \frac{21}{8} & 0 \ldots & 0 \\ \frac{37}{3} & \frac{103}{4} & \frac{37}{3} & \frac{21}{8} \ldots & 0 \\ \cdots & \cdots & \cdots & \cdots & \cdots \\ 0 \ldots \frac{21}{8} & \frac{37}{3} & \frac{103}{4} & & \frac{37}{3} \\ 0 \ldots 0 & \frac{21}{8} & \frac{37}{3} & & \frac{103}{4} \end{bmatrix}$$

$$T_1 = \begin{bmatrix} 6 & -2 & -1 & 0 \ldots & 0 \\ -2 & 6 & -2 & -1 \ldots & 0 \\ \cdots & \cdots & \cdots & \cdots & \cdots \\ 0 \ldots -1 & -2 & 6 & & -2 \\ 0 \ldots 0 & -1 & -2 & & 6 \end{bmatrix}$$

Elements of the matrix K_m are determined from the following relationships

$$K_{\ell i} = A_{\ell+1,i} + B_{\ell i} + C_{\ell-1,i}$$

where

$$A_{\ell i} = \int_{r_1'}^{r_2''} K(r,\lambda_i)\tilde{a}_\ell(r)dr$$

$$B_{\ell i} = \int_{r_1'}^{r_1''} K(r,\lambda_i)\tilde{b}_\ell(r)dr$$

$$C_{\ell i} = \int_{r_1'}^{r_1''} K(r,\lambda_i)\tilde{c}_\ell(r)dr$$

$$A_{m+1,i} = B_{mi} = C_{oi} = 0$$

and

$$\tilde{a}_\ell(r) = [r_\ell r_{\ell+1}(r_{\ell+1} - r_\ell) - (r_{\ell+1}^2 - r_\ell^2)r + (r_{\ell+1} - r_\ell)r^2]/\Delta_\ell$$

$$\tilde{b}_\ell(r) = [-r_{\ell-1}r_{\ell+1}(r_{\ell+1} - r_{\ell-1}) + (r_{\ell+1}^2 - r_{\ell-1}^2)r - (r_{\ell+1} - r_{\ell-1})r^2]/\Delta_\ell$$

$$\tilde{c}_\ell(r) = [r_\ell r_{\ell-1}(r_\ell - r_{\ell-1}) - (r_\ell^2 - r_{\ell-1}^2)r + (r_\ell - r_{\ell-1})r^2]/\Delta_\ell$$

$$\Delta_\ell = r_\ell r_{\ell+1}(r_{\ell+1} - r_\ell) - r_{\ell-1}r_{\ell+1}(r_{\ell+1} - r_{\ell-1}) + r_{\ell-1}r_\ell(r_\ell - r_{\ell-1}) \quad .$$

The choice of h is determined by the dimension m of the solution vector $\mathbf{s}$, and must be made in accordance with the value of σ. The dimensionality required depends, in turn, upon the behavior of the density function $s_0(r)$ on the size interval R and should be larger for more complicated size spectra of the media sounded. One should keep in mind, however, that an increase of m requires a corresponding decrease of the errors, and hence, lowering the acceptable upper limit of the regularizing parameter α. One interesting aspect of the matrix operator $(K_m^*K_m + \alpha H)$ used in the method of histograms concerned with the choice of required values of α and m should be noted. It can readily be seen that the determinant of this matrix differs from zero even if the dimensionality of the $\mathbf{s}$ vector exceeds the number of optical measurements, i.e., when $m > n$. The equation $(K_m^*K_m + \alpha H)\mathbf{s} = K_m^*\boldsymbol{\beta}$ differs essentially in this respect from the equation $K_m^*K_m\mathbf{s} = K_m^*\boldsymbol{\beta}$, which is connected with direct minimization of the optical discrepancy $\|K_m\mathbf{s} - \boldsymbol{\beta}\|_{L_2}^2$. Of course, the increase of m to exceed n does not increase the information content of the vector $\mathbf{s}$ in a given inverse problem. This only indicates the fact that the number m becomes merely a formal parameter in the method of smoothing functional. The fact that the quadratic form $\Omega_m(\mathbf{s})$ is minimized on ψ_m in this functional in addition to minimizing the discrepancy $\rho(K_m\mathbf{s},\boldsymbol{\beta})$ is equivalent to imposing some additional relations on the components of the vector $\mathbf{s}$. One can easily find that the vector $\mathbf{s}$ minimizing $\Omega_m(\mathbf{s})$ obeys the equation $H\mathbf{s} = 0$, which imposes additional linear relationships between the components s_ℓ of the solution vector. This of course makes the choice of m exceeding n essentially useless, especially when measurement errors are too large.

To complete a short overview of the regularization methods for the equation $Ks = \beta$ based on a variational principle, let us give two more numerical schemes, which have been proved to be effective in processing the data of optical sounding of atmospheric aerosols. Thus, in particular, the minimization of Tikhonov's functional $T_\alpha(s)$ can be performed directly in Φ, without an Euler equation (2.61). Using quadratures for polydispersed integrals [2.10] one can transform the equation $Ks = \beta(s \in \Phi)$ into the linear system $K_m \mathbf{s} = \boldsymbol{\beta}(\mathbf{s} \in \Psi_m, \boldsymbol{\beta} \in B_n)$. It is not difficult to show that, in this case, the quadratic form $T_\alpha(s)$ defined on Φ can be made to correspond to the functional $T_{m\alpha}(\mathbf{s})$ defined on Ψ_m. In this case, s can be determined numerically by minimizing $T_{m\alpha}(\mathbf{s})$ on ψ_m with the use of standard algorithms for searching for the minimum of a function of m variables in R provided certain limiting conditions exist. Such algorithms are efficient for use in the inverse problems of light scattering and the results of their applications are presented in [2.33,34]. It appears that under otherwise equal conditions the use of these algorithms enables one to obtain a solution s_α by minimizing $T_\alpha(s)$ on Φ which is less smoothed than that obtained following the scheme (2.62), since formally the differentiability of the derivative $s'(r)$ is not required because it is sufficient for its second power to be summable. However, it should be noted that regularizing algorithms based on the minimization in R_m are a little more cumbersome for practical use as compared with schemes like (2.62). The next scheme of a regularizing algorithm useful for processing the optical information can be written as follows

$$s_\alpha = (GK^*K + \alpha I)^{-1} GK^*\beta \quad , \tag{2.67}$$

where G is the integral operator corresponding to the Green's function for the operator D_2 in (2.62). As opposed to (2.62), the scheme (2.67) requires no numerical representation of the D_2 operator which is inherently unstable on the set of approximate data. In addition, the use of (2.67) is justified when the kernels of the equation $Ks = \beta$ are known approximately. The operator G possesses a certain smoothing effect which is useful in some cases as, e.g., when atmospheric haze is sounded since a weak dependence of the aerosol substance index of refraction upon λ may occur in this case. Since as a rule the function $m(\lambda)$ is unknown, a situation may occur in which it is necessary to smooth the wavelength behaviors of optical measurement data and of the kernel of the initial equation. This is performed by introducing the operator G into the regularizing algorithm. In addition, the contruction of the operator G requires the boundary conditions to be imposed on the size-distribution function s(r) sought; when these are imposed, they are equivalent, in the inversion scheme, to an a priori supplement defining the problem to be solved. The latter circumstance provides an additional stabilizing effect on the solution, especially near the points R_1 and R_2, where $s_0(r)$ is, as a rule, small in value and, as a consequence, it is too weak to be revealed by a given optical method. The above discussion characterizes the scheme (2.67) as an efficient one for practical use in optical sounding of aerosols, especially for poor information support and high noise level. Corresponding formulas for calculations were given in [2.35].

2.5.4 An Example of the Inversion of the Spectral Optical Depth of an Aerosol Atmosphere

In order to illustrate the above schemes of regularizing algorithms as well as the optical methods for determining microstructures of atmospheric aerosols, let us consider, an example, of an experimental study based on spectral measurements of the optical depth of atmospheric aerosols in the region from 0.3 to 2.5µm [2.36] and on the microstructure analysis of the aerosol aloft [2.37]. Let us discuss briefly this complex experiment and its results.

Measurements in the spectral region 0.3µm - 0.5µm were carried out using a grating spectrometer with a PMT photoreceiver. It took about 30 s to make measurements over this region. The spectral resolution of these measurements was 0.8 nm.

Spectral measurements in the visual range 0.4µm to 0.85µm were made with a spectral photometer having a resolution of 1nm to 4nm.

In the IR (0.6µm - 2.5µm) a prism spectrometer with a non-cooled PbS photodetector was used [2.38]. The spectral resolution in this region was about 4nm; this permitted one to select intervals within the transmission microwindows free of molecular absorption. The optical depth of the atmosphere was evaluated according to the Bouger law. The mean error ε_τ was found to be less than 20%. It is important to note that measurements were made during periods when the optical properties of the atmospheres were most stable.

The aerosol part of the investigation comprised measurements of the number density of the aerosol particles and of their size-distribution functions at different altitudes and at several different times. The sampling of aerosol particles was made using filters. The analysis of the results obtained showed that the normalized size-distribution function of aerosol particles was very stable. To a first approximation, it could be described by a monotonically decreasing (with size increase) function that can be approximated satisfactorily by a power law. Some instability of the number density of particles with the sizes 0.25µm to 0.35µm was observed.

The diurnal stability of the size-distribution function allowed one to assume that a time-lag between the measurements of optical depth and aerosol sampling did not significantly affect the results of comparison of the inverted data with those of the direct microstructure analysis. In this respect, the complex experiment presented here is of particular interest for estimating the efficiency of optical methods for aerosol studies.

As to the absolute measurements, diurnal variations of the aerosol number density were observed with maxima and minima revealed on evenings and mornings, respectively. A small relative increase in the concentration of large particles during evenings was also observed. As the chemical analysis showed, the main fraction of aerosol ensembles was formed by quartz particles. For example, only about $100 \mu g m^{-3}$ of $250 \mu g m^{-3}$ of the total mass density of aerosol particles were formed by non-quartz particles. Under these conditions the refractive index of the aerosol substance could be assumed to be close to 1.65.

The vertical structure of the aerosols was studied using a photoelectric particle counter enabling one to measure the number density of particles with $r > 0.2 \mu m$, and the size-distribution function (10 fractions). An impactor for studying the particles with $r \geq 0.1 \mu m$ and filters for making chemical and microstructures analysis of particles with $r \leq 0.02 \mu m$ were also used. Measurements were made at several horizontal levels up to a height of 8 km.

The results obtained showed that the normalized size-distribution function $\varphi_N(r)$ was quite stable with increasing altitude, and that the number density decreased exponentially with the height. In this altitude region, the aerosol particles had mostly oval shapes. This circumstance allowed one to expect the errors of the data on the inversion of spectral behaviors $\tau^{(a)}(\lambda)$ due to particles non-sphericity to be acceptably small. The determination of microstructure from such an optical experiment is generally reduced to solving the integral equation

$$c \int_R K_{ext}(r,\lambda)\varphi(r)dr = \tau^{(a)}(\lambda) \quad , \tag{2.68}$$

where $\varphi(r)$ is the normalized size-distribution function sought and c is an unknown constant.

The function $\varphi(r)$ bears information only on the shape of the distribution sought. An example of its behavior obtained from (2.68) according to (2.67) is presented in Fig.2.6 (solid curve). Let this distribution be denoted as φ_α. The dashed curve in this figure represents a solution of (2.68) in the class of "too smooth" functions, such as, e.g., the family of single-mode distribution Φ_M, described by a gamma distribution function. Note that the

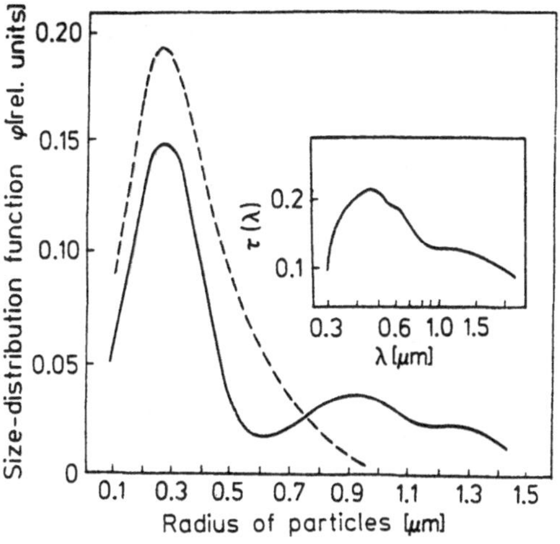

Fig.2.6. The inversion of $\tau^{(a)}(\lambda)$ according to (2.67) (——) and by the method of model assessments, scheme (2.80) (---)

discrepancies $\rho(K\varphi_\alpha, \tau^{(a)})$ and $\rho(K\varphi_M,\tau^{(a)})$ have approximately the same values as those corresponding to $\sigma \simeq 0.1\ \tau^{(a)}_{max}$ and, hence, both functions φ_α and φ_M are formally solutions of (2.68) for a given accuracy of optical measurements.

The above example shows the important role played by the choice of a reference set of solutions (or, equivalently, numerical method) for a given inverse problem. It should be noted, however, that the spectral interval of soundings Λ was wide enough and the number of $\tau^{(a)}(\lambda)$ measurements in it (about twenty) was sufficient for more generalized inversion schemes to be used instead of the fitting single-mode model.

The size-distribution function φ_α distinctly reveals the existence of two fractions in the ensemble of particles with mean radii $r_{s1} \simeq 0.25\mu m$ and $r_{s2} \simeq 0.9\mu m$, respectively. When proceeding to the density function $n(r)$, it happens that the appropriate mean radii r_{n_1} and r_{n_2} will move to the left, becoming 0.1?μm and 0.6μm, respectively; this agrees well with existing assumptions on the atmospheric aerosol microstructure.

It was interesting to obtain data on the function $n(r)$ characterizing the distribution of particles number with respect to their sizes when taking the distribution φ_α as a reference for further estimations of these or other parameters of aerosol microstructure. In Tables 2.1 and 2.2 we give estimated data on histogram $N(r_i < r < r_{i+1})/N(r > r_1)$ where the intervals $(r_i, r_{i+1})(i = 1,2,\dots,)$ correspond to the aerosol fractions discovered by a direct microstructure analysis using sampling methods [2.37].

Table 2.1. Relative values of number density of aerosol particles measured directly and obtained by inverting the spectral behavior of the optical depth [2.36]

Measuring technique	Radius r[μm] 0.25	0.25-0.30	0.3-0.6	0.6-1.20	>1.20
Direct measurements					
Impactor [2.37])	0.639	0.178	0.132	0.026	0.030
Inversion of $\tau^{(a)}(\lambda)$ [Scheme (2.67)]	0.704	0.104	0.164	0.023	0.006

Table 2.2. Relative values of the number density of aerosol particles obtained from direct measurements and by inverting the spectral behavior of the optical depth [2.36]

Measuring technique	Radius r[μm] 0.20-0.25	0.25-0.30	0.30-0.40	0.40-0.80	0.80-2.0	2.0-3.0
Direct measurements Filter [2.37]	0.479	0.282	0.138	0.053	0.037	0.011
Inversion of $\tau^{(a)}(\lambda)$ [Scheme (2.67)]	0.382	0.254	0.235	0.098	0.031	-

A comparison of corresponding values in the tables shows good agreement between them, especially if one takes into account that the experiment was carried out under the conditions of a real atmosphere with the inherent temporal instability and spatial inhomogeneity. One should also take into account the fact that methods for direct microstructure analysis based on the use of impactors and similar devices cannot be reckoned as etalon ones. Leaving the discussion of the data presented in the above tables aside, note that the information about aerosol microstructure can be obtained more easily and at lower cost by inverting $\tau^{(a)}(\lambda)$ rather than by direct methods for microstructure analysis. The example considered above also demonstrates the feasibility of the optical sounding method for aerosol microstructure studies.

2.5.5 Assessment of the Admissible Number of Independent Measurements for Soundings in Spectral Intervals

Summarizing the discussion of the numerical construction of K_α^{-1} for problems of optical sounding in polydispersed media, one should pay attention to a very important peculiarity of the inversion schemes based on the variational principle. It is easily seen from (2.61) that the term $\alpha D_2 s$ is related to the function $\Omega(s)$ and determines mainly the stability of the family of regularized solutions with respect to the noise in the initial data. The term K^*K of this equation is related to the initial operator K. The smaller the value of α, the weaker is the influence of the information introduced a priori into (2.61) and the solution s_α is conditioned to a greater degree by the measured function $\beta(\lambda)$ and by the initial operator K. At the same time, it is obvious that for small α, (2.61) is by no means equivalent to the initial equation $Ks = \beta$.

In fact, when $\alpha \to 0$ (2.61) takes on the form $K^*Ks = K^*\beta$, which differs from the initial equation $Ks = \beta$. This is also an integral equation of the first kind, but it has a quite different kernel and integral operator. The operator K^*K is symmetric and positively defined on Φ, while the same cannot be said about the initial operator K. The kernel of the equation $\int_\Lambda K(r,\lambda)K^*(\eta,\lambda)d\lambda$ is defined on the region $[R \times R]$ and possesses a higher order of smoothness as compared to that of the initial kernel $K(r,\lambda)$. Therefore, in the sense discussed above (Sect.2.4), the information potentialities of the equation $K^*Ks = K^*\beta$, are naturally lower than those of the initial equation $Ks = \beta$.

Thus, the regularization of solutions of the equation $Ks = \beta$ is necessarily connected, within the framework of the variational approach, with nontrivial transformations which lead, as a consequence, to lowering the information potentialities of the method of integral equations. It is not difficult to see that the equation obtained is analytically a simpler functional relation between the functions s and λ. In particular, the methods of spectral analysis can be applied to the operator K^*K while the functions s and β may, in principle, be expanded as a series $\sum c_k(s)v_k(r)$ and $\sum c_k(\beta)u_k(\lambda)$, where v_k and u_k are the eigenfunctions of the operators K^*K and KK^*, respectively. The main use of these is the fact that

$$c_k(\beta) = \mu_k^2 c_k(s) \quad , \tag{2.69}$$

where μ_k^2 are the eigenvalues of the operator K^*K. Such analytical relationships cannot be written for the initial equation $Ks = \beta$. All this makes the idea of using operators of the type K^*K in studies of properties of the initial operators K quite useful, but, of course, only to the degree that the former reflect the properties of the latter. In particular, the smoothness of the kernel corresponding to the operator K^*K determines the rate at which the spectrum of eigenvalues $\{\mu_k^{(2)}\}$ $(k = 1,2,\ldots)$ of the operator K^*K diminishes. Since the analytical properties of this kernel depend, in turn, upon the smoothness of the function $K(r,\lambda)$ on the interval $[R \times \Lambda]$, then the spectrum $\{\mu_k^{(2)}\}$ $(k = 1,2,\ldots)$ characterizes the measure to which the initial equation $Ks = \beta$ is conditioned. Let us consider briefly, in this connection, the spectral analysis of the operators K^*K and KK^*, especially taking into account the fact that this gives one the possibility to estimate the admissible number of independent optical measurements over the spectral interval Λ. Let us write the coefficients of the Fourier expansion of measured characteristic $\beta(\lambda)$ as follows

$$c_k(\beta) = \int_\Lambda \beta(\lambda)u_k(\lambda)d\lambda/m(\Lambda) \quad ,$$

where $m(\Lambda) = \lambda_{max} - \lambda_{min}$ and $u_k(\lambda)$ are the eigenfunctions of the integral operator KK^* whose kernel is $\int_R K(r,\lambda), K(r,\xi)dr/m(R)$. The following estimation is valid for $c_k(\beta)$:

$$|c_k(\beta)| \leq m^{1/2}(R) \cdot \|s\| \cdot \mu_k \quad .$$

The components $c_k(\beta)u_k(\lambda)$ of the expansion $\sum c_k(\beta)u_k(\lambda)$ are independent, and decrease when $k \to \infty$ due to the fact that the norm $\|s\|$ is bounded and μ_k vanishes. If an error in the measurement of $\beta(\lambda)$ does not exceed σ, on the average over the interval Λ, then the above series should be truncated at the qth term of the expansion. The number q can be found from the condition

$$m^{1/2}(R)\|s\|\mu_q = \eta\sigma \quad , \tag{2.70}$$

where $\eta \geq 1$ (the reliance coefficient). Equation (2.70) may be of use provided that the norm $\|s\|$ and $\mu_k (k = 1,2,...)$ are known. In inverse problems of light scattering by polydispersed aerosols, it is more convenient to use the value $S = \int_R s(r)dr$ instead of the norm $\|s\|$. Using the obvious inequality relation $\|s\| > S/m^{1/2}(R)$, one can rewrite (2.70) as follows

$$\mu_q = \eta\sigma S^{-1} \quad . \tag{2.71}$$

Finally, introducing the value of minimum relative error as $\varepsilon_{min} = \sigma[\max_\lambda \beta(\lambda)]^{-1}$, one has

$$\mu_q = \eta\varepsilon_{min}\bar{K}_{max} \quad , \tag{2.72}$$

where K_{max} is the maximum value of the polydispersed efficiency factor $\bar{K}(\lambda) = S^{-1}\beta(\lambda)$. The value $\bar{K}_{max}$ satisfies the relation $\bar{K}_{max} \leq K_{max}$ for any system of particles, where K_{max} is the maximum value of the monodispersed efficiency factor $K(r,\lambda)$ over the variables r and λ. For any given error ε_{min}, the formulas given above enable one to estimate the admissible number of independent measurements of an optical characteristic $\beta(\lambda)$ to be made within the interval Λ. As to the properties of the subsequence $\{\mu_k\}$, we shall only give Table 2.3 as an illustration.

Table 2.3. Eigenvalues of the matrix analogs for the operators $K^*_{sc}K^*_{sc}$ and $K^*_{\pi}K_{\pi}$ and different dimension of the measurement vector β

$K^*_{sc}\,K_{sc}$		$K^*_{\pi}\,K_{\pi}$	
n = 7	n = 11	n = 7	n = 11
7.13	8.97	13.40	17.00
1.24	1.52	4.20	4.47
0.64	0.76	2.90	3.65
0.33	0.37	2.68	3.39
0.16	0.16	2.20	2.86
0.15	0.15	2.10	2.71
0.15	0.15	1.55	2.44
	0.14		2.41
	0.14		1.71
	0.14		1.30
	0.1		1.16

The reference data used in the calculations of the matrix analogs of the operators presented in Table 2.3 were: index of refraction 1.56-0i; spectral interval Λ = [0.4μm, 0.7μm]; R = [0.05μm, 1μm]. The wavelengths $\lambda_i (i = 1,2,..,n)$ were set in an increment $\Delta\lambda$ = 0.05μm in the first case, and 0.03μm in the second case. Using (2.72) and Table 2.3,one can easily show that for $\varepsilon \sim 10\%$ in the above spectral interval, not more than 4 or 5 independent measurements of the aerosol scattering coefficient ($\bar{K}_{max} \leq 3$) can be made. In the case of the backscattering coefficient the number q does not exceed 6 or 7. The estimation of the maximum value of the polydispersed efficiency factor was performed, in the last case, by proceeding from the maximum value of a smoothed monodispersed efficiency factor (Sect.3.3).

In conclusion, note that the series $\sum c_k(s)v_k(r)$ may, in principle, be used for the construction of the size-distributions s(r), if one takes into account (2.54). Unfortunately, but for understandable reasons, these series converge poorly and become quite irregular when noise occurs in measurements. It is for this reason that their use in practice is possible only for the creation of corresponding methods of stable summation.

2.6 Determination of Microphysical Characteristics of Dispersed Media by the Method of Model Assessments from the Data of Optical Sounding

2.6.1 Parametric Families of Size-Distributions and Optical Models

The interpretation of optical data in the studies of microstructures of polydispersed media by the method of optical sounding can often be carried out using a simpler mathematical formalism than that above. For example, in certain cases one can refrain from searching for the general analytical form of the density function s(r) and restrict oneself to the estimation of one or more of its parameters. It is obvious that such an approach aims only at investigations of the medium microstructure as a whole, when there is no basis for the determination of local peculiarities of the size spectrum. The formulation of such problems, in practical atmospheric studies, is conditioned by the restriction of the information potentialities of the sounding systems used, when neither the accuracy of the measurements nor their bulk permit one to invert the data by solving integral equations. In addition, the form of the size-distribution function can sometimes be estimated a priori. Let us consider briefly the mathematical formulation of the problems of optical sounding in this approach.

The parametric characterization of the microstructures of dispersed media, using, e.g., an array of parameters $a_1, a_2, a_3, \ldots, a_m$ is, as a rule, connected with an a priori assignment of some analytical form to a model size-distribution function $s_M(r, a_1, a_2, \ldots, a_m)$. The array of parameters $\{a_i\}$ $(i = 1,2,\ldots,m)$ may be regarded as a point in a parametric space P, the elements of which are $p = \{a_1, a_2, \ldots, a_m\}$. If p takes values from a limited region $\Omega \in P$, then any given size-distribution law $s_M(r, a_1, a_2, \ldots, a_m)$ generates a parametric family of functions Φ_M. The size-distribution functions from this family will be denoted below simply as s_M. If, for the initial operator K, the solution of an inverse problem of light scattering is being sought in the class Φ_M then mathematically this problem is equivalent to the determination of the parameters $a_1, a_2, \ldots, a_m$ from the following system of equations

$$\beta_m(\lambda_i, a_1, a_2, \ldots, a_m) = \beta_i \quad , \quad i = 1,2,\ldots,n \geq m \quad , \tag{2.73}$$

where $\beta_m(\lambda_i,p) = \int_R K(r,\lambda_i)s_M(r,p)dr$. Usually, $\beta_m(\lambda_i)$ are called the model of optical characteristics. Since the initial inverse problem $Ks = \beta$ is "ill-posed", the choice of the region Ω should satisfy the condition that Φ_M is a compact set for any $p \in \Omega$. If this condition is fulfilled, then the minimization of $\rho^2(\beta(p),\beta)$ in the Ω region can serve as a basis for constructing the algorithm for seeking the solution s_M. That means, equivalently, that system (2.73) is solved by the variational method. Besides the above condition,a necessary requirement is for the point p^*, which gives the minimum to $\rho^2(p)$, to be the only such point in Ω. Otherwise, the solution of the inverse problem will not be unique. Unfortunately, the last requirement is difficult to satisfy in practice because of the nonlinear character, with respect to the variables $a_1, a_2, \ldots, a_m$ of the model size-distributions used. For example, some of the most widely used models for the size spectra of aerosols are:

$$n(r) = \begin{cases} a_1 & r_1 \le r < r_2 \\ a_2 r^{-\nu} & r_2 \le r \le r_3 \end{cases} \qquad (2.74)$$

$$n(r) = ar^{\alpha}\exp(-r^{\gamma}b) \quad , \quad (0 \le r < \infty) \qquad (2.75)$$

$$n(r) = (N\delta/\pi^{1/2}r)\exp(-\delta^2 \ln^2 r/r_0) \quad , \quad 0 \le r < \infty \; . \qquad (2.76)$$

It should be stressed here that the nonlinear character of the dependence of $\beta_M(\lambda_i p)$ on the parameters $\{a_i\}$ makes the task of seeking the minimum of the discrepancies for the system (2.73) considerably less simple than it might seem at first (a nonlinear case of the least squares method is meant). With these very general remarks,let us return to the essence of the method for interpretation of optical data under consideration (method of model assessments [2.3]).

If the values a_i^* $(i = 1,2,\ldots,m)$ obtained by numerically solving (2.73), are to describe the microstructure of a realistic medium, within the limits of estimated errors, the size-distribution function s_M^* should not differ too greatly, in the interval R_1, from the exact one s_0. In other words, it is necessary to require that $\|s_0 - s_M^*\| \le \delta(\sigma)$, where σ characterizes the errors of optical measurements. Situations may occur for which this requirement is not met in practice, since the models (2.74-76) describe only some statistically averaged size-distributions, while the true size distributions have, as a rule, more complicated characters [2.39]. Thus, the discrepancies between the model size-distribution $s_M(r)$ and the exact one $s_0(r)$ within the interval R yield the methodological errors in the estimations of the parameters $\{a_i\}$, which can be significant. In this connection, the choice of a size-distribution law to be used as a basis for the system (2.73), should obey certain requirements when interpreting the vector β.

Investigations [2.3,40,41] have shown that the efficiency of the scheme (2.73) depends not only upon the model $s_M(r)$ and its accordance with the real size-distribution $s_0(r)$, but also upon the number of parameters to be estimated and their characters. Here the following principle can be formulated. If it is known a priori that the model s_M used in (2.73) differs in an important way from the real size-distribution within the interval R, then the integral parameters of the microstructure must be estimated first of all. The latter include the full measure s of the size-distribution, the median, the mean radius $\bar{r}$, etc. (to say generally, all the moments of s_0). All these parameters, as is easily seen, have quite definite physical meanings and are not a result of any analytical features of the models. The integral parameters characterize microstructures regardless of the ways in which they are described, including the case of numerical description in the form of histograms. And if just these parameters are chosen as unknowns in the system (2.75) then one may hope that corresponding estimates will depend only weakly on the analytical form of the models s_M used.

For example, let the mean radius r_0 of the size-distribution $s_0(r)$ be estimated. The corresponding error can be written as follows

$$(\bar{r} - \bar{r}_M^*) = \int_R r[s_0(r) - s_M(r)]dr \quad . \tag{2.77}$$

It is not difficult to see that for the condition $|\bar{r}_0 - \bar{r}_M^*| \leq \delta(\sigma)$ to be fulfilled,the value of $||s_0(r) - s_M(r)||$ should be required to be small enough. The estimate $\bar{r}_M^*$ of the value $\bar{r}_0$ may be quite valid even if the value of $||s_0(r) - s_M(r)||$ reaches large magnitudes at some points of the interval R. It is for this reason that the efficiency of such estimations can be quite acceptable for solving these or other applied problems.

Concerning the practical use of the scheme (2.73), the principle formulated can be considered as a particular case of the general approach to the problem of optical data interpretation when a significant a priori uncertainty occurs. Indeed, the integral characteristics of the microstructures may be regarded as functionals of the corresponding size-distributions. Thus, the full measure $S = \int_R s\,dr$ of a polydispersed ensemble of particles is a linear functional of s(r) defined on Φ. The same can be said also about r, the median μ_s, etc. It is known that in an appropriate metric system, the sequence of functionals $F_j = F(s_j)(j = 1,2,\ldots)$ can converge , to F_0 even if the reference sequence $\{s_j\}$ diverges in Φ. The so-called weak convergence (or topology) is meant here. It is found to be widely used in solving these or other purely physical problems [2.42]. Also of interest is the application of this concept to the interpretation of optical measurements obtained under real atmospheric conditions.

Since in the above studies of the microstructures of the scattering media the integral characteristics were introduced into (2.73) as functionals of s, such a technique can be called the method of integral assessments for the interpretation of optical data. Applications of this method to the solution of purely optical problems can be of certain interest. It often occurs in practice that the optical characteristics of light scattering by aerosols being sought cannot be measured directly. Such problems can be solved, within the framework of this approach, by numerical methods based on the data of measurements of the other optical characteristics or, simply, as often occurs, on the characteristics of optical signals (Sect.2.5). The idea of such solutions is to estimate the function s_M^* from the measured vector β_1 and then to restore the vector β_2 using the size-distribution found. Denoting the ith component of the vector β_2 as $\beta_{2i} = \beta_2(\lambda_i)$ one can write the expression for the error of restoration as follows

$$\beta_{2i}^{(0)} - \beta_{2i}^{(m)} = \int_R K(r,\lambda_i)[s_0(r) - s_M(r)]dr \quad . \qquad (2.78)$$

This expression is analogous to (2.77) and again the condition that the value $|\beta_{2i}^{(0)} - \beta_{2i}^{(M)}|$ must be small is not necessarily connected with the requirement for the norm $\|s_0 - s_M\|$ to be also small. In this scheme the optical characteristic $\beta_2(\lambda)$ is nothing but the sought functional of size-distributions of the aerosol particles.

It is not difficult to see that the above method for the determination of one optical characteristic of dispersed media in terms of the others is a particular case of the more general method called the method of the inverse operator. In this case,the implicit form for constructing the regularizing operator for the equation $Ks = \beta$ is meant. The application of the method of integral assessments to the solution of such a problem as a restoration of spectral transmittance of the atmospheric ground layer from the data of laser sounding at three wavelengths has been presented in [2.43]. Similar studies on the determination of optical and microphysical parameters of stratospheric aerosols were also carried out [2.44,45]. Two and three-wavelengths lidars were used in these investigations. In addition to these applications, the method of integral assessments, as well as the method of inverse operator, can be used for the construction of optical models for the scattering atmosphere based on optical sounding data. This method does not require such a vast amount of input data on the aerosol microphysics, in contrast with the situation generally occurring in such problems [2.18].

2.6.2 Scheme of the Method of Model Assessments. An Example of Inverting the Spectral Transmission

Let us proceed now to concrete examples of the use of the method of integral assessments in optical sounding. One of the examples has already been discussed above (Fig.2.6). The dashed curve in this figure presents the data obtained by inverting $\tau^{(a)}(\lambda)$ according to (2.73) and using the model (2.75). The parameters sought were the total geometrical cross-section S of particles in a unit scattering volume and the mode r_s of the size-distribution $s(r)$. Introduction of these parameters requires the transformation of the reference model $S(r)$ to the form

$$s_M(r) = S\varphi(r,r_s,\alpha,\gamma) \quad ,$$

$$\int_R \varphi(r,r_s,\alpha,\gamma)dr = 1 \quad . \tag{2.79}$$

The remaining parameters of the model, i.e., α and γ, may be set a priori by using, for example, recommendations given in [2.18]. Naturally, a large number of measurements of the characteristic $\tau^{(a)}(\lambda)$ is not required for the determination of two parameters. It is for this reason that we have used only three points of $\tau(\lambda_i)$; $i = 1,2,3$ in the example shown in Fig.2.6. Here λ_i are from the visible region. The value of the discrepancy $\rho(Ks_M,\tau^{(a)})$ for the system (2.73) was comparable with the measurements errors, being of the order of 10% of the measured values of $\tau^{(a)}(\lambda)$. The solution of (2.73) had been sought according to the scheme which included the estimation of r_s^* from the condition

$$\min_{r_s \in [r_{s1},r_{s2}]} \sum_{\substack{i=1\\ i\neq m}}^{n} \left[\frac{\tau^{(a)}(\lambda_i)}{\tau^{(a)}(\lambda_m)} - \frac{\int_0^\infty K_{ext}(r,\lambda_i)\ \varphi(r,r_s)dr}{\int_0^\infty K_{ext}(r,\lambda_m)\ \varphi(r,r_s)dr}\right] \tag{2.80a}$$

and S from the formula

$$S^* = \sum_{i=1}^{n} \tau^{(a)}(\lambda_i) \int_0^\infty K_{ext}(r,\lambda_i)\ \varphi(r,r_s^*)dr \Big/ \sum_{j=1}^{n} [\tau^{(a)}(\lambda_j)]^2 \quad . \tag{2.80b}$$

The value λ_m in (2.80a) corresponds to the index i for which the value of $\tau^{(a)}(\lambda_i)$ reaches a maximum ($i = 1,2,\dots$). The value S^* found according to (2.80) is the normalizing constant for the measured values $\tau^{(a)}(\lambda_i)$. Recall that in this example the estimation of the parameter S with the accuracy of a constant multiplier c is meant, see (2.71).

Let us discuss briefly the values of S^* and r_s^* obtained when considering the solution s_α (s: solid curve in Fig.2.6) to be, in this case, the exact size-distribution $s_0(r)$. It should be noted, first of all, that the deviation of S^* from $S_\alpha = \int_R s(r)dr$ does not exceed 10% of their values. Thus, this estimation of the value of S can be regarded as quite satisfactory for the given accuracy of the measurements. But as to the estimation of the mode, the value r_s^* found should be assigned to the absolute maximum, since s_α is a multimode size-distribution. It should be noted here that the initial formulation of the inverse problem, where one of the unknown parameters has been chosen to be r_s, proceeds from a priori assumption on the existence of a sharp absolute maximum in the size-distribution sought.

The value of r_s^* obtained presents quite a satisfactory estimate of the position of the absolute maximum of the function s(r) in the interval R (Fig.2.6). If one were to transform the value r_s^* into the value of r_n^*, i.e., to find the mode of the distribution of the number density of particles with respect to their sizes n(r), then one would find that the absolute maximum of s_α is caused, out of the whole size-distribution, by the fraction of small particles with sizes near 0.1μm. It is known that the vicinity of this point is the most stable in the size spectrum of particles of the background atmospheric aerosol where, under the conditions of a slightly turbid atmosphere, the fraction of small particles is localized near 0.1μm.

Comparing the values of $s_M(r_s^*)$ and max $s_\alpha(r)$, one can easily see that the former exceeds the latter significantly. It is for this reason that the estimation of S^* has been performed using $\tau^{(a)}(\lambda)$ values measured in the region $0.3\mu m \leq \lambda \leq 0.7\mu m$, where the contribution of the fraction of large particles to the light scattering can be considered to be constant in a first approximation. Just this "bias" yielded the overestimation of the value of $s_M(r_s^*)$ as compared with max $s_0(r)$. It should be emphasized here that the knowledge of the value of max $s_0(r)$ is very important. In this connection, the scheme (2.80) is of great importance since it enables one to make an approximate evaluation of this value based on data from only two or three measurements. To a first approximation these are the possibilities of the method of integral assessments and of the calculation scheme (2.80) as applied to the studies of aerosol microstructures based on measurements of spectral behavior of the aerosol optical depth $\tau^{(a)}(\lambda)$.

Recently, problems dealing with the inversion of the data obtained from lidar measurements became of special interest along with the use of photometric data to restore parameters of the aerosol microstructures. Below we give an example of the inversion of the data of multi-wavelengths lidar sounding of stratospheric aerosols. These questions will be discussed in more detail in Chap.3.

2.6.3 Examples of Inversion of Lidar Measurements

The reference information for these examples are the values of the backscattering coefficient $\beta_\pi(\lambda)$, measured at three wavelengths 0.53μm, 0.69μm and 1.06μm. The measurements were made in Tomsk at the test field of the Institute of Atmospheric Optics on May 14, 1975 [2.46]. The experiment was carried out to determine the parameters of microstructure of the Junge aerosol layer in the lower stratosphere using a ground-based multi-wavelengths lidar. The results of the inversion are shown in Fig.2.7. The real $\bar{m}_0$ and imaginary $\varkappa_0$ parts of the index of refraction of the aerosol substance were assumed to be 1.43 and 0, respectively. The histogram was obtained by the direct minimization of $T_\alpha(s)$ (Sect.2.5), while the solid curve was obtained according to the scheme (2.80) for the model (2.75). This histogram illustrates the applications of multi-wavelengths lidars to the study of microstructures of aerosols as well as the mathematical methods developed for the inversion of optical data. It is not difficult to see that the histogram characterizes the microstructure of the aerosol sounded more informatively than the second distribution does, although it is quite clear that three measurements cannot be sufficient to identify amplitudes of the fractions. In our case, the histogram $[\Delta(N)/\Delta(r)]$ is closer in form to the typical size spectrum obtained with the use of impactors at the same altitudes. A corresponding example is presented in Fig.2.8, where a size-distribution function $n(r)$ obtained from measurements with impactors (Rylsk, 5 August, 1975 [2.47]) is presented in arbitrary units.

In addition to qualitative comparisons, quantitative assessments of these or other characteristics of microstructures are possible from the inversion of experimental results. Let us proceed to the integral size-distribution of the number density of particles $N(r \geq a)$ to improve quantitative estimates, where a is some fixed size of particles. Such a representation is also interesting since it allows one to compare the inversion results of optical measurements with those obtained from the direct microstructure analysis. The function $N(r \geq a)$ is presented in Table 2.4. The first column of Table 2.4 corresponds to the histogram from Fig.2.7, while the second column corresponds to the model size-distribution $n_M(r)$.

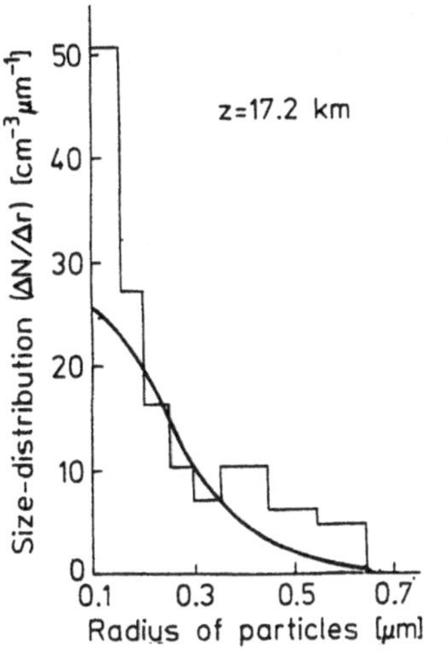

Fig.2.7. Size spectra of stratospheric aerosols, obtained from the data of three-frequency lidar sounding; the stepwise curve corresponds to the method of histograms (Sect. 2.5) while the smooth line presents the inversion results obtained by the method of model assessments

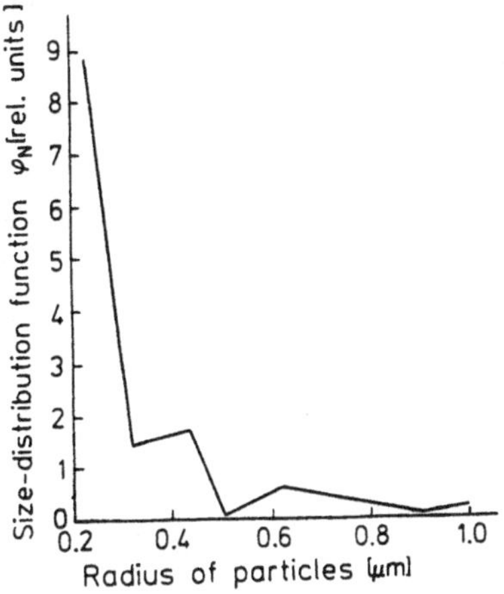

Fig.2.8. Typical size-distribution function of aerosols in the Junge layer according to impactor sampling [2.47]

Table 2.4. Parameters of a stratospheric aerosol in the Junge layer (z = 17.2 km) obtained by inverting three-frequency lidar sounding data

a [µm]	$S(r \geq a)$ [km^{-1}] 1	2	$N(r \geq a)$ [cm^{-3}] 1	2
0.1	$1.40 \cdot 10^{-3}$	$1.30 \cdot 10^{-3}$	6.7	5.0
0.15	$1.30 \cdot 10^{-3}$	$1.20 \cdot 10^{-3}$	4.2	3.8
0.2	$1.20 \cdot 10^{-3}$	$1.10 \cdot 10^{-3}$	2.8	2.80
0.25	$1.0 \cdot 10^{-3}$	$1.0 \cdot 10^{-3}$	2.0	1.9

It is easily seen that, on the whole, the integral size-distribution functions and, in particular, $S(r \geq a)$ obtained by different inversion methods are close enough to each other. This means that methodological errors caused by the use of a particular inversion method for determining size-distribution functions $S(r \geq a)$ are unimportant. This in turn substantiates the statement that the method of integral assessments provided quite an adequate information on the microstructure of the aerosol sounded.

2.6.4 Determination of Optical Constants of the Aerosol Substance by the Method of Integral Assessments

The last remark to be made here is required by the fact that the system (2.73) is not defined if the refractive index of the aerosol substance is unknown. To overcome this difficulty one can assume the values $\bar{m}$ and $\varkappa$ to be the unknown parameters. The most suitable procedure for this purpose is the scheme (2.80). The expression (2.80) may be regarded as a normalized discrepancy $\tilde{\rho}(a_i,\bar{m},\varkappa)$ for the system (2.73). If, as frequently occurs, the values $\partial\ln\tilde{\rho}/\partial\bar{m}$ and $\partial\ln\tilde{\rho}/\partial\varkappa$ significantly exceed corresponding derivatives with respect to microstructure parameters, then the determination of the latter becomes meaningless unless $\bar{m}_0$ and $\varkappa_0$ are known. In such situations,it is expedient that microstructure characteristics should be set a priori, while the optical constants be found by minimizing $\tilde{\rho}$ with respect to $\bar{m}$ and $\varkappa$. Corresponding examples of the interpretation of characteristics of stratospheric aerosols were given in [2.48].

It is necessary to emphasize here that the polarization of the scattered radiation, in this case, depends essentially on the optical constants of the aerosol substance or finally on the chemical composition of the particulate matter. This, in turn, allows one to solve corresponding inverse problems. Thus, an imaginary part of the refractive index of the aerosol substance can be estimated from the values of such characteristics of light scattering as the degree of linear polarization, the tangent of the angle of elliptic polarization and so on [2.48,49]. The schemes (2.73) and (2.80) can serve as a mathematical basis in this case.

Let us use the experimental data given in [2.50] to illustrate the above mentioned possibility. Measurements of the angular behavior of the tangent of the angle of elliptic polarization are made in this work for aerosols at the visual range $S_M \geq 40$ km in the atmosphere. Assuming, to a first approximation, the real part of the refractive index to be $\bar{m} \cong 1.56$ and using the model (2.75) one can construct the discrepancy $\rho_{tg}(r_s,\varkappa)$ according to (2.80a). Fig.2.9 shows the behavior of this function in the region Ω, whose boundaries are determined by the conditions that $0.06 \leq r_s \leq 0.1\mu$m and $0 \leq \varkappa \leq 0.06$. Since optical measurements have errors, then the direct minimization of $\rho_{tg}(r_s,\varkappa)$ in Ω is meaningless and therefore the point $(r_s^*,\varkappa^*)$ should not be taken as a solution of the inverse problem of light scattering, but rather some region Ω^*, whose points satisfy the condition

$$\rho_{tg}^2(r_s,\varkappa) \leq \delta(\sigma^2) \quad .$$

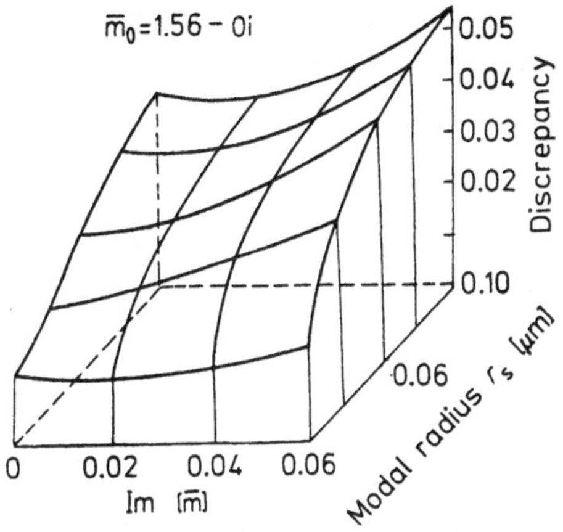

Fig.2.9. The discrepancies behaviour in the method of model assessments applied to the interpretation of elliptical polarization. Im{$\bar{m}$} denotes the imaginary part of the refractive index

Proceeding from the error analysis it is possible to assume that $\delta(\sigma^2) \lesssim 0.02$ and, hence, the estimations r_s and $\varkappa^*$ are the following $0.06 \le r_s^* \le 0.07\mu m$, $0.01 \le \varkappa^* \le 0.03$. It should be noted that the function $\rho_{tg}^2(r_s,\varkappa)$ has no absolute maximum in this region Ω. This is caused, as the numerical study shows, by the fact that not all the initial assumptions are fulfilled. The estimation of the real part of the index of refraction, in this case, is very approximate and also it is difficult to say to what degree the assumption on the particle sphericity is valid. However, on the whole, the values of r_s^* and $\varkappa^*$ obtained here do not contradict the data on optical constants and parameters of microstructures of aerosols at these altitudes.

We should like to note in conclusion that more rigorous methods for the inversion of data of optical sounding aiming at inference of the information about optical constants of aerosol substance will be considered in Chap.3.

3. Determination of the Microphysical Parameters of Aerosol Ensembles by the Method of Multifrequency Laser Sensing

The method of multifrequency laser sensing for remotely determining the microphysical parameters of atmospheric aerosols considered in this chapter is an application of the general theory of optical sounding of polydispersed aerosol systems. Solutions of the corresponding inverse problems of light scattering by polydispersed systems are given here within the framework of a single scattering approximation.

The basis of the theory of multifrequency laser sensing considered in the following are the operators for light scattering by polydispersed systems. These operators allow a simple derivation of methods and algorithms for the numerical solution of systems of the transfer equations for the locating signals which propagate through the atmosphere. In addition to the purely theoretical aspects of the method, the questions of its practical realization are presented in detail. Furthermore, the information potentialities of the multifrequency lidars are investigated and assessments of their accuracy are given. Particular attention is paid to methods for solving inverse problems of multifrequency sounding where not only the microstructures of the ensembles but also the refractive index of the aerosol substance are to be determined. A scheme is described for the numerical processing of the optical data that are to be used for the solution of practical problems, such as monitoring the aerosol pollutions of the atmospheric ground layer. The spatial resolution inherent in pulsed laser sounding makes it possible, in principle, to investigate the structure and dynamics of inhomogeneities of the aerosol atmospheric formations. Of course, the realization of these capabilities requires an appropriate methodological treatment. It is therefore necessary to develop methods for solving the inverse problems of light scattering by polydispersed media in the case of the broken size-distribution functions that adequately describe the microstructures of aerosol ensembles in small volumes. Also, it is necessary to have methods for inverting the optical data when the size limits of the size-distributions are unknown. The latter peculiarity of the inverse problem is of particular importance when the aerosols of the atmospheric ground layer are sounded. The peculiarities of the remote determination of microstructures of the low stratospheric aerosols using lidars are also considered in detail. Some examples concerning the inversion of the results of two- and three-wavelength lidar soundings are also presented here. At the end of this chapter,questions of multifrequency sounding of clouds are discussed and assessments of the effect of multiple light scattering on the accuracy of the inversion of optical data are given.

Multifrequency lidars are very complicated optical systems, and therefore their use in the practice of atmospheric studies should be carefully considered. We present the following material on the methodology of multifrequency laser sounding with this goal in mind. This methodology is expected to stimulate, to a certain degree, the development of lidar facilities for multifrequency sounding and to extend the use of these methods in atmospheric physics, atmospheric optics and also in practical meteorology.

3.1 General Theory of Multi-Frequency Laser Sensing of Polydispersed Systems in the Single-Scattering Approximation

The basis of the theory of multifrequency laser sounding as an optical method for remotely determining the microphysical parameters of the atmospheric aerosols is given by the transfer equations for laser radiation propagating through the atmosphere and by the operators for light scattering due to the systems of particles considered in Chap.2. In the theory of multifrequency sounding in the single scattering approximation the basic equation is the lidar equation that relates the amplitudes of the lidar returns to the optical characteristics of the scattering components of the atmosphere (2.61). For simplicity, the contribution of the molecular component to the total backscattered signal will be considered either to be negligible as compared with the aerosol component or to be known to within an adequate accuracy. In this case (2.61) can be written as follows

$$P(z,\lambda_i) = P_0(\lambda_i)z^{-2}\beta_\pi(z,\lambda_i)\exp\left[-2\int_{z_0}^{z} \beta_{ext}(z',\lambda_i)dz'\right] \quad , \qquad (3.1)$$

where λ_i, $(i = 1,2,\ldots,n)$ are the wavelengths at which the lidar operates, while β_π and β_{ext} are the optical characteristics of the aerosol defined in Chap.1. Note that the discussion below will deal mainly with questions of aerosol sensing in the ground atmospheric layer, where light scattering by aerosols is predominant. Before deriving the methods that yield information on the aerosol microstructure from the optical data $\{P(z,\lambda_i)\}(i = 1,2,...n)$, one should prove that the system (3.1) provides information about the size-distribution function $s(r)$, provided that the proper choice of sounding wavelengths is made. Let us consider, for this purpose, the operators K_π and K_{ext} and present the optical characteristics $\beta_\pi(\lambda)$ and $\beta_{ext}(\lambda)$ in the form $(K_\pi s)(\lambda)$ and $(K_{ext}s)(\lambda)$, respectively. As before, these designations are equivalent to the polydispersed integrals $\int_R K_\pi(r,\lambda)s(r)dr$ and $\int_R K_{ext}(r,\lambda)$ $s(r)dr$. Since the optical measurements are usually made at a discrete number of wavelengths within the sounding interval Λ, it is convenient to use the vector form of the basic relations. We shall therefore denote the values of $\beta_\pi(\lambda)$ and $\beta_{ext}(\lambda)$ at the fixed wavelengths λ_i $(i = 1,2,...,n)$ by the vectors $\boldsymbol{\beta}_\pi$ and $\boldsymbol{\beta}_{ext}$, respectively. The components of these vectors are: $\beta_{\pi i} = \beta_\pi(\lambda_i)$ and $\beta_{ext,i} = \beta_{ext}(\lambda_i)$. Using an implicit form of the lidar equation, one can write the reference system (3.1) for the case when polydispersed aerosols are sounded as follows:

$$F(P_i,\beta_{\pi i},\beta_{ext,i},z,\lambda_i) = 0$$

$$K_\pi \mathbf{s} = \boldsymbol{\beta}_\pi \tag{3.2}$$

$$K_{ext}\mathbf{s} = \boldsymbol{\beta}_{ext} \qquad i = 1,2,...,n \quad .$$

Here K_π and K_{ext} are meant to be the matrix analogs of the initial integral operators. It is obvious that this system of equations is quite well determined with respect to the vector $\mathbf{s}$, which characterizes approximately the size-distribution $s(r)$ being sought. Indeed, there are two equations for finding the vector $\mathbf{s}$, whose right-hand sides are related to each other through the lidar equations although neither is known directly. Denoting the components $\beta_{\pi i}$ and $\beta_{ext,i}$ in terms of $(K_\pi \mathbf{s})_i$ and $(K_{ext}\mathbf{s})_i$, respectively, and substituting them into the corresponding equations, one can, in principle, reduce the system (3.2) to the following form

$$F(P_i,(K_\pi \mathbf{s})_i, (K_{ext}\mathbf{s})_i, (z,\lambda_i) = 0, \quad i = 1,2,...,n \quad . \tag{3.3}$$

The system (3.2) is completely defined with respect to the vector $\mathbf{s}$ sought at any range z along the sounding path. This demonstrates the consistency of multifrequency sounding as a remote method for determining the microstructures of the atmospheric aerosols.

The system (3.3) can be used when the refractive index of the aerosol substance is known and an assumption about the shape of the aerosol particles is made. The discussion presented in this chapter concerns only the case of systems of spherical particles, which enables one to use the Mie theory as a model for interpreting the lidar sounding data for the atmospheric aerosols.

Let us now consider the questions concerning the development of numerical methods for solving systems like (3.2) and (3.3). For greater clarity and simplicity, one can use the operator W introduced into the theory of optical sounding of the polydispersed aerosol systems. In this case, (3.2) can be rewritten as follows

$$F(P_i, \beta_{\pi i}, \beta_{ext,i}, z, \lambda_i) = 0$$

$$\boldsymbol{\beta}_{ext} = W\boldsymbol{\beta}_\pi \quad , \quad i = 1,2,\dots,n \quad . \tag{3.4}$$

Let us first write (3.4) in discrete form. It will be assumed, for this purpose, that for all $i = 1,2,\dots,n$ the sequence of equidistant values $P_{ki} = P(z_k, \lambda_i)$ is chosen with the step $\Delta z = z_{k+1} - z_k$ $(k = 1,2,\dots)$ along the z axis. Then, for every k, one will have the system of k equations

$$\bar{S}_{ki} = \beta_{\pi,ki} \exp\left(-2\Delta z \sum_{j=1}^{k} \omega_j(\beta_{ext,ji})\right) , \tag{3.5}$$

where $\bar{S}_{ki} = P_{ki} z_k^2 / P_{0i} B_i$ and ω_j are quadrature coefficients (with no loss in generality, one may use the quadrature coefficients of the trapezoid formula). If (3.5) is solved successively, beginning from $k = 1$, then all the values of the optical characteristics including $\beta_{\pi,k-1,i}$ and $\beta_{ext,k-1,i}$ will be known when z_k is reached. In this case, a more convenient form of (3.5) is

$$\beta_{\pi ki} \exp(-2\tau_{ki}) = \bar{S}_{ki}/T_{k-1,i} \qquad i = 1,2,\dots,n \quad , \tag{3.6}$$

where

$$T_{k-1,i} = \exp\left(-2\Delta z \sum_{j=1}^{k-1} \omega_j \beta_{ext,ij}\right), \quad k \geq 2, \quad T_{1i} = 1 \quad ,$$

$$\tau_{ki} = \Delta z(\beta_{ext,k-1,i} + \beta_{ext,ki})/2 \quad .$$

The value τ_{ki} is the optical depth of the kth layer of the atmosphere located between the points z_{k-1} and z_k. Here the index i denotes the number of wavelengths λ_i at which the lidar operates. An iteration algorithm for solving the system (3.6) jointly with the equation $\boldsymbol{\beta}_{ext} = W\boldsymbol{\beta}_\pi$ has been suggested and verified mathematically in [3.1]. The scheme of this algorithm can be presented in the analytical form

$$\beta_{\pi ki}^{(m)} \exp(-2\tau_{ki}^{(m-1)}) = \bar{S}_{ki}/T_{k-1,i}$$

$$\boldsymbol{\beta}_{ext\ k}^{(m-1)} = W\boldsymbol{\beta}_{\pi k}^{(m-1)} \quad , \quad i = 1,2,\dots,n \quad ; \quad k = 1,2,\dots, \tag{3.7}$$

where $\boldsymbol{\beta}_{\pi k}$ and $\boldsymbol{\beta}_{ext\ k}$ are n-dimensional vectors with components $\beta_{\pi ki}$ and $\beta_{ext\ ki}$, m is the number of iterations, and the operator W is the matrix analog of the operator $K_{ext}K_{\pi\alpha}^{-1}$. The system (3.7) does not formally depend on the distribution s(r) and is a concrete numerical realization of the general functional equation (2.54).

If the particles are assumed to be spherical, the matrix operator W can be calculated by using the Mie theory. The absolute values of its elements are determined by the values of the efficiency factors $K_{\pi}(r,\lambda)$ and $K_{ext}(r,\lambda)$ and they also depend upon the disposition of nodes $\{r_e\}$ on the size interval $R = [R_1, R_2]$, as well as on the choice of operating wavelengths within the sounding interval Λ. It should be noted that the action of the operator W, on the set of reference data B_{π} is more regular than that of the operator $K_{\pi\alpha}^{-1}$. Regularity is understood here as the more stable behavior of the values $(W\boldsymbol{\beta}_{\pi})$ with respect to errors in the vector $\boldsymbol{\beta}_{\pi}$, as compared with the corresponding spread of vectors $(K_{\pi\alpha}^{-1}\boldsymbol{\beta}_{\pi})$. This feature makes the scheme for inverting the optical data $\{P_i\}$ efficient, particularly when the optical characteristics of atmospheric aerosols are to be determined.

As the numerical analysis shows, at least five sounding wavelengths are usually required in the visual range in order to make a satisfactory reconstruction of the microstructures of atmospheric hazes (models H, L, M according to [3.2]) from lidar measurements which are accurate to about 10%, on the average. At the same time, it is sufficient to restrict the number of sensing wavelengths to $n \cong 3$ in order to determine only the optical characteristics of the aerosols. In this case,the errors of reconstruction for β_{π} and β_{ext} will not exceed 20% provided that the errors of measurements $\sigma_{meas.}$ are less than 10%.

Thus, two aspects of the use of multiwavelengths lidars can be considered. First, they are instruments for determining the optical properties of the aerosol component of the atmosphere (e.g., spectral transmission). The second aspect concerns studies of aerosol microstructures. In this case,the lidar systems must be more sophisticated, since more experimental information is required.

We now return to (3.7) and consider the convergence of the iteration algorithm. Omitting a detailed and rigorous derivation of the necessary and sufficient conditions (for this case see, e.g., [3.1]) one can write the inequality

$$\|\boldsymbol{\tau}_k^{(m)}\| = \max_i \tau_{ki}^{(m)} < 1 \tag{3.8}$$

which, if satisfied for all m and k, ensures the convergence of the scheme (3.7). The condition (3.8) means that the choice of the Δz value, when discretizing $P(z,\lambda)$, should be made such that the maximum optical depth of any of the layers does not exceed unity over the whole spectral region Λ. It is obvious that this condition can easily be satisfied in the case of hazes and slightly turbid atmospheres. However, when optically dense aerosol media, such as fogs and clouds, are sounded it becomes difficult to fulfil condition (3.8). One should also take into account that the speed of convergence of the iteration algorithm is not higher than that of the geometrical progression with the denominator $\|\boldsymbol{\tau}\|$ [3.1]. Therefore, the smaller the value of $\|\boldsymbol{\tau}\|$, the faster the scheme (3.8) converges. Taking this property of the iteration method into account, let us examine in greater detail the properties of the lidar equation when τ is comparable to or larger than unity. For simplicity, we consider the case of one-frequency sensing (n = 1). In order that (3.1) be defined with respect to the optical characteristics of aerosols, one should be able to assign a priori any value to the lidar ratio b(z) along the direction of sounding. The iteration scheme (2.7), for n = 1, will be written as follows

$$\beta_{ext,k}^{(m)} = (\bar{S}_k/T_{k-1}b_k)\exp(2\tau_k^{(m-1)}) \quad . \tag{3.9}$$

Mathematically (3.9) is equivalent to the iteration method for solving the transcendental equation $c = \beta\exp(-\Delta z\beta)$ where c is any number and β is the unknown value sought. This equation has two roots, one of which lies to the left of the point $\beta = 1/\Delta z$, and the other to the right of it. At the point β_0 the function $\beta\exp(-\Delta z\beta)$ has a maximum. The iteration scheme (3.9) determines the root of this equation in the region of small β values for a given value of the range increment along the sounding path.

Since the spatial resolution of the pulsed laser systems is high enough, the value of Δz can be chosen, in principle, to be small (of course, the sensitivity of the measuring tract should be taken into account) and in this case the iteration schemes suggested above will converge rapidly. In order to find the second root of the equation $c = \beta\exp(-\Delta z\beta)$, it is necessary to write it as follows $\beta = (\ln c\beta^{-1})/\Delta z$ and only then to make use of the iteration method for the numerical solution. The corresponding scheme for (3.1) is written in the form

$$\tau_k^{(m)} = \ln\left(\beta_{ext,k}^{(m-1)}\bar{S}_k^{-1}T_{k-1}b_k\right) \quad . \tag{3.10}$$

The scheme (3.10) is useful when τ is large ($\tau > 1$); this can occur when dense scattering media (e.g., mists and clouds) are sounded. We note, however, that although this scheme leads to a formal solution of (3.1) for $\tau > 1$, its use in practice is very difficult. This happens because,as τ increases, the contribution of multiple scattering to the total backscattered signal increases, and as a consequence, the measured values of $\bar{S}_k$ exceed those that would correspond to single scattering. It is easily seen that (3.10) is meaningful only for such values of $\beta_{ext,k}$, $\bar{S}_k$, T_{k-1} that satisfy the condition

$$\beta_{ext,k}\, T_{k-1}b_k\bar{S}_k^{-1} > 1 \tag{3.11}$$

for all k along the sounding path. Overestimating the $\bar{S}_k$ values can invalidate condition (3.11) for given T_{k-1} and b_k.

It should be noted that it becomes more difficult to satisfy condition (3.11) with increasing τ_k because of the fact that T_{k-1} vanishes rapidly. In this connection, condition (3.11) can be regarded as a criterion of applicability of (3.1) for interpreting the data of lidar sensing of optically dense media.

Omitting a detailed discussion of the above questions, we note that an iteration scheme for (3.1) was suggested in [3.1], for which the fulfilment of the convergence condition depends directly on the degree to which the theory of single scattering is applicable to a given optical experiment. Therefore, we can state that the use of iteration methods for solving the systems of lidar equations, when multifrequency lidar sensing is used, allows one not only to obtain the solutions of inverse problems but also to assess, to a certain extent, the applicability of the initial functional relations.

In conclusion, we should like to emphasize one point. All the above analytical constructions are based on the use of the operator $W = K_{ext}K_{\pi\alpha}^{-1}$, which describes the transition from the optical characteristic $\beta_\pi(\lambda)$ of a polydispersed system to the characteristic $\beta_{ext}(\lambda)$ (within spectral interval Λ). Of course, certain analytical results can be obtained using the operator $\tilde{W} = K_\pi K_{ext,\alpha}^{-1}$ which may be regarded as an inverse operator with respect to W. A corresponding theory can be useful in the case of absorbing dispersed media, since the aerosol extinction coefficient $\beta_{ext}(\lambda)$ strongly affects the amplitudes of lidar returns in such media. However, such problems are of practical use only when the complex refractive index of the aerosol substance is known. Therefore, the use of the operator $\tilde{W}$ is also advisable in those inverse problems for which the refractive index is to be evaluated along with the aerosol microstructure parameters. Such an example will be considered in Sect.3.4.

3.2 Single-Frequency Lidars in Studies of the Optical Properties of the Atmosphere

3.2.1 The Choice of the Lidar Ratio for the Accuracy of the Solution

Single-frequency lidars are devices used widely in atmospheric studies, although it goes without saying that the problem of determining the aerosol microstructure parameters cannot be solved with them. Single-frequency lidars can obtain only qualitative estimates of the optical characteristics of aerosols. Nevertheless, it is interesting to discuss, as rigorously as the single scattering approximation allows, the interpretation of the lidar returns $P(z,\lambda)$ obtained at a fixed wavelength λ. Our principal concern will be to assess of the accuracy of determining of the aerosol optical characteristics. At the same time, some peculiarities of laser sounding, and of the remote optical method for atmospheric researches, will be considered as well.

Three optical characteristics of aerosols $\beta_\pi(z)$, $\beta_{ext}(z)$ and $p(z) = T^{1/2}(z)$ can be determined, in principle, from the lidar equation (3.1). Here P(z) characterizes the profile of the aerosol transmission coefficient along the sounding path between the points z_1 and z at a given wavelength λ.

The profile of the lidar ratio $b(z) = \beta_\pi(z)/\beta_{ext}(z)$ should be set arbitrarily. It is important to assess the effect of deviations from the (arbitrarily chosen) b(z) on the accuracy of the solution of the lidar equation.

Let the profile b(z) be assumed to vary insignificantly, within the distance interval from z_1 to z_2 of the sensing path, so that it can be approximated by a constant b^* that obeys the condition

$$\max_{z\in[z_1,z_2]} |b(z) - b^*| < \eta(\overline{\Delta^2 b})^{1/2} \quad , \tag{3.12}$$

where η is a certain coefficient of reliability ($\eta > 1$), and $(\overline{\Delta^2 b})^{1/2}$ characterizes the standard deviation of the solution of (3.1) caused by the choice of b^* on the path interval $[z_1,z_2]$. Now the question can arise: with what accuracy can the above optical characteristics be determined from (3.1) at a given value of $(\overline{\Delta^2 b})^{1/2}$?

If the lidar ratio b is assumed to be constant on the interval $[z_1, z_2]$, then the following relations can be obtained from (3.1)

$$T(z) = T(z_1) - 2I(z)(b^*)^{-1}$$

$$\beta_\pi(z) = \bar{S}(z)T^{-1}(z)$$

$$\beta_{ext}(z) = \beta_\pi(z)(b^*)^{-1} \quad , \qquad (3.13)$$

where $I(z) = \int_{z_1}^{z_2} \bar{S}(z')dz'$ and $z_1 \le z \le z_2$. It is convenient, for further discussion, to introduce the relative errors ε_B, ε_T and so on, assuming that $\varepsilon_B = (\overline{\Delta^2 b})^{1/2}/b^*$, $\varepsilon_T = (\overline{\Delta^2 T})^{1/2}/T$ and so on. It is not difficult to show the validity of the following relationships

$$\varepsilon_T = (1 - T)\varepsilon_B/2T$$

$$\varepsilon_\pi = (1 - T)\varepsilon_B/T$$

$$\varepsilon_{ext} = \varepsilon_B/T \quad , \qquad (3.14)$$

which characterize the expected values of the errors of the optical characteristics $T(z)$, β_π and β_{ext}, respectively. These optical parameters are determined by solving the lidar equation, which involves a value b, set with errors. In (3.14) we assumed $T(z_1) = 1$ with no loss of generality.

Using the relations (3.14),one can assess the information content of single-frequency lidars in the remote determination of optical characteristics of atmospheric aerosols. The corresponding assessments will be related, first of all, to the optical depth τ of the layer sounded. The value of τ is determined not only by the value of the aerosol extinction coefficient β_{ext} but also by the geometrical length of the sensing path. The latter plays an important role in the theory of optical sounding. Recall that $\tau(z) = \int_{z_1}^{z} \beta_{ext}(z')dz'$ and, hence, $T(z) = \exp[-2\tau(z)]$.

In the case of a slightly turbid atmosphere, when the optical depth is small, the value of T(z) is close to unity and one can find from (3.14) that $\varepsilon_T \cong 0$, $\varepsilon_\pi \cong 0$ and $\varepsilon_{ext} \cong \varepsilon_B$ which means that the errors of the choice of the lidar ratio affect only the value of the aerosol extinction coefficient β_{ext}. We recall that here the values of β_{ext} are those determined from (3.13). As for β_π and T, the errors $(\overline{\Delta^2 b})^{1/2}$ due to the choice of b^* do not affect essentially the values determined from (3.15) for small τ.

However, when optically dense aerosol layers of the atmosphere are sensed, the situation changes drastically. Let $\tau \to \infty$ approach infinity. Then $T(\tau) \to 0$ and, as a consequence, the values $(1-T)/T$ and T^{-1} increase continuously. Therefore, the requirements on the accuracy of the a priori choice of the lidar ratio b become more stringent with the increase of the optical depth of the layer sounded. This result is also valid when the data of multi-wavelengths sounding are interpreted. Indeed, in single-frequency sounding, the value b^{-1} is equivalent to the operator W, since the transition from β_π into β_{ext} is performed according to the relation $\beta_{ext} = b^{-1}\beta_\pi$.

It follows from the above that the requirements on the accuracy in assigning of the value of the refractive index of the aerosol substance should become more stringent with the increase of $\max\tau(\lambda)$, in the multi-wavelengths variant. This is also relevant to the probable deviations of the shapes of particles from spheres.

The above analysis of how the errors of an a priori choice of some parameters affect the accuracy of solutions of the lidar equation is instructive in the sense that analogous conclusions are nearly impossible to obtain in the case of systems like (3.4) with the operator W set approximately, due to analytical difficulties.

3.2.2 On the Choice of the Lidar Ratio in the Case of Single-Frequency Sensing

When seeking a proper choice of admissible values of $(\overline{\Delta^2 b})^{1/2}$ one should proceed, in any particular case, from the values of errors of the optical measurements. Let these errors be characterized by the value $\varepsilon = (\overline{\Delta^2 I})^{1/2}/I$. It can easily be shown that the error in the determination of T, which is caused by ε and ε_B, is given, to a first approximation, by

$$\varepsilon_T \cong (1 - T)(\varepsilon^2 + \varepsilon_B^2)^{1/2}/T \quad . \tag{3.15}$$

Hence, according to (3.15), the accuracy of the a priori choice of b^* may be regarded satisfactory if ε_B does not exceed ε. In this case, the uncertainty introduced in the value of T sought, when solving the lidar equation with the approximate b, will be comparable with that due to the errors of the optical measurements. Therefore, the requirement $\varepsilon_B \leq \varepsilon$ is necessary when interpreting the data of single-frequency lidar sensing. This condition will be used below in the form

$$(\overline{\Delta^2 b})^{1/2} \lesssim \varepsilon b_0 \quad ,$$

where b_0 is the exact value of the lidar ratio in a given optical experiment. Since b_0 is unknown, one can use an arbitrarily chosen value b^* that permits, at least, in the first approximation, an estimation of the value of $(\overline{\Delta^2 b})^{1/2}$.

Consider now the question of the choice of a value for b . There are three possible approaches, in the most general case. The first is based on the use of optical models of the atmosphere. Examples of such models representing these or other optical situations in the atmosphere can be found in [3.3,4]. Since the lidar ratio b is determined by the microstructure parameters of aerosol ensembles and by the index of refraction of aerosol substance, the corresponding optical models for $b(z,\lambda)$ are, naturally, based on models of these microstructures.

The assumption that b is constant within the layer $[z_1, z_2]$ is equivalent, in fact, to the assumption that the microstructure of an aerosol ensemble is constant within this layer. Of course, this assumption becomes less realistic for larger values of $(z_2 - z_1)$. As it was emphasized above, only qualitative methods for interpreting the lidar measurements could be discussed within the framework of this approach. It should be stressed that the use of multi-frequency lidar sounding enables one to decrease significantly the uncertainty in the interpretation of sensing results which is caused by the a priori choice of the size-distribution function and by the assumption of its constancy along the sensing path as well. The choice of b^* is made, in the case under consideration, by numerical calculation of the characteristics $\beta_{\pi M}$ and $\beta_{ext,M}$. Thus, if the model size-distribution function $n(r) \cong ar^{-\nu}$ is chosen, then the following expression can be shown to be valid

$$\varepsilon_B \leq 0.2\,(\overline{\Delta^2\nu})^{1/2} \quad ,$$

where $(\overline{\Delta^2\nu})^{1/2}$ is the rms deviation of the a priori assessment of the microstructure parameter ν. The above assessment has been obtained for $1.4 \leq \bar{m} \leq 1.5$ and $\varkappa \leq 0.01$. Since the most probable values of ν for the hazes in the ground layer of the atmosphere are in the interval (3.4), we may assume $(\overline{\Delta^2\nu})^{1/2} = 0.5$. As a consequence, the error ε_B, caused by the a priori choice of the lidar ratio, can be safely assumed to be at the level of 15% or 20% provided that the size spectra of atmospheric hazes are close to a power law [3.5].

A second approach for giving a supplementary definition of the lidar equations for sounding aerosols is also based on optical models, but on those which are constructed with the use of data from appropriate optical measurements. In other words, some preliminary experiments have to be

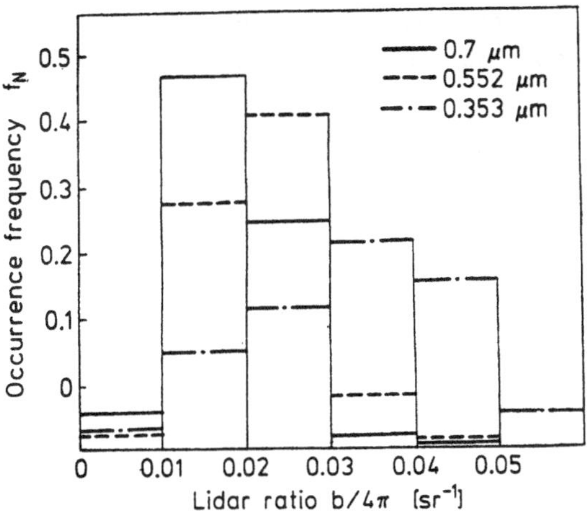

Fig. 3.1. Histogram of the occurrence frequencies for the lidar ratios at different wavelengths according to [3.6]

carried out, in which the values of the lidar ratio b (or some other characteristics of light scattering) are directly measured in various optical situations in the atmosphere. Then using the experimental data, models are constructed. In this case, the model is understood as an interval of values of the observed parameter and the distribution of corresponding probabilities. As an example, we shall consider the probability model of the lidar ratio b, which is constructed using the experimental data presented in [3.6]. The value estimated in the experiment was $\xi = D_{11}(180^\circ)/\beta_{sc}$ where $D_{11}(180^\circ)$ is the direction scattering coefficient $D_{11}(\theta)$ for $\theta = 180^\circ$. The measurements were made during autumn in a semi-desert area when the visual range S_{Met} was longer than 7 km. The measurements of $D_{11}(\theta)$ were made for θ varying from 170° to 178° in 2° increments. The value of $D_{11}(180^\circ)$ was obtained from these measurements by extrapolation. The ratio ξ gives the value of b up to a constant factor 4π.

Figure 3.1 presents the histogram of the occurrence frequency for ξ values. The intervals, within which the value of ξ varies are shown along the abscissa. As is seen from Fig.3.1, the tendency of the most probable values of ξ to increase as λ decreases is observed. Here the occurrence frequency is understood as the probability of occurrence of the corresponding values of b^* in certain optical situations. When using the probability models of the lidar ratio, one can take for $\bar{b}$ its mean or most probable values. The values of the expected error $(\overline{\Delta^2 b})^{1/2}$ are estimated by calculating corresponding variances.

A third way to choose the values of b^* is to perform suitable optical experiments. The interpretation of the data from such experiments should provide an estimate of the optical values sought, including the lidar ratio. This approach is one of the most effective and well-founded ways to overcome ambiguity in the initial functional relations being used in practice for optical sensing of the atmosphere. In Chap.4 we shall give such an example, in which the lidar ratio b is evaluated from the data of the inversion of the scattering phase function $\mu(\theta)$ measured within the angular interval from 10 to 170 degrees. Since this method is based on the inversion of optical data, it can be called the inverse operator method. The above theory of multi-frequency laser sounding of aerosols is an example of a concrete application of the inverse operator method to the solution of problems of atmospheric optics.

3.2.3 Examples of the Interpretation of Single-Frequency Sensings Data

The iteration scheme (3.7) for i = 1 can be written if the profile b(z) is known, as follows

$$\beta_{ext,k}^{(m)} = \bar{S}_k \exp(2_k^{(m-1)})/T_{k-1}b_k$$

$$T_{k-1} = \exp\left(-2h \sum_{j=1}^{k-1} \omega_j \beta_{ext,j}\right) \qquad (3.16)$$

$$\tau_k = \Delta z(\beta_{ext,k-1} + \beta_{ext\ k})/2 \qquad k = 1,2,\ldots, \qquad .$$

With no loss of generality, the coefficients of the trapezoidal quadrature formula can be taken as ω_j. The system (3.16) is solved in succession beginning from k = 1. The transmission T_0 along the path between the lidar and the initial point of the path is assumed to be equal to 1. As mentioned above, the iteration scheme (3.7) and, hence, the scheme (3.16) converges for those indices k and m, for which $\tau < 1$.

In Fig.3.2,an example of the profile of an aerosol extinction coefficient $\beta_{ext}(z)$, reconstructed according to the scheme (3.16), is presented. The lidar sensings were carried out during the night time under conditions of good visibility (the test field of the Institute of Atmospheric Optics, October, 1973 [3.7]). Thin strata were observed at altitudes from 1 to 2 km which were readily revealed in the height profiles of $\beta_{ext}(z)$. The lidar used had a receiving mirror of 0.3 m in diameter and a pulsed ruby laser with 8 MW peak power. This lidar provided reliable observations of the spatial and temporal variations in the lidar returns up to the altitudes of 3 km. In the interpretation of sensing data the lidar ratio was assumed constant

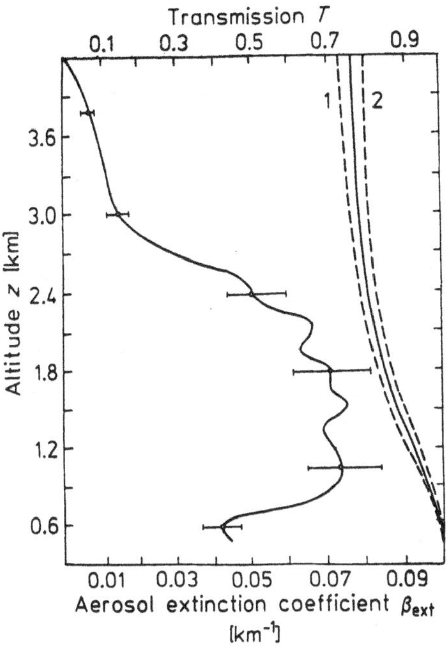

Fig.3.2. Profiles of β_{ext} and T inferred from the single-frequency lidar measurements. The lidar ratio was chosen a priori to have the rms deviation $(\overline{\Delta^2 b})^{1/2} = 0.05$

($b^* = 0.41$) within the above layer. The value was chosen in accordance with the probability model of ξ which was constructed for a given area and the season and wavelength $\lambda = 0.69\mu m$. The most probable value ξ^*, in the model used, lies within the interval 0.02; 0.03. The calculated variances of the observed values of near ξ gave the estimation $(\overline{\Delta^2 b})^{1/2} < 0.05$.

The bars in Fig.3.2 show the possible intervals of the solutions of (3.1) which correspond to the variation of b determined by the relation $|b^* - b| \leq (\overline{\Delta^2 b})^{1/2}$. If b(z) is assumed to be constant along the sensing path then the error $(\overline{\Delta^2 b})^{1/2}$ yields the systematic shift of the profile T(z), as is illustrated by the dashed curves in Fig.3.2 (Curve 1 corresponds to b = 0.36 and the Curve 2 to b = 0.46). Thus, a given error in b produces different errors in the profiles $\beta_{ext}(z)$ and T(z) which is quite understandable since the last profile is the integral optical characteristic and it is natural that the errors in the lidar ratio have less effect on such a characteristic along the sensing path. The example considered here confirms the conclusions of the analysis of (3.14).

The values T(z) and b are related to each other by (3.13) if b(z) is assumed to be constant for $z_1 \leq z \leq z_2$. Therefore, if it is possible in an optical experiment to measure $T(z_2)$, then the value of b^* can be estimated according to the simple relation

$$b^* \simeq 2I(z_2)[T(z_1) - T(z_2)]^{-1} \quad . \tag{3.17}$$

If the aerosols are sounded at night, then $T(z_2)$ can be measured with the use of a star photometer. Since the attenuation of light by the atmospheric aerosols is mainly due to the lower layers of the troposphere, then $\lim_{z\to\infty} T(z) \cong T(z_2)$, where $z_2 \leq 3$ or 5 km. As a consequence, the value of b^* found from (3.17) should be related to this layer.

Consider now some other examples of the interpretation of single-frequency sensing data performed using the scheme (3.16). Figure 3.3a presents the profiles of the lidar returns $P(z)$, obtained during three nights in October 1973 at the test field of the Institute of Atmospheric Optics using the same lidar as in the experiments discussed just previously. Corresponding results of the interpretation are shown in Fig.3.3b and c as the profiles of $\beta_{ext}(z)$ and $T(z)$. These profiles obtained with a spatial resolution of 75 m demonstrate the behavior of the aerosol extinction coefficient $\beta_{ext}(z)$ as a function of altitude. The same curves demonstrate also the spatial inhomogeneities of the aerosol backscattering coefficient. The studies of these inhomogeneities are of great importance in investigations of gaseous components of the atmosphere, where the aerosol is used as a tracer (Chap.5).

As a conclusion to this section, it is interesting to compare the efficiency of the method considered here and that of the method of spectral transmittance.

Let the spectral photometer with the base y measure the transmittance of the atmosphere $P(y)$ at a wavelength λ with an error ε_p. The initial functional relation, in this method, is written as $p(y) = \exp(-\beta_{ext}\, y)$, where β_{ext} is some mean value of the aerosol extinction coefficient averaged over the sensing path of length y. In the case of an aerosol layer with thickness $y = z_2 - z_1$, the analogous relation has the form $\bar{S}(y) = b\, \beta_{ext} \exp(-2\beta_{ext}\, y)$, and the error of measurements are ε_p and ε_s, respectively.

Let us estimate the effects of the measurement errors ε_p and ε_s on the accuracy of the determination of b in order to compare the efficiencies of the methods. Applying the formula of finite increments to the cases considered, one can obtain that $(\overline{\Delta^2\beta_{ext}})^{1/2} \leq \varepsilon_p y^{-1}$ and $(\overline{\Delta^2\beta_{ext}})^{1/2} < \varepsilon_s\beta_{ext}/(1 - \beta_{ext}y)$, respectively). If the optical thickness τ is used, then these relationships will take the form $(\overline{\Delta^2\tau})^{1/2} \leq \varepsilon_s$ and $(\overline{\Delta^2\tau})^{1/2} \leq \varepsilon_s\tau/(1 - \tau)$.

One can easily see that the method of spectral transmittance becomes ineffective for small τ. The value ε_p in this method characterizes, in fact, the maximum sensitivity of the method to variations $\Delta\tau$. If sensings are made when τ is small then $(\overline{\Delta^2\tau})^{1/2} \lesssim \varepsilon\tau$ and, hence, the relative error

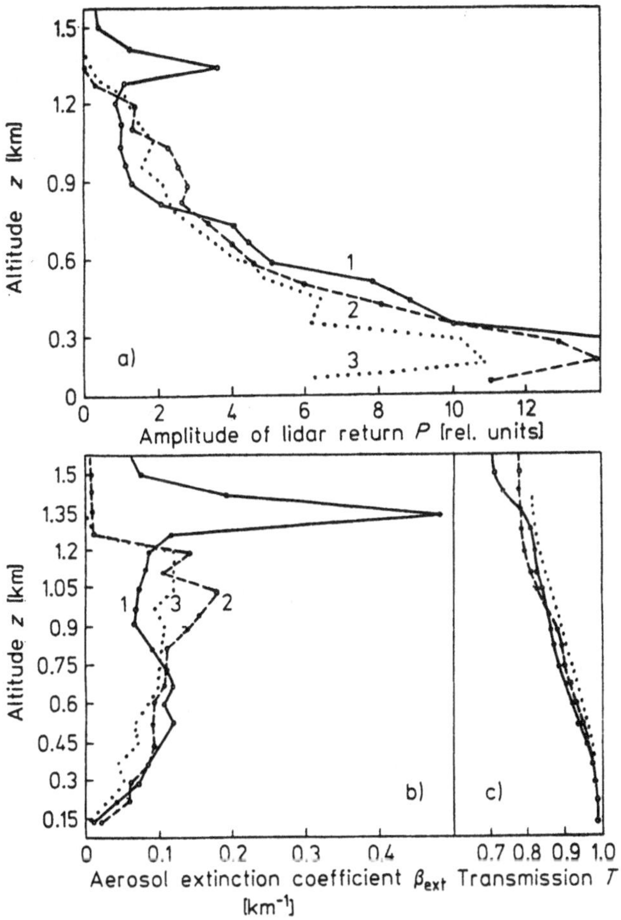

Fig.3.3. Examples illustrating the interpretation of lidar returns using the iteration scheme (3.16)

of the determination of $P(z)z^2/P_0B$ value is, at the same time, the relative error of the determination of τ. This causes the information potentialities of the lidar sensing to be higher as compared with those of the method of spectral transmittance, under the conditions of a slightly turbid atmosphere. It is for this reason that even the first experiments on the lidar sensing of the atmosphere showed that lidars can be used effectively to study the atmospheric aerosol stratification under clear air conditions [3.8,9]. This situation is explained by the fact that the amplitudes of lidar returns are proportional to the number of scatterers in the volume illuminated and, therefore, the functional equation, which relates the values $P(z)$ and $\beta_{ext}(z)$, is well conditioned (in the sense defined in Sect.2.6). At least, in the case of $b(z)$ = const, lidar sensing is a direct method for determining the optical characteristics β_π and β_{ext}. As for the method of spectral transmittance, the relation between $p(z)$ and $\beta_{ext}(z)$ is not as well conditioned and for small τ becomes useless for determining β_{ext}. The above advantages of the lidar method decrease as τ increases and it becomes necessary, in this case, to use qualitatively new relations between $P(z)$ and $\beta_{ext}(z)$ that would enable one to take into account the contribution of the multiply scattered light to the total lidar backscatter. These questions will be discussed at the end of this chapter.

3.3 Qualitative Methods for Data Interpretation in Multi-Frequency Sensing

3.3.1 The Concept of an Optically Active Interval of Sensing

In addition to rigorous methods for solving the functional equation (3.1), based on the use of the integral operators of light scattering by polydispersed systems of particles, it is possible in some cases, to use less rigorous techniques which have a simpler interpretation. Since we are discussing the determination of microstructures of aerosol ensembles from the data of multi-frequency lidar sensings $P(\lambda_i,z)$, methods for data interpretation and their peculiarities, in the case of the optical characteristics $\beta_{sc}(\lambda)$ and $\beta_\pi(\lambda)$ measured in the spectral interval Λ, should be considered. In turn, the interpretation of optical characteristics is based on the investigation of the relation between the peculiarities of the density function $s(r)$ and those of the spectral behavior of $\beta(\lambda)$ within the spectral interval. Let us consider a known example from atmospheric optics.

Assume that one can use the function $ar^{-\nu}$, [see (2.74)], to describe, as a first approximation, the size spectrum of the aerosol ensemble of the volume sounded. It can be shown that the integral $\int_{R_1}^{R_2} K(r,\lambda)\pi r^2 n(r)dr$ in this case may be written in the form $(\lambda/2\pi)^{-\nu+3} \int_{x_1}^{x_2} K(x)\pi a x^{-\nu+2}dx$ where $x_1 = 2\pi R_1\lambda^{-1}$ and $x_2 = 2\pi R_2\lambda^{-1}$.

When $R_2 \to \infty (R_1 \ll R_2)$, the last integral depends weakly on λ and consequently, the first approximation to $\beta(\lambda)$ is given by

$$\beta(\lambda) \cong c\lambda^{-\nu+3} \quad , \tag{3.18}$$

where c is some constant independent of λ. This relation is often used to interpret the spectral component of the atmospheric transmittance owing to the attenuation by aerosols. The interpretation procedure is very simple and consists of the following steps. Let us assume that the spectral behavior or $\beta_{ext}(\lambda)$, within the spectral interval Λ, can be approximated, in the first approach, by a power law with the exponential coefficient η, then ν can be determined using a simple relation, viz.

$$\nu^* \cong \eta + 3 \quad , \tag{3.19}$$

which follows from (3.18). This simple method is widely used in problems of applied optics (examples may be found elsewhere in [3.10]). Nevertheless, it should be noted that the use of (3.19) is justified when interpreting the spectral behavior of $\beta_{ext}(\lambda)$ only if the sought size spectrum n(r) has a shape described by a power law.

Some examples can be found, when the measured characteristic $\beta(\lambda)$ diminishes monotonically in the spectral interval Λ such that its behavior can be described approximately by the function $c\lambda^{-\eta}$ even though the size-distribution function n(r) has a very complicated form. As an example, Fig.3.4 presents the spectral behavior of the scattering efficiency factors $\bar{K}_{sc}(\lambda) = S^{-1}\beta_{sc}(\lambda)$ and $\bar{K}_{\pi}(\lambda) = S^{-1}\beta_{\pi}(\lambda)$ of a system of spherical particles with the microstructure s(r), which can in no way be approximated within the size interval R by a power law. If the spectral interval to the right from $\lambda = 0.7\mu m$ is considered, then the function $\beta_{sc}(\lambda)$ diminishes monotonically as $c\lambda^{-\eta}$ with η close to unity, but it does not follow from this that the size-distribution function is approximated by the function ar^{-4}. In this example the size-distribution function s(r) [equivalently n(r)] has two modes in the vicinities of points $r_s \cong 0.2$ and $r_s \cong 0.6\mu m$, i.e., it is bimodal.

The monotonic decrease of $\beta_{sc}(\lambda)$ in the interval to the right of $\lambda = 0.7\mu m$ can be explained by the fact that the behavior of optical characteristics $\beta(\lambda)$ is determined not only by the peculiarities of the density function s(r) but also by the behavior of the kernel $K_{sc}(r,\lambda)$ of the initial polydispersed integral. For clarity, let us consider the scattering efficiency factor $K_{sc}(x)$ shown in Fig.3.5. This curve reaches its maximum at a value of x between 4 and 5. This is in a good agreement with the known approximation for the abscissa of the first diffraction maximum of the factor $K_{sc}(x)$ in the form $x_m \cong 2/(\bar{m} - 1)$.

It is easily seen that if

$$2\pi R_2 x_m^{-1} = \lambda^* < \lambda \quad , \tag{3.20}$$

a particular factor $K(r,\lambda)$ will be a non-decreasing function within the interval $R = [R_1\ R_2]$ for $r \to R_2$. In this case, the following inequality is obviously valid

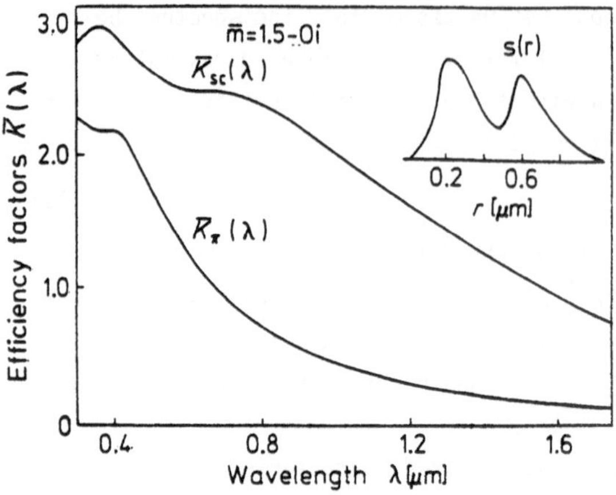

Fig.3.4

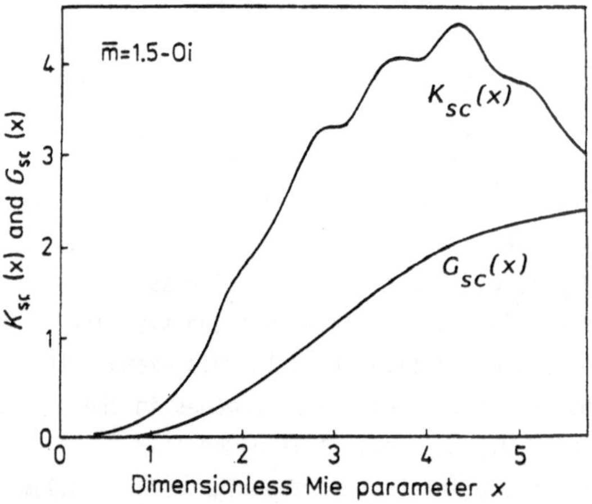

Fig.3.5

Fig.3.4. Spectral behavior of the polydispersed efficiency factors $\bar{K}_{sc}$ and $\bar{K}_{\pi}$ of an aerosol ensemble with the two-modes size-distribution

Fig.3.5. The scattering efficiency factor $K_{sc}(x)$ and corresponding function $G_{sc}(x) = x^{-1} \int_0^x K_{sc}(x')dx'$ appearing in (3.26)

$$\int_R K_{sc}(r,\lambda)s(r)dr \leq \max_{r \in R} K_{sc}(r,\lambda)S = K_{sc}(R_2,\lambda)S \quad , \tag{3.21}$$

where $\lambda > \lambda^*$. For this reason the optical characteristic $\beta_{sc}(\lambda)$ will be a decreasing function when λ increases above λ^*, since in this case the value $K_{sc}(R_2,\lambda)$ decreases monotonically as λ increases. It should be emphasized that this analytical property of the spectral characteristic $\beta_{sc}(\lambda)$ is not caused by any of the assumptions on the shape of the size spectrum n(r) of the dispersed media. Therefore, the decrease of $\beta_{sc}(\lambda)$ as $\bar{\lambda}$ increases above λ^*, is a general analytical property of this family of optical characteristics.

The estimation of λ^* gives a value close to 1μm, in good agreement with that obtained above from the spectral behavior of $\beta_{sc}(\lambda)$ (Fig.3.4). Thus it becomes clear that the estimations of the power ν in the size-distribution function $n(r) \cong ar^{-\nu}$ in terms of the value of η are valid only if the interval of optical sounding Λ, where the value of η is sought, is located to the right of λ^*. Otherwise, the value of η is by no means related to the parameter of microstructure ν of the medium sounded, and characterizes only the rate of decrease of the kernel $K(x)$ when $x \to 0$. Therefore, one should be very careful when using estimations like (3.19).

As the analysis made in [3.11] has shown, the characteristic feature of the sounding interval to the left of λ^* is the correlation between the spectral behavior of $\beta_{sc}(\lambda)$ and the function $s(r)$, i.e., if $s(r)$ is a diminishing function in the interval R, then $\beta_{sc}(\lambda)$ is also diminishing with respect to λ from $\Lambda^*(\lambda < \lambda^*)$. If $s(r)$ is convex, then $\beta_{sc}(\lambda)$ is also a convex function and so forth. In the case of unimodal size-distribution functions,the following approximate relationship can be used

$$r_s \leq \lambda_m/\pi(\bar{m} - 1) \quad , \tag{3.22}$$

where r_s is the mode of the size-distribution $s(r)$; λ_m is the point, in the interval of wavelengths, where the characteristic $\beta_{sc}(\lambda)$ reaches its absolute maximum. According to the peculiarity of the spectral interval $\Lambda^*(\lambda \leq \lambda^*)$ considered above, it was suggested in [3.11] that this interval be called the optically active interval of sensing.

The spectral measurements made in this interval are the most informative with respect to the function $s(r)$, especially in the task of revealing the local peculiarities of its analytical behavior. This result is valid regardless of the way the optical measurements are inverted. The spectral interval $\Lambda(\lambda > \lambda^*)$ is of certain interest, particularly for problems in which the optical constants of the aerosol substance are to be estimated using the methods of optical sensing. Indeed, in this case the behavior of $\lambda_{sc}(\lambda)$ is determined, as follows from (3.21), by the behavior of the kernel $K_{sc}(x)$ in the region of small x. If the decrease of $\beta_{sc}(\lambda)$ with increasing λ is not uniform, then this fact can serve as an indicator of the spectral dependence of the index of refraction of the aerosol substance.

The relation (3.22) can be used as a basis for a variety of approximate techniques to interpret spectral optical measurements in the IR region, in the case when atmospheric hazes are sounded.

The above relationships have outlined qualitative methods of the theory of optical sensing within spectral intervals that provide approximate assessments of the aerosol microstructure parameters to be made directly from corresponding spectral measurements. In terms of the number of optical measurements, these assessments can be obtained using quite a limited amount of information.

The formulation of inverse problems of light scattering in a more rigorous form (e.g., in the form of integral equations) requires, naturally, more measurements in broad spectral intervals, especially when the size interval R is large. It should be noted that the value of λ^* answers the question on the maximum value λ_{max} required for sensing the microstructure of a dispersed medium whose particle sizes do not exceed the value R_2 see[(3.20)].

3.3.2 Some Peculiarities of the Spectral Behavior of the Backscattering Coefficient

Although the qualitative approach developed above deals with the interpretation of the spectral behaviors of $\beta_{sc}(\lambda)$ and $\beta_{ext}(\lambda)$, it may also be generalized for making an analogous analysis of other optical characteristics, in particular of $\beta_\pi(\lambda)$. Difficulties may arise as a result of a rapidly oscillating form of the kernel $K_\pi(r,\lambda)$ in the corresponding polydispersed integrals $\int_R K_\pi(r,\lambda)s(r)dr$. However, if the size interval occupied by the aerosol particle is assumed to be wide enough, then the oscillating component of $K_\pi(r,\lambda)$ cannot markedly affect the spectral behavior of $\beta_\pi(\lambda)$, and a smoothed model of the efficiency factor can be used for making rough estimations. Obviously, the simplest method of removing the oscillating component from $K_\pi(r,\lambda)$ is the so-called local smoothing, performed with the use of the integral (2.46).

Figure 3.6 presents examples of such smoothed kernels, which will be denoted in the following by $\tilde{K}_\pi(x)$. The interval of local averaging has been chosen so that the explicitly oscillating component of $K_\pi(x)$ is removed. As is seen, e.g., from Fig.2.1, for $0 < x < 20$ and $\bar{m} = 1.65$, this oscillating component does not deviate much on the average from a harmonic oscillation with period Δx of about 0.8. The value of Δx increases slightly as the real part $\bar{m}$ of the refractive index of the aerosol substance decreases. The appearance of the imaginary part $\varkappa$ does not affect the period of this oscillation very much and results only in a sharp decrease of its amplitude. The values 1.5 and 1.65 of the real part of the refractive index are chosen for calculations because of their usefulness

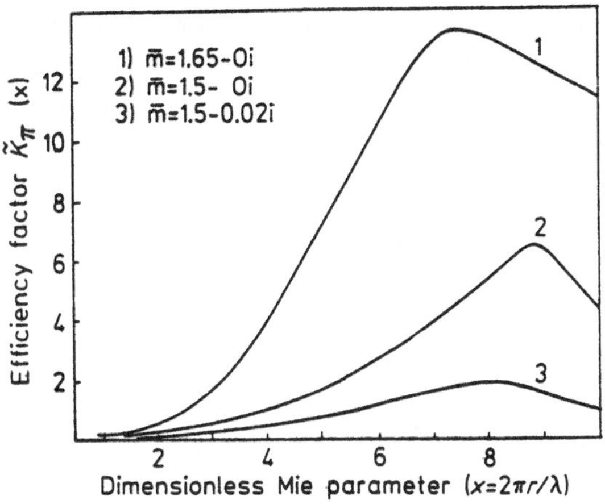

Fig.3.6. Examples of smoothed backscattering efficiency factors $K_\pi(r,\lambda)$ [see (2.46)]

in optical sensing of atmospheric hazes. A rapid increase of the smoothed efficiency factors $\tilde{K}_\pi(x)$ is observed as the real part of the refractive index increases, as is seen from the curves presented. An analogous increase occurs also for the polydispersed efficiency factors $\bar{K}_\pi(\lambda)$. Therefore, the more (optically) "tough" particles, i.e., those with larger values of $|\bar{m} - 1|$, produce stronger backscattering. In contrast, an increase of the imaginary part $\varkappa$ of the refractive index results in a reduced backscattered signal. As is seen from Fig.3.6, the absolute maxima of the function $\tilde{K}_\pi(x)$ are at values x_π^* between 7 and 9, which exceeds the location of the maximum for the efficiency factor $K_{sc}(x)$ (Fig.3.4) by a factor of approximately two. Thus, by analogy with the relation $x_{sc} = 2/(\bar{m} - 1)$ which is valid for the scattering efficiency factor, the relation $x_\pi^* \cong 4/(\bar{m} - 1)$ should hold for the locally smoothed efficiency factor $\tilde{K}_\pi(x)$. The corresponding value λ^* will be approximately half as large for a given R, as that for the coefficient $\beta_{sc}(\lambda)$. Therefore, the optically active interval, in the method of multifrequency lidar sensing, is shifted to the region of short wavelengths, as compared with that in the method of spectral transmittance where the values of $\beta_{ext}(\lambda)$ are measured.

Indeed, as can be seen from Fig.3.4, the value λ^* at which the efficiency factor $\tilde{K}_\pi(x)$ starts decreasing with increasing λ, is close to 0.4μm, i.e., it is approximately half as large as that for the factor $\bar{K}_{sc}(\lambda)$. This is very important for the theory of optical sensing of polydispersed aerosols. In particular, it follows that the interval of optical sensing Λ should be chosen to the left of that in which $\beta_{sc}(\lambda)$ is measured in order for the function s(r) to be reconstructed with an error not exceeding, for example the value σ, within the given size interval $R = [R_1, R_2]$. If one inverts the optical characteristics β_π and β_{sc} simultaneously on the same size interval, then the approximate inverse solutions $(K^{-1}_{\pi\alpha}\beta_\pi)(r)$ and $(K^{-1}_{sc\alpha}\beta_{sc})(r)$ will differ markedly from each other in the region of small r, all other conditions being equal. The first solution is more correct, in the sense that it describes the size-distribution $s_0(r)$ better in the region of large sizes, while the second one is more correct in the region of small sizes. This peculiarity of the inverse problem for the equation $K_\pi s = \beta_\pi$ should be taken into account in the analysis of multifrequency sensing of aerosols, especially when τ is small. Indeed, the value $T(z,\lambda) = \exp[-2\tau(z,\lambda)]$ is in this case close to unity and, as a consequence, the characteristic $\beta_{ext}(z,\lambda)$ is not necesarily revealed from the behavior of the inverted function $P(z,\lambda)$. Therefore, in the case of small τ values, the backscatter characteristic $\beta_\pi(\tau)$ contains, for practical purposes, almost all the information on the microstructures of the aerosols sounded.

3.3.3 The Logarithmic Derivative Method Applied to the Interpretation of Spectral Behaviors of Optical Characteristics

Let us now consider a more rigorous approach to the study of the analytical properties of the polydispersed integrals β_{sc} and β_π. Appropriate methods have been suggested and developed for application to the theory of optical sensing in [3.11-13]. These methods allow one to introduce some interesting relations between the parameters characterizing the analytical behavior of the size-distributions sought, on the one hand, and the functions $\beta_{sc}(\lambda)$ and $\beta_\pi(\lambda)$ measured experimentally, on the other. A single important functional property of the efficiency factors $K_{sc}(r,\lambda)$ and $K_\pi(r,\lambda)$ is the basis for deriving these relationships. These factors are the kernels of corresponding polydispersed integrals (polydispersed operators) and the above property is that the factors $K_{sc}(r,\lambda)$ and $K_\pi(r,\lambda)$ are uniform functions, of power zero, with respect to the variables r and λ, i.e., their values are determined by the value of the Mie parameter $x = 2\pi r\lambda^{-1}$. Incidentally, this holds for all the monodispersed efficiency factors from the theory of scattering of electromagnetic waves by particles.

For the polydispersed integral β, with a uniform kernel, the following relationship is valid

$$d\ \ln\beta/d\ \ln\lambda - 1 = \left[\int_R K(r,\lambda)s(r)(d\ \ln s/d\ \ln r)dr - c(\lambda)\right]/\beta \qquad (3.23)$$

where $c(\lambda) = R_2 s(R_2)K(R_2,\lambda) - R_1 s(R_1)K(R_1,\lambda)$.

Since (3.23) allows the calculation of the derivative of $\beta(\lambda)$ at any point λ of the spectral interval Λ, it can be regarded as a differentiation formula for spectral optical characteristics. The fact that the derivatives of $K(r,\lambda)$ with respect to λ are not necessary for this procedure is very important. It is known from the Mie theory that the efficiency factors $K(r,\lambda)$ are represented by a weakly converging series, which is not differentiable. Therefore, differentiation according to (3.23) gives a possibility for investigating the analytical properties of the spectral optical characteristics of polydispersed aerosol systems. Such investigations are useful for solving inverse problems of light scattering. In particular, they suggest development of new methods for the qualitative interpretation of optical measurements.

If the size-distribution $s(r)$ satisfies zero boundary conditions, i.e., $s(R_1) = s(R_2)$, then $c(\lambda)$ is equal to zero for all values of λ. It must be noted that the term $c(\lambda)$ is, as a rule, small in magnitude, even if the size-distribution $s(r)$ does not satisfy the above zero boundary conditions. Thus, for example, in the case of a power law size-distribution, $n(r) = ar^{-\nu}$ $(R_1 \le r \le R_2)$ the value $s(R_1)$ is not itself small, although the factor $K(r,\lambda)$ can be small provided that λ is chosen properly. Therefore, to a first approximation $c(\lambda)$ can be considered small, especially if the size-distribution function $s(r)$ (in the integrals) decreases for $r \to R_1$, and $r \to R_2$. Equation (3.23) relates the logarithmic derivatives of the characteristics $\beta(\lambda)$ and those of the size-distributions $s(r)$. It follows, in particular, from (3.23) that if the function $n(r)$ is described approximately by a power law $ar^{-\nu}$, the approximate relationship $d\ln\beta/d\ln\lambda - 1 \cong -\nu + 2$ is valid. As a consequence, the behavior of $\beta(\lambda)$ is also close to that described by a power law. Thus we have obtained (3.19) but in a slightly different way. Now, using this more general analytical method for investigating spectral characteristics, one can obtain from (3.23) some new relationships which are of interest for the interpretation of spectral measurements. In particular, it follows that

$$\max_{\lambda\in\Lambda}\left|\, (d\ln\beta/d\ln\lambda - 1\right| \le \max_{r\in R}\left|\, d\ln s/d\ln r\right| \quad . \tag{3.24}$$

If the smoothness of the functions $\beta(\lambda)$, $s(r)$ is characterized by the moduli of their logarithmic derivatives, it then follows from (3.24) that the spectral behavior of the optical characteristics $\beta(\lambda)$ will always be less smooth than that of the size-distribution functions of the aerosol ensembles. This peculiarity of the function $\beta(\lambda)$ follows from the basic property of the operators of light scattering by polydispersed systems, which is due to the

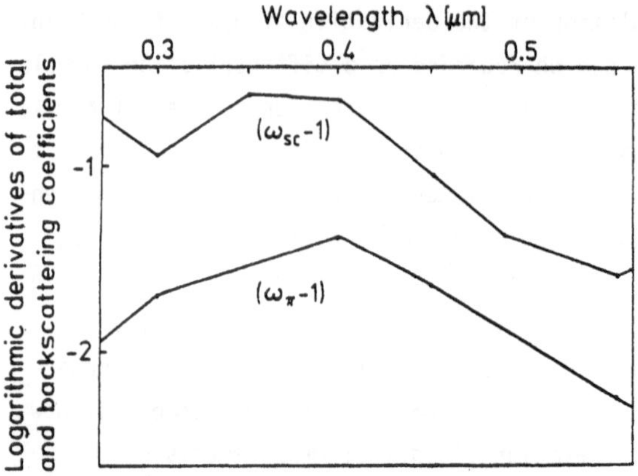

Fig.3.7. Spectral behavior of the function $[\omega(\lambda)-1]$ for the optical characteristics β_{sc} and β_π of the aerosol ensembly with the bimode size distribution from Fig.3.8

fact that the transformation $K: \Phi \to B$ smoothes the size-distribution functions. By comparing the values of $|d\ln\beta/d\ln\lambda|$ and $|d\ln s/d\ln r|$, one can estimate the smoothing effect of the integral operator K. In Sect.3.6 these aspects will be discussed in connection with estimating the acceptable smoothness of the regularized solutions of the equation $K_\pi s = \beta_\pi$. Unfortunately, the direct application of (3.24) appears to be useless in estimating the moduli of the logarithmic derivatives of the size-distributions s(r), due to the very smooth spectral behaviors of the light scattering characteristics as compared with those of the aerosol size-distribution functions used. As an example, the curves $\omega(\lambda) - 1 = (d\ln\beta/d\ln\lambda - 1)$ are presented in Fig. 3.7. These curves correspond to the characteristics shown in Fig.3.4. The maximum value of $\nu(r) = d\ln s/d\ln r$ reaches 12, in the case of the reference size-distribution, while the value $\max_\lambda|\omega(\lambda) - 1|$ does not exceed 3. One can use (3.23) to strengthen the inequality (3.24) [for $c(\lambda) \approx 0$] as follows

$$|\omega(\lambda) - 1| \leq (1/\beta) \int_R K(r\lambda)s(r)|d\ln s/d\ln r|\,dr \quad .$$

Let the known Chebyshev inequality for definite integrals be applied to this integral. One obtains, in this case,

$$\int_{R_1}^{R_2} K(r,\lambda)s(r)|\nu(r)|dr < \int_{R_1}^{R_2} K(r,\lambda)s(r)dr \int_{R_1}^{R_2}|\nu(r)|dr/(R_2 - R_1) \quad .$$

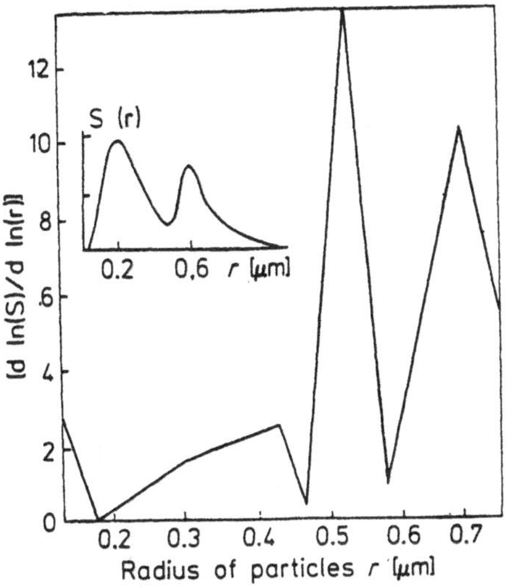

Fig.3.8. Double-peaked model size-distribution function for the atmospheric haze and modulus of its logarithmic derivative

Designating the mean modulus of the logarithmic derivative of s(r) as $<|\nu|>$ one obtains

$$|\omega(\lambda) - 1| \leq <|\nu|> \quad ,$$

which is valid for all values of λ, including that at which $|\omega(\lambda) - 1|$ reaches a maximum. Hence, one has finally the estimation

$$\max_{\lambda \in \Lambda} |\omega(\lambda) - 1| \leq <|\nu|> \quad . \tag{3.25}$$

In (3.25) the mean value of the logarithmic derivative $\nu(r)$ is used and, as a consequence, its estimation in terms of $|\omega(\lambda) - 1|$ is more reliable. Figure 3.8 presents an example, in which $<|\nu|>$ reaches a value of about 4. This value is closer to the values of $\max|\omega(\lambda) - 1|$ (Fig.3.7).

For the two-modes size-distribution function used in previous examples as a model, the mean modulus of the logarithmic derivative $<|\nu|>$ does not exceed 4 (Fig.3.8). It can easily be shown from the spectral behavior of $(\omega - 1)$ for β_π and β_{sc} (Fig.3.7) that the value of $\max|\omega(\lambda) - 1|$ for this model is between 2 and 3. As a consequence, one can state that (3.25) is more efficient than (3.24) for estimating the parameters of the size-distribution s(r). However, the latter case deals only with the estimation of mean values. It should be noted, in conclusion, that $(\omega_\pi - 1)$ is larger in absolute value than $(\omega_{sc} - 1)$ within the spectral interval presented in Fig.3.7. This indicates a weaker smoothing effect of the operator K_π on the size-distributions belonging to Φ, as compared with that of K_{sc}. As a consequence, the inverse problems for the characteristic β_π are better conditioned than corresponding problems for β_{sc}.

3.3.4 Methods for Determining Boundaries of the Size-Spectra from Spectral Optical Measurements

The problem considered here arises from the fact that the integral representation $\beta(\lambda) = \int_R K(r,\lambda)s(r)dr$ is defined only in the case when the limits R_1 and R_2 of the size interval R are known. If atmospheric hazes are sounded then, as a rule, $R_1 \ll R_2$ and R_1 can be considered close to zero as a first approximation. Usually, in calculations of optical characteristics, these values are of the order of 0.02 - 0.05µm. Actually, the uncertainty in the choice of R_1 does not, at least in the visual range, affect the magnitude of the polydispersed integral $\beta(\lambda)$ significantly since in this case the values of the efficiency factor $K(r,\lambda)$ are small. On the other hand, the same cannot be said about the value of R_2 because the efficiency factor $K(r,\lambda)$ is large for r near R_2.

The formulation ot the inverse problem of light scattering becomes indefinite if the limit R_2 is unknown. An incorrect choice of R_2 can yield significant errors in the solutions obtained. The a priori assessments of R_2 values cannot always be made with an admissible accuracy. In this connection, it may be advisable to use the measured spectral variation of the optical characteristics to find a reasonable estimate of the value of R_2. In this method, the value R_2 is regarded as an unknown parameter of the microstructure. In order to find it, a spectral subinterval is to be selected from the sensing interval Λ, in which the behavior of $\beta(\lambda)$ is determined mainly by the particles having sizes close to the limiting value R_2.

It is obvious that the interval Λ' extending to the right of λ^*, mentioned above in connection with (3.20), can serve as such a subinterval, since it contains λ that, as a rule, exceed λ^*. Therefore, for a given polydispersed ensemble of particles, the interval Λ' can be regarded as an interval of optical sensing in the region of large λ. Using experimental data for $\beta(\lambda)$ in the interval Λ', one can easily construct an approximate estimation R_2^* for the R_2 value sought [3.14]. Such an estimation is based on the fact that the parameter R_2 determines, via $K(R_2,\lambda)$ in (3.21), the rate at which $\beta(\lambda)$ falls as λ increases. Note that this property, i.e., the dependence of the slope of $\beta_{ext}(\lambda)$ on the value of the effective boundary R_2 was known earlier [3.15,16]. In the following discussion we shall try to substantiate this fact more rigorously and derive a number of relationships for use in the approximate evaluation of R_2.

Let us introduce the normalizing function $\varphi(r) = S^{-1}s(r)$ and consider the polydispersed efficiency factors $\bar{K}(\lambda) = \int_R K(r,\lambda)\varphi(r)dr$. In addition to $\bar{K}(\lambda)$, introduce the integral $\int_R K(r,\lambda)\psi(r)dr$ where $\psi(r) = (R_2 - R_1)^{-1}$ is the uniform size-distribution function ($R_1 \le R \le R_2$). By partial integration, one can obtain for these integrals

$$\int_{R_1}^{R_2} K(r,\lambda)\varphi(r)dr = K(R_2\lambda) - \int_{R_1}^{R_2} q_\varphi(r)K_r'(r,\lambda)dr$$

$$\int_{R_1}^{R_2} K(r,\lambda)\psi(r)dr = K(R_2\lambda) - \int_{R_1}^{R_2} q_\psi(r)K_r'(r,\lambda)dr \quad ,$$

where $q_\varphi(r) = \int_{R_1}^{r} \varphi(r)dr$ and $q_\psi(r) = \int_{R_1}^{r} \psi(r)dr(q_\varphi = q_\psi = 0)$ at the point $r = R_1$ and $q_\varphi(r) = q_\psi(r) = 1$ when $r = R_2$. If the function $q_\varphi(r)$ is convex from below, then it lies above its chord $q_\psi(r) = rR_1(R_2 - R_1)$, and, as a consequence, $q_\varphi(r) \ge q_\psi(r)$ everywhere in the interval R. Since the subinterval Λ' is characterized by decreasing efficiency factors $K(r,\lambda)$ as $r \to R_1$ for all λ, then $K'(r, \ ) \ge 0$ everywhere in R, and, consequently,

$$\int_{R_1}^{R_2} q_\varphi K_r'(r,\lambda)dr \ge \int_{R_1}^{R_2} q_\psi K_r'(r,\lambda)dr \quad ,$$

from which the inequality

$$\int_{R_1}^{R_2} K(r,\lambda)dq_\varphi(r) < \int_{R_1}^{R_2} K(r,\lambda)dq_\psi(r) \qquad (\lambda \in \Lambda')$$

follows, which leads to the relationship

$$\bar{K}(\lambda) < \int_{R_1}^{R_2} K(r,\lambda)dr/(R_2 - R_1) \quad . \tag{3.26}$$

If the polydispersed factor $\bar{K}(\lambda)$ is assumed to be known, at least for a certain λ in Λ', then, designating the right-hand side of (3.26) as $G(R_2)$, one obtains the following equation for estimating R_2

$$\bar{K}(\lambda) = G(R_2^*,\lambda) \quad . \tag{3.27}$$

Since the function $G(R_2^*,\lambda)$ does not decrease as R_2 increases for λ in Λ' (an example of the kernel $K_{sc}(\lambda)$ is shown in Fig.3.5), the solution R_2^* of (3.27) is smaller than the value of R_2 sought for a given polydispersed ensemble. Taking this into account, one can see that R_2^* is a lower bound for the upper limit R_2 of the aerosol size-distribution.

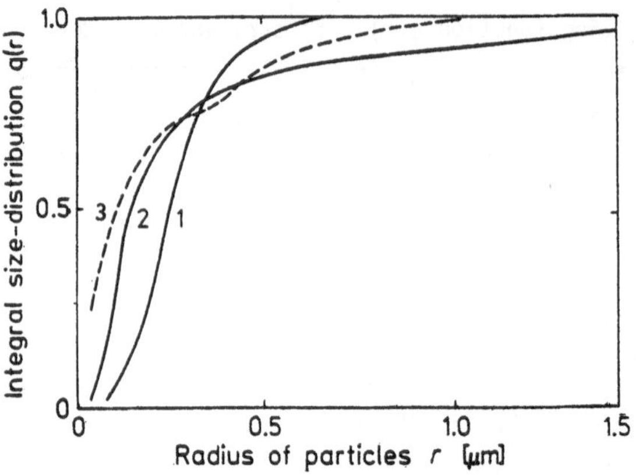

Fig.3.9. Normalized integral size-distribution functions q(r) for the haze H [3.2] (Curve 1); Curve 2 is the Junge size-distribution function with $\nu = 4$, and Curve 3 represents the experimental data of [3.5]

When deriving (3.26), we assumed $q_\varphi(r)$ to be a convex (from below) function. In practice, this is not always valid. As an example, several size-distributions are presented in Fig.3.9, which are usually used as models for atmospheric hazes. Curve 1 is a size-distribution one like (2.75) with $r_s = 0.2\mu m$ (haze H [3.2]). The second curve corresponds to the model (2.75) with $\nu = 4$ and $r_2 = 0.1\mu m$. The dashed curve is plotted according to the data of direct sampling of aerosol particles from the air of the ground layer using filters [3.5]. In the case of the first model, one can see that there is an interval of r values, near the left boundary $R_1 = 0$, for which $q_\varphi(r)$ lies below the chord $q_\psi(r)$, which passes through the points $q(R_1) = 0$ and $q(0.7) \cong 1$. However, in this interval the values of the efficiency factor $K(r,\lambda)$ are small for all λ from Λ and hence this fact does not violate the inequality (3.26).

Integral size-distributions q(r), for which the condition (3.26) is valid, can be called on the average convex, size-distributions with a given kernel $K(r,\lambda)$. In the case of such functions q(r), the size-distribution density $\varphi(r)$[or equivalently s(r)] should decrease on the average as $r \to R_2$. The size intervals, in which the density function increases, are mainly local.

Let us consider briefly a technique for determining R_2^* from (3.27). Assume that the spectral interval of optical sounding $\Lambda = [\lambda_{min}, \lambda_{max}]$ is sufficiently wide so that it includes wavelengths both larger than and smaller than λ^*. In this case, according to (3.26),

$$\bar{K}(\lambda_{max}) \leq G(R_2,\lambda_{max}) \quad .$$

In turn, $\bar{K}(\lambda_{max}) = S^{-1}\beta(\lambda_{max})$, where the total geometrical cross-section of particles per unit volume can be estimated by using a maximum value of $\beta(\lambda)$ on the interval Λ. Let the point of the absolute maximum of the measured characteristic $\beta(\lambda)$ be in the interval Λ. In this case, $S = \beta_m/\bar{K}_m$, where β_m is the value of absolute maximum of $\beta(\lambda)$ and $\bar{K}_m$ is the maximum value of the efficiency factor $\bar{K}(\lambda)$. Of course, the value of $\bar{K}_m$ depends upon the normalized size-distribution $\varphi(r)$ and it is practically impossible to estimate it a priori. However, for a kernel such as $K_{sc}(r,\lambda)$, which is a simpler function as compared with $K_\pi(r,\lambda)$, one can succeed in obtaining a reasonable estimate of $\bar{K}_m$. In particular, as calculations have shown for typical model size-distributions (Sect.2.6), the value $\bar{K}_m$ lies within the interval from 2.6 to 3.3 (when the real part of the refractive index of the aerosol substance is $1.4 \leq \bar{m} \leq 1.6$).

The narrower the size-distribution s(r) the larger is the value of K_m (i.e., the narrower is the interval R of averaging in corresponding polydispersed integrals). Thus, for example, the value K_m is close to 3.3 for the haze H as it is the narrowest of the model size-distributions M, L and H. The broad size-distribution s(r) of model M yields a strong averaging of the monodispersed efficiency factor $K_{sc}(r,\lambda)$, when the characteristic $\beta_{sc}(\lambda)$ is calculated, and, as a consequence, the corresponding values of $\bar{K}_m$ are close to 2.6. It should be noted, in this connection, that the ratio $K_m/\bar{K}_m$ can be regarded, to some extent, as a characteristic of the polydispersedness of the ensemble of particles sounded. Unfortunately, such assessments cannot be carried out for the factors $\bar{K}_\pi(\lambda)$, since, on the one hand, their values depend strongly upon the refractive index of the aerosol substance and, on the other hand, there is no distinct absolute maximum of diffraction in this case. Using the estimation $\bar{K}_m$ and the last inequality, one can obtain the following expression for determining R_2^*

$$\alpha_\beta \bar{K}_m = G(R_2^*,\lambda_{max}) \quad . \tag{3.28}$$

The value $\alpha_\beta = \beta(\lambda_{max})/\beta_m$ in (3.28) characterizes the mean slope of the measured optical characteristic $\beta(\lambda)$ within the limits of the spectral interval Λ. It is obvious that the value of α_β cannot be smaller than the value of the smallest relative error of optical measurements $\varepsilon_{min} = \sigma/\beta_m$. If, in (3.28) one assumes $\alpha_\beta = \varepsilon_{min}$, then the solution R_2^* of (3.28) is the maximum resolvable value of the upper limit of the size-distribution s(r),

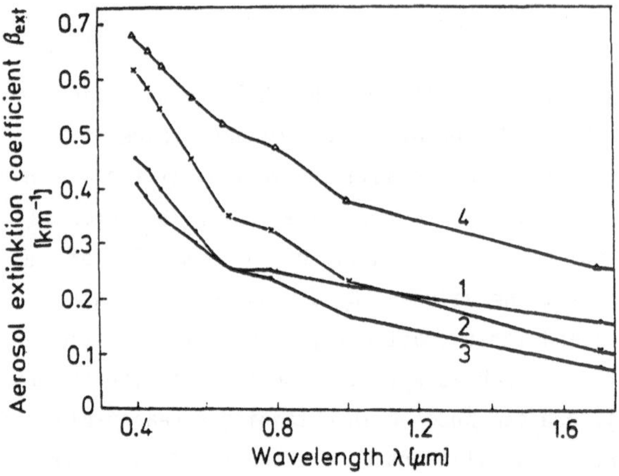

Fig.3.10. Spectral behavior of the atmospheric haze extinction coefficient $\beta_{ext}(\lambda)$ [3.15]. Curve 1 represents the data obtained on April 28, Curves 2 and 3 on April 29 at 11 and 13 h. LT, respectively, and Curve 4 represents the data obtained on April 30

in a given optical experiment (i.e., for given λ_{max} and ε_{min} [3.17,18]). It is important that if R_2 is known, in principle, to determine a required value of λ_{max} as the upper limit of the sensing interval Λ from (3.28). To illustrate this technique, we consider some examples.

In Fig.3.10 the curves $\beta_{ext}(\lambda)$ plotted according to the data from [3.15] are presented. The measurements were carried out in the spectral region of 0.4 - 1.8μm over a period of several days. In the order in which the curves are numbered, the α_β values are: 0.372; 0.165; 0.175; 0.405. Assuming $\bar{m} = 1.5$, on the average, [the corresponding function G(x) is presented in Fig.3.5], one can find from (3.28) that the values of R_2 are: 0.86; 0.59; 0.63; 0.94μm, respectively. Recall that these values are estimations of the maximum sizes R_2 of particles of the ensemble from below, that means that real values of R_2 are somewhat larger than these values. Of course, only the effective maximum size of particles is meant here; thus it is possible that particles with sizes $r > R_2$ are present in the ensemble. However, their optical contribution to the scattering process does not exceed the errors of optical measurements. In this connection, a question arises on the more exact definition of the concept of the effective upper limit of the size-spectrum of a polydispersed ensemble.

Let us consider, according to [3.17], the value of R_2 satisfying the condition

$$1 - q(R_2) = \alpha_2 \leq \eta\varepsilon_{min} \quad ,$$

as such a boundary, where α_2 is the quantiple of the size-distribution, ε_{min} is the smallest relative error of the optical measurements in the spectral interval Λ, and $\eta \geq 1$ is the reliability coefficient. Using this relationship,one can assess the effectiveness of the values R_2 found in this example.

Assume that the size spectra of the atmospheric hazes, to which the curves presented in Fig.3.10 correspond, can be represented by the curves shown in Fig.3.9. These size-distributions can be used for estimating the value of α_2. It follows from the curves shown in Fig.3.9 that the values of the quantiple α_2 are, even in the worst case, not smaller than 10% or 15%.

Let us consider also the asymptotic behavior of the R_2 assessments. The Mie theory predicts that the efficiency factors K(x) have the property that $K(x) \to Ax^n$, as x vanishes (or $\lambda \to \infty$). Here A is some constant that depends on the refractive index of the aerosol substance. In particular, for the case of $K_{sc}(x)$ $n \simeq 4$, as is shown in [3.14] $\beta(\lambda) \to K(\sqrt[n]{\overline{r^n}}\lambda)S$ and $R^* \to \sqrt{(n+1)\overline{r^n}}$ for $\lambda \to \infty$, where $\overline{r^n}$ is the nth moment of the normalized size-distribution $\varphi(r)$. Note that if the kernel of a polydispersed integral approaches a function described by a power law of degree n, then $\beta(\lambda)\cdot\lambda^n \to$ const. Of course, realistic optical characteristics have more complicated behavior due to the influence of a spectral dependence of the refractive index of the aerosol substance. Such a situation naturally complicates the practical use of the above methods for the interpretation of spectral optical measurements, since it is then necessary to take into account the dependences $\bar{m}(\lambda)$ and $\varkappa(\lambda)$ [3.19, 20].

3.3.5 On the Interpretation of the Spectral Behavior of Lidar Signals

All the above relates to the optical characteristics β_π, β_{sc} and has nothing to do with the lidar signals $P(z,\lambda)$ obtained in spectral intervals Λ. However, it is obvious that the spectral behaviors $\beta_\pi(\lambda)$ and $\beta_{ext}(\lambda)$ will affect the lidar signals $P(z,\lambda)$. These questions will be discussed in more detail here. Let us assume that the characteristics $\beta_\pi(\lambda)$ and $\beta_{ext}(\lambda)$, in the spectral interval Λ, can be approximated within the experimental accuracy by the power law functions $c_\pi\lambda^{-\eta_2}$ and $c_{ext}\lambda^{-\eta_1}$, respectively. Choose two wavelengths from that spectral interval, viz. λ_1 and λ_2. It can easily be shown that the functions $T(z,\lambda_1)$ and $T(z,\lambda_2)$ are related by

$$T(z,\lambda_2) = [T(z,\lambda_1)]^{\lambda_1^{\eta_1}\lambda_2^{-\eta_1}} \quad . \tag{3.29}$$

Denote, as before, the value $P(z,\lambda)z^2/B(\lambda)P_0(\lambda)$, as $\bar{S}(z,\lambda)$, where $P(z,\lambda)$ is the amplitude of the lidar return received from the scattering volume at a distance z from the receiver, $B(\lambda)$ is an instrumental constant and $P_0(\lambda)$ is the power of a sensing pulse of light. It is not difficult to show that for any λ_1 and λ_2 from Λ the following expression can be derived

$$T(z,\lambda_1) = \left[\bar{S}(z,\lambda_2)\lambda_2^{\eta_2}/\bar{S}(z,\lambda_1)\lambda_1^{\eta_2}\right]^{1/(\lambda_1^{\eta_1}\lambda_2^{-\eta_1}-1)} \quad . \tag{3.30}$$

This expression can serve as a basis for the development of methods for estimating the spectral transmittance based on the data of two-frequency (and/or multifrequency) lidar sensings. In order to make use of this formula, one should choose, in some manner, estimates for the values η_1 and η_2. As to the value of η_1, vast material on its determination can be found elsewhere in the literature [3.21-23]. Unfortunately, investigations of the spectral behavior of $\beta_\pi(\lambda)$ have not so far been systematic. It is possible that the wider use of multifrequency lidars in the practice of atmospheric researches will make such systematic investigations available.

Let us use the estimations of η_1 and η_2 made in [3.22]. Corresponding optical measurements have been carried out with the use of nephelometers in the spectral region [0.3μm, 0.7μm]. The sites of measurements involve different geographical regions. The mean values of η_1 and η_2 as estimated from the measurements made at the seaside (near Odessa) are 0.57 and 1.02, respectively. In the highlands, near Alma-Ata, these values were: $\eta_1 = 1.42$ and $\eta_2 = 1.80$. It is seen that in all cases $\eta_1 < \eta_2$, although such estimations are, to say generally, relative.

The problem is that the measured functions $\beta_{sc}(\lambda)$ and $\beta_\pi(\lambda)$ are not always diminishing monotonically with increasing λ in the visible range and, as a consequence, the power law approximations $\beta_{sc}(\lambda) \cong c\lambda^{-\eta}$ are not always valid. The measurements made in [3.22] have shown that the characteristics $\beta_{sc}(\lambda)$ and $\beta_\pi(\lambda)$ diminished monotonically with increasing λ (in the region from 0.3 to 0.7μm) only in 50%-60% of the total number of observations, while in about 20% of the cases uniform spectral behavior was observed. With approximately the same frequency of occurrence, convex (from below) characteristics are observed and in about 5%-10% of cases one can observe $\beta_{sc}(\lambda)$ and $\beta_\pi(\lambda)$ increasing as λ increases.

Thus the interpretation methods based on power law approximations to the spectral behavior of the above characteristics are essentially qualitative and one should be very careful when using them in practice. Therefore, it is preferable to estimate η_1 and η_2 from the data of accompanying measurements of optical characteristics and, above all, of $T(z,\lambda)$. In doing so, one should proceed from the fact that if $\beta_\pi(\lambda)$ and $\beta_{ext}(\lambda)$ have certain characters in the spectral interval Λ, then the optical signals $P(z,\lambda_1)$ and $P(z,\lambda_2)$ are also definite functions of the wavelength. Indeed, since $T(z,\lambda) < 1$ for all λ and z values, one has for any z, λ_1 and λ_2 that

$$\bar{S}(z,\lambda_2)\lambda_2^{\eta_2}/\bar{S}(z,\lambda_1)\lambda^{\eta_2} < 1 \quad .$$

This inequality leads to the condition

$$P(z,\lambda_2)\lambda_2^{\eta_2} < P(z,\lambda_1)\lambda_1^{\eta_2} \tag{3.31}$$

which means that the value of η_2 used in the interpretation of two given profiles $P(z,\lambda_1)$ and $P(z,\lambda_2)$ cannot be arbitrary. Thus, use of (3.31) enables one to avoid major errors when choosing η_2 a priori. If, during an optical experiment, one can measure $T(z^*,\lambda_1)$ and $T(z^*,\lambda_2)$ (z^* is a point on the sensing path) then it is possible to estimate η_1 and η_2 from (3.29) and (3.30).

Note, in conclusion, that the closer to each other are λ_1 and λ_2 the better can be the choice of the values of η_1 and η_2 provided that appropriate accuracy of the corresponding optical measurements can be achieved, since, in this case, the dependences (3.29) and (3.30) are not as well conditioned (in the sense of Sect.3.2). Therefore, one should be extremely careful when interpreting such optical data.

3.4 Determination of the Aerosol Microstructure and Index of Refraction Using Methods of Multi-Frequency Sensing

3.4.1 The Method of Optical Operators (Variational Approach)

Remote sensing is considered for the determination of the microstructure and the refractive index of the aerosol substance from the data of lidar sensings along horizontal and slanted paths. Such problems arise in situations for which the aerosol refractive index is unknown and, moreover, cannot be set a priori with an admissible accuracy. The aerosol backscattering coefficient $\beta_\pi(\lambda)$ strongly depends, in the visible region, upon the index of refraction $\bar{m}$ of the aerosol substance and, hence, the solutions $s_\alpha = K_{\pi\alpha}^{-1}\beta_\pi$ are very sensitive to the error $\Delta\bar{m}$. It should be noted, in this connection, that in the case of natural and well studied aerosols, it is possible to set such parameters beforehand, while in the case of aerosols of artificial origin such a priori setting is problematic. Moreover, the purpose of lidar sensing is to determine $\bar{m}$.

It is obvious that if one must determine not only the microstructure of aerosols but also the index of refraction of the particulate matter, the amount of measured information should be correspondingly enlarged; as a rule, this is possible only at the expense of additional measuring tools. In this case, the question arises as to the best way to make a supplementary definition of the data of multi-frequency sensings. Since the uncertainties in the refractive index of the aerosol substance are important in this case, then, in supplementary measurements, one should measure a characteristic of light scattering which depends less strongly on variations $\Delta\bar{m}$ than $\beta_\pi(\lambda)$.

Since the lidar returns depend primarily on $\beta_\pi(\lambda)$ (when the optical depth of the atmosphere is small enough), one should choose an optical characteristic $\beta(\lambda)$ which obeys the condition

$$\| \partial \ln \beta(\bar{m},\lambda)/\partial\bar{m} \| \ll \| \partial \ln \beta_\pi(\bar{m},\lambda)/\partial\bar{m} \| \qquad (3.32)$$

at every wavelength at which the multi-frequency lidar operates. As a numerical analysis of the properties of the aerosol optical characteristics has shown, the aerosol scattering (extinction) coefficient can be taken as such a characteristic in the visible, since its dependence on $\bar{m}$ and, hence, the variations $\Delta\beta_{ext}$ caused by $\Delta\bar{m}$, are much weaker than those of β_π. It is for this reason that the idea of making a supplementary definition of the data of multi-frequency lidar sensings by using data on the spectral transmittance of the atmosphere arises.

Such experiments are possible when the sensings are made along horizontal or slant paths. Since $\beta_{ext}(\lambda)$ enters into the lidar equation (3.1), it can be inferred from the lidar signals $P(\lambda)$ provided, of course, that certain conditions are fulfilled. Let the calibrated targets be placed at distances z' and z" along the sensing path and have reflection coefficients $\beta_{\pi M}$. In this case one can easily find the values $T(z',\lambda)$ and $T(z'',\lambda)$ which allow the estimation of the mean extinction coefficient $\beta_{ext}(\lambda)$ over the path portion between z' and z". Using this, one can find the value of $\beta_{ext}(\lambda)$ from (3.1) and then use the array of data $\{\beta_\pi(\lambda_i), \beta_{ext}(\lambda_i)\}$ $(i = 1,2,\ldots,n)$ as reference information. A corresponding inverse problem is formulated, in this case, in the form of two integral equations

$$\int_R K_\pi(\bar{m}(\lambda_i), r, \lambda_i)s(r)dr = \beta_\pi(\lambda_i)$$

$$\int_R K_{ext}(\bar{m}(\lambda_i),r,\lambda_i)s(r)dr = \beta_{ext}(\lambda_i) \quad i = 1,2,\ldots,n \tag{3.33}$$

in which the unknown functions are $s(r)$ and $\bar{m}(\lambda)$. This inverse problem will be written in vector form:

$$K_\pi(\bar{\mathbf{m}})\mathbf{s} = \boldsymbol{\beta}_\pi$$

$$K_{ext}(\bar{\mathbf{m}})\mathbf{s} = \boldsymbol{\beta}_{ext} \quad . \tag{3.34}$$

In order to find the vectors $\bar{\mathbf{m}}$ and $\mathbf{s}$ from (3.34), one can construct operators analogous to the operator W, which has already been introduced into the theory of optical sensing of dispersed media. It is advisable, in this case, to choose the operator $\hat{W} = K_\pi K^{-1}_{ext\alpha}$, which, for every fixed $\bar{\mathbf{m}}$, is the operator which maps the vector $\boldsymbol{\beta}_{ext}$ to $\boldsymbol{\beta}_\pi$ defined on the sensing interval Λ. If the particles sphericity is assumed, the above operators (or matrices) can be constructed using the Mie formulas. Assuming the operator W to be a continuous function of the index of refraction, one can determine the vector $\bar{\mathbf{m}}$ by solving the following variational problem

$$\min_{m \in \Omega_m} \|\hat{W}(\bar{\mathbf{m}})\, \beta_{ext} - \beta_\pi\|_{L_2} \tag{3.35}$$

where Ω_m is the region where the variables $\bar{m}(\lambda_i)$ are defined. As a first approximation, one can consider $\bar{m}(\lambda)$ to be constant within the limits of the interval Λ. In this case, (3.35) takes the simple form

$$\min_{\bar{m} \in [m_1,m_2]} \|\hat{W}(\bar{m})\beta_{ext} - \beta_\pi\|_{L_2} \quad . \tag{3.36}$$

In this problem the value $\bar{m}^*$ obtained is a mean value over the interval Λ. The formulation of the variational problem (3.36) is useful when the interval Λ is not too wide and significant selective absorption is present.

After determining $\bar{\mathbf{m}}$ from (3.35), one can proceed to the estimation of the microstructure of the aerosol. The corresponding approximate solution is written as

$$\mathbf{s}_\alpha = K^{-1}_{ext\alpha}(\bar{\mathbf{m}})\, \boldsymbol{\beta}_{ext} \quad . \tag{3.37}$$

From (3.37) one can see why the operator $\hat{W}$ was used instead of $W = K_{ext}K^{-1}_{\pi\alpha}$ (Sect.3.1), when solving (3.34). Since the characteristic $\beta_{ext}(\lambda)$ is less affected by variations of the refractive index, then the set of approximate solutions $\mathbf{s}_\alpha$ which are generated by the operator $K^{-1}_{ext\alpha}$ is more stable with respect to uncertainties in $\bar{m}(\lambda)$. This means that the operator $K^{-1}_{ext\alpha}$ is more useful in the determination of microstructures of dispersed media than $K_{\pi\alpha}^{-1}$ when uncertainties in the optical constants of the aerosol substance are present, although the first equation of (3.34) may be conditioned better than the second one (Sect.2.5).

3.4.2 The Iteration Method and Inversion of the Scattering Coefficient in the Visible

The application of this general method for solving the inverse problem of light scattering in the form (3.33), which uses the direct construction of the inverse and direct operators of the theory of light scattering by polydispersed ensembles of spherical particles, is difficult, since it is practically impossible to assess the effect of the experimental inaccuracies on the correctness of the interpretation by using only an elementary mathematical apparatus. For this region,it seems to be advisable to discuss inversion schemes which are less general in form but more simple and illustrative in application. The iteration method can be considered as one such scheme for solving (3.34). The vector $\mathbf{s}$ is sought in the same manner as the vector $\bar{\mathbf{m}}$ is, i.e., by solving the appropriate variational problem. The idea of the method is the following. Let the zero approximation of $\bar{\mathbf{m}}$ be set according to the a priori estimations. Then, using this vector denoted as $\bar{\mathbf{m}}^{(0)}$, one can find the first approximation of the vector $\mathbf{s}$ minimizing the functional

$$T_\alpha(\mathbf{s},\bar{\mathbf{m}}^{(0)}) = \rho^2(K_{sc}(\bar{\mathbf{m}}_\lambda^{(0)})\mathbf{s}, \boldsymbol{\beta}_{ext}) + \alpha\Omega^2(\mathbf{s}) \tag{3.38}$$

in the space $\boldsymbol{\Phi}_m$. The solution $s^{(1)}$ thus obtained can be used further to correct the value of the refractive index of the aerosol substance for a given spectral interval Λ. The first approximation of $\bar{\mathbf{m}}$ can be found with the use of the variational method for the quadratic functional

$$\rho^2(K_\pi(\bar{\mathbf{m}})\mathbf{s}^{(1)}, \boldsymbol{\beta}_\pi) = \sum_{i=1}^{n}\left[\sum_{\ell=1}^{n} G_{\pi i\ell}s_\ell - \beta_\pi(\lambda_i)\right]^2 \tag{3.39}$$

where the matrix elements $\{G_{\pi i\ell}\}$ are determined by the quadrature formulas of the method of histograms (Sect.2.3) and are continuous functions of the values $\bar{m}_i$ sought, i.e., of the components of the vector $\bar{\mathbf{m}}$. The convergence of the iteration method is determined with the conditions $\|\bar{\mathbf{m}}^{(q+1)} - \bar{\mathbf{m}}^{(q)}\|_{\ell_2} \le \delta(\sigma_\pi)$ and $\|\mathbf{s}^{(q+1)} - \mathbf{s}^{(q)}\|_{\ell_2} \le \delta(\sigma_{ext})$ where q is the number of the iteration, and σ_π and σ_{ext} are the measurement errors in $\boldsymbol{\beta}_\pi$ and $\boldsymbol{\beta}_{ext}$, respectively.It should be noted that a rigorous analysis of the convergence conditions is very difficult in this case [3.18]. Some specific features of

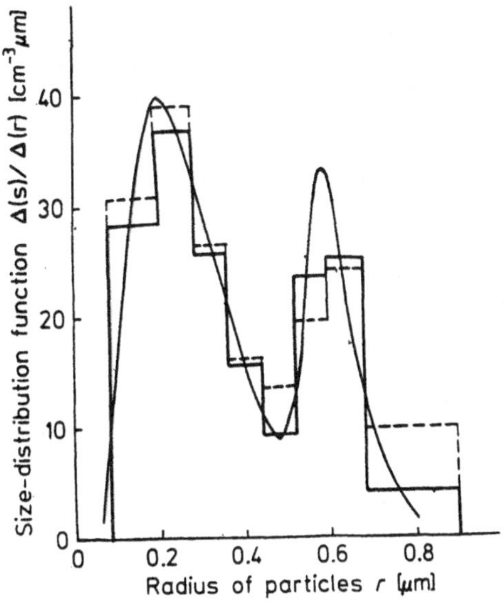

Fig.3.11. Double-peaked size-distribution function s(r) obtained by inverting $\beta_{sc}(\lambda)$ using the method of histograms (m = 8, n = 6)

the problem will be discussed below; let us now consider certain peculiarities inherent in solving the variational problem for the functional (3.38) when the spectral optical characteristics are inverted. Note, first of all, that the determination of the microstructure of the medium sounded is connected with the construction of the inverse operator $K^{-1}_{ext\alpha}$ when solving the system (3.34). This operator has been found explicitly when constructing the operator $\hat{W}$ and implicitly when solving the variational problem for $T_\alpha(\mathbf{s},\bar{\mathbf{m}})$. It is interesting, in this connection, to consider separately the peculiarities of the inversion procedure for inverting the spectral extinction coefficient $\beta_{ext}(\lambda)$ in the visible and IR range. This problem is of interest, not only in connection with the problem of lidar sensing of aerosols, but also because such inversion methods are widely used in optical investigations of the atmosphere for inverting numerically the data on spectral transmittance.

In the following,the efficiency of the method for inverting the spectral extinction coefficient $\beta_{ext}(\lambda)$ by the direct minimization of the quadratic form $T_\alpha(\mathbf{s},\bar{\mathbf{m}})$ with respect to the vectors $\mathbf{s} \in \Phi_m$ is illustrated. A continuous size-distribution function $s_0(r)$ is taken as a reference model. Recall that in this case a piecewise continuous size-distribution function $s_0(r)$ can be set, corresponding uniquely to every vector $\mathbf{s}$, and this function can be interpreted as an approach to the size-distribution function sought. An example of such a solution is presented in Fig.3.11. Here the bimodal size-distribution $s_0(r)$ which is a composition of two lognormal size-distributions with the nodes $r_{s_1} = 0.2$ and $r_{s_2} = 0.6\mu m$, is taken as a reference.

The total integral $S_0 = \int_{0.05}^{1.0} s_0(r)dr$ is assumed to be equal to 0.096 km^{-1} which corresponds approximately to the total geometrical cross-section of particles per unit volume of the haze H" from [3.2]. The reference size-distribution function is shown in Fig.3.11 by the solid curve. A histogram $\Delta(S)/\Delta(r)$ can be plotted for this size-distribution function if the nodes $\{r_0\}$ are fixed. Such a histogram is presented in the figure as a stepwise function plotted with a solid line. The dashed curve in Fig.3.11 represents the histogram inverted from the data on $\beta_{sc}(\lambda)$ measured at the wavelengths: 0.353; 0.380; 0.405; 0.446; 0.542; 0.700μm (the spectral behavior of the corresponding polydispersed efficiency factor $K_{sc}(\lambda) = S_0^{-1}\beta_{sc}(\lambda)$ is shown in Fig.3.4). The level of errors in β_{sc} is at most 2%. The quadratic form $T_\alpha(s,\bar{m})$ minimized is taken in a simpler form than that in the method of histograms [3.24], viz.,

$$\Omega(s) = p_0 h \sum_{\ell=1}^{m} s_\ell^2 + p_1 h^{-1} \sum_{\ell=1}^{m} (s_{\ell+1} - s_\ell)^2 \quad .$$

The value of the refractive index of the aerosol substance was taken for calculations to be 1.5, and the dimensionality of the solution vector was 8. The result obtained can be considered good, on the whole, if one requires only the reproduction of the histogram $\Delta(S)/\Delta(r)$, but the errors in the histogram $\Delta(N)/\Delta(r)$ are naturally larger.

In order to illustrate some general peculiarities of inverse problems for spectral optical characteristics, it is advisable to analyze this example in detail. It should be noted, first of all, that the error in the reconstructed function $S_0(r)$, in the vicinity of the second maximum, is much larger than that near the first one. It is obvious that the measurements of β_{sc} at short wavelengths are of great importance for the reconstruction of the size-spectrum in the region of small sizes, while, for larger sizes, the behavior of β_{ext} in the wavelength interval near the right-hand boundary of the sensing interval is more important. Taking this into account, one concludes that the measurements of $\beta_{sc}(\lambda)$ in an interval of small λ are more informative than those made at larger wavelengths. In fact, it is not difficult to show that the scattering efficiency factors $K_{sc}(r/\lambda)$ become smoother functions, with respect to the variable r, as λ increases. This can be seen from the following obvious relation

$$dK(r,\lambda)/dr = K'(x)2\pi/\lambda \quad . \tag{3.40}$$

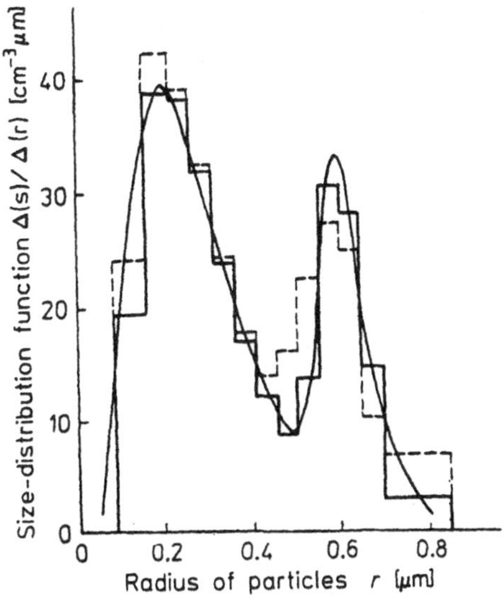

Fig.3.12. The same as Fig.3.11 but for m = 13, n = 9

Since the derivative of the factor K(x) with respect to the diffraction parameter x is restricted in magnitude by a constant whose value is independent of λ, then, for $\lambda \to \infty$ the modulus of the derivative in the left side of (3.40) decreases continuously. All this makes the optical measurements of $\beta_{sc}(\lambda)$ at large λ less informative for the reconstruction of the microstructure as compared with measurements at short wavelengths. Naturally, the terms short and long wavelengths are relative and should be defined in every particular case, according to the boundaries of the sensing interval R. In the above case, the effective upper boundary of the size-spectrum does not exceed 0.8μm. To improve the efficiency of the methods for reconstructing the size-distribution function in the vicinity of its second maximum, one should improve the accuracy of the measurements or increase the number of measuring points for the characteristic $\beta_{sc}(\lambda)$. The usefulness of the latter procedure is illustrated by Fig.3.12 where an example of such a reconstruction is shown. In addition to the above wavelengths used in spectral measurements of $\beta_{sc}(\lambda)$, four others are included in this case, viz. (λ = 0.300 and 0.345μm) in the range of short wavelengths and (λ = 1.06 and 1.450μm) in the long-wavelength region. The increase of the dimensionality of the vector $\boldsymbol{\beta}_{sc}$ allows one to increase also the dimensionality of the solution vector $\mathbf{s}$, which in this case is equal to 13.

Some remarks are necessary concerning the choice of the model size-distribution function $s_0(r)$ in the above examples. Unimodal size-distributions are used in most works on numerical methods for restoring the size-spectra of aerosol ensembles from the data of optical measurements. Unfortunately, in this case it is not always possible to illustrate fully the capabilities of the methods to resolve fine details in the microstructures of the dispersed media sounded, especially if sufficiently sophisticated inversion methods are applied.

In the above numerical experiment we aimed at demonstrating the information potentials of the method of histograms, presented in Chap.2, by applying it to the inversion of the spectral behavior of the aerosol extinction coefficient. The construction of unimodal size-distributions can be made quite satisfactorily using the method of model assessments. In this case, neither a large amount of experimental data nor complicated (and cumbersome) calculation schemes are required. In this connection, the authors prefer the above model for use as a reference in all examples concerning numerical experiments on the inversion of optical data. This size-distribution function is defined in the region $R_2 \leq 1\mu m$ and hence it can be quite useful for the description of microstructures of the atmospheric hazes.

3.4.3 Method of Logarithmic Derivative for the Assessment of the Regularization Parameter. Sensing of Hazes in the UV Spectrum

In the above examples, the inverse problem was solved by minimizing smoothing functionals and therefore some remarks on the choice of a regularization parameter α should be made. General aspects were discussed in Sect.2.4. Let us consider here one method for estimating α, which can be useful for inverting spectral optical characteristics, since it explores the peculiarities of the analytical behavior of these characteristics in the region of short wavelengths.

As is seen from (3.40), the kernels $K(r,\lambda)$ of the integral equations (3.33) become less smooth as λ decreases and, as a consequence, inverse problems become better conditioned. In particular, this means that the behavior of the measured function $\beta(\lambda)$ in the region of short wavelengths (i.e., near the left-hand boundary of the sensing interval) is conditioned to a greater extent by the behavior of the size-distribution function $s(r)$. This situation is clearly demonstrated with the curves $\bar{K}_{\pi}(\lambda)$ and $\bar{K}_{sc}(\lambda)$ presented in Fig.3.4. This peculiarity of the behavior of $\beta(\lambda)$ can be used for deriving some quantitative assessments. For this purpose, consider the logarithmic derivatives $[\omega(\lambda) - 1]$ and $\nu(r)$ related to each other by the inequality (3.24).

From the numerical results presented in Figs.3.7 and 3.8, one can see that these quantities are closest to each other in the region of short λ and small r. In other words, it is advisable to compare the values $[\omega(\lambda)-1]$ and $\nu(r)$ in the intervals Λ' and R' that are located near λ_{min} and R_1, respectively. As to the examples presented above, such intervals can be chosen in the regions of $\lambda \leq 0.5\mu m$ and $r \leq 0.4\mu m$. These are the intervals in which the inequality (3.24) provides the most effective estimation of the logarithmic derivative $\nu(r)$ from the spectral behavior of $\beta(\lambda)$ in the region of small λ. In the special case in which the behavior of the characteristic $\beta_{sc}(\lambda)$, in the vicinity of a given point λ, is determined by the position of the absolute diffraction maximum of the kernel, it is possible to estimate the boundaries of the interval R' provided by a given interval Λ'. Thus, for example, if the points λ_1' and λ_2' are the boundaries of the interval Λ', one can recommend the following relationships for assessing R_1' and R_2': $R_1' = \lambda_1'[\pi(\bar{m}-1)]^{-1}$ and $R_2' = \lambda_2'[\pi(\bar{m}-1)]^{-1}$. In order to avoid oversmoothing of the solution, one should choose a value of the regularization parameter α, such that

$$\max_{r\in R'} |\nu_\alpha(r)| \geq \max_{\lambda\in\Lambda'} |\omega(\lambda)-1| \quad . \tag{3.41}$$

Since in this case the logarithmic derivatives are compared, the method for estimating α may be called the method of logarithmic derivatives. The histograms presented in Figs.3.11 and 3.12 correspond to the value $\alpha=\alpha^*$, which illustrates the usefulness of the method suggested. The choice of α in this method depends directly on the information content of the spectral measurements, i.e., on the complexity of spectral behavior of the characteristic $\beta(\lambda)$ measured in the region of short wavelengths. In the case of atmospheric hazes this region is close to the ultraviolet.

We have seen that optical sensing of atmospheric hazes is undoubtedly of practical interest. In this case, one can apply less general but more effective numerical methods for interpreting the data. Let us consider the example of the inversion of the spectral behavior $\beta_{ext}(\lambda)$ based on the data of [3.15]. In Fig.3.13, the spectral behavior of $\beta_{ext}(\lambda)$ in the region [0.4μm, 2μm] and the appropriate histogram $\Delta(S)/\Delta(r)$ are given. The interval Λ' was chosen to the left of the point $\lambda = 0.55\mu m$ and the corresponding size-interval R' included $r \leq 0.4\mu m$. In the calculations, inequality (3.41) was written in terms of the components of the vectors $\mathbf{s}_\alpha$ and $\boldsymbol{\beta}$ max

$$\max_{\ell\leq m'} |r_\ell(s_{\alpha,\ell+1}-s_{\alpha,\ell})/s_{\alpha\ell}(r_{\ell+1}-r_\ell)| > \max_{i\leq n'} |(\beta_{i+1}\lambda_i-\beta_i\lambda_{i+1})/\beta(\lambda_{i+1}-\lambda_i)| \quad , \tag{3.42}$$

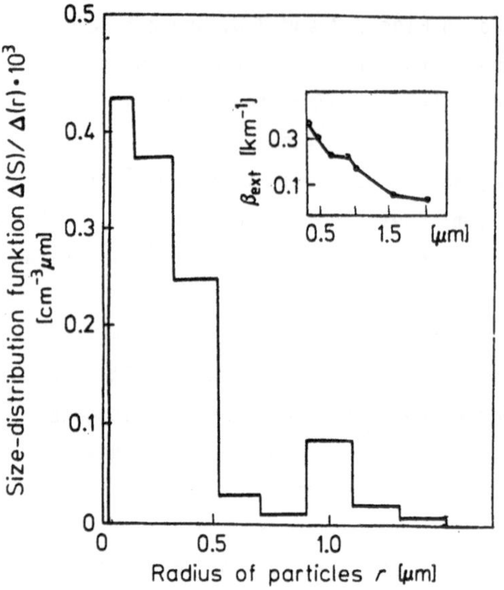

Fig. 3.13. The results of inverting the experimentally measured spectral behavior $\beta_{ext}(\lambda)$ [3.15] using the method of histograms

where m' and n' are conditioned by the choice of the boundaries of the intervals R' and Λ', respectively. The histogram represents the solution defined on the interval [0.05μm, 2μm] with the refractive index of the aerosol substance being taken to be 1.46. Of course, the estimation of α according to (3.41) is possible only if the errors of the optical measurements are small enough; otherwise the calculations of discrete analogs for the logarithmic derivatives are difficult. In addition, the number of points at which the spectral characteristic $\beta(\lambda)$ is measured must be sufficiently large and the spectral distances between them should decrease with decreasing λ. As is shown in [3.20], the smallest distance between the points at which $\beta(\lambda)$ is measured should obey the condition

$$\Delta\lambda \leq c\varepsilon\lambda \ , \tag{3.43}$$

where c is a constant whose value depends on the microphysical parameters of the aerosol sounded and ε is the relative error in the measurements of $\beta(\lambda)$. Condition (3.43) as well as (3.41) is based, to a great extent, on the relation (3.40). In the case of atmospheric hazes whose size-spectrum in the region of small r satisfies the condition $d \ln n(r)/d \ln r \leq \nu^*$, the value of the constant C is approximately equal to $(\nu^* + 3)^{-1}$ [3.20]. Condition (3.43) indicates, at the same time, that the distance between the data points can be larger when $\beta(\lambda)$ is measured in the region of longer wavelengths, as compared with that for short wavelengths. A larger number of measurements, in this case, does not increase the information content since individual measurements are not independent.

3.4.4 The Parametric Form of the Iteration Method

Let us assume for simplicity that the size-distribution function of the aerosol ensemble does not differ significantly from the power law $n(r) \cong ar^{-\nu}$ $(R_1 \le r \le R_2)$. Also let the values of $\beta_\pi(\lambda)$ and $\beta_{ext}(\lambda)$ be measured at the lidar wavelength λ_0. The parameter ν of the power law is to be determined from these data as well as the imaginary part of the refractive index of the aerosol substance. The initial system of equations can be written as follows

$$\beta_{ext}(\nu,\varkappa) = \tilde{\beta}_{ext}$$
$$\beta_\pi(\nu,\varkappa) = \tilde{\beta}_\pi \quad . \tag{3.44}$$

This system is the simplest version of the inverse problem (3.34). The values of ν and $\varkappa$ from (3.44) can be determined using the iteration method described above. For this method to be convergent, the Jacobian of the system (3.44)

$$I(\beta_{ext},\beta_\pi) = \beta_{ext}\beta_\pi(Q_\varkappa^{(ext)}Q_\nu^{(\pi)} - Q_\varkappa^{(\pi)}Q_\nu^{(ext)}) \tag{3.45}$$

must differ from zero at the point λ_0. As before (Sect.2.6) the quantities Q denote the logarithmic derivatives of β_{ext} and β_π at the point λ_0. The above condition is fulfilled if

$$Q_\varkappa^{(ext)}/Q_\nu^{(ext)} \neq Q_\varkappa^{(\pi)}/Q_\nu^{(\pi)} \quad . \tag{3.46}$$

The parameters Q characterize the variability of the corresponding optical characteristics with respect to variations of $\varkappa$ and ν. In particular, if the inequalities

$$|\partial\beta_\pi/\partial\varkappa| > |\partial\beta_{ext}/\partial\varkappa|$$
$$|\partial\beta_\pi/\partial\nu| < |\partial\beta_{ext}/\partial\nu| \tag{3.47}$$

are valid, then (3.46) is fulfilled and the iteration method converges when applied to the system (3.44). It is very important that the larger the value of $Q_\varkappa^{(ext)}/Q_\nu^{(ext)} - Q_\varkappa^{(\pi)}/Q_\nu^{(\pi)}$, the faster the method converges.

Figure 3.14 presents the values $Q_\varkappa^{(\pi)}/Q_\nu^{(\pi)}$, $Q_{\bar{m}}^{(\pi)}/Q_\nu^{(\pi)}$, $Q_{\bar{m}}^{(ext)}/Q_\nu^{(ext)}$ and $Q_\varkappa^{(ext)}/Q_\nu^{(ext)}$ (in the sequence of the curves' enumeration) for the visual range. It was assumed in the calculations that $\nu_0 = 4$; $\bar{m}_0 = 1.5$; $\varkappa_0 = 0$. As is seen from the figure, the curves $Q_\varkappa^{(\pi)}/Q_\nu^{(\pi)}$ and $Q_\varkappa^{(ext)}/Q_\nu^{(ext)}$ do not intersect in the spectral interval considered and, consequently, the system (3.44)

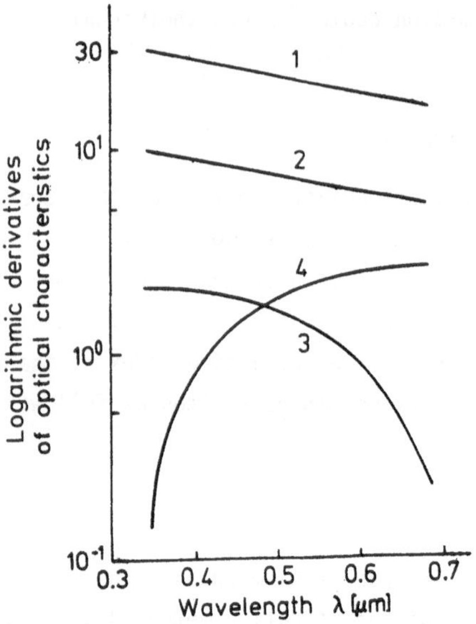

Fig.3.14. Spectral behavior of the ratios: $(\ln\beta_\pi)'_\varkappa/(\ln\beta_\pi)'_\nu$: (Curve 1); $(\ln\beta_\pi)'_{\bar{m}}/(\ln\beta_\pi)'_\nu$: (Curve 2); $(\ln\beta_{ext})'_\varkappa/(\ln\beta_{ext})'_\nu$: (Curve 3); $(\ln\beta_{ext})'_{\bar{m}}/(\ln\beta_{ext})'_\nu$ for the case of the Junge size-distribution function

is consistent for estimating $\varkappa$ and ν. The analogous result can be obtained for the parameters $\bar{m}$ and ν since the values $Q_{\bar{m}}^{(\pi)}/Q_\nu^{(\pi)}$ and $Q_{\bar{m}}^{(ext)}/Q_\nu^{(ext)}$ differ from each other at every λ. However, this conclusion is incorrect in the IR range, since here the Curves 2 and 3 come close to each other and the condition (3.46) is not fulfilled.

To a first approximation, the values $|Q_\varkappa^{(\pi)}/Q_\nu^{(\pi)} - Q_\varkappa^{(ext)}/Q_\nu^{(ext)}|$ and $|Q_{\bar{m}}^{(\pi)}/Q_\nu^{(\pi)} - Q_{\bar{m}}^{(ext)}/Q_\nu^{(ext)}|$ can be regarded as qualitative characteristics of the degree to which the inverse problems of light scattering written in the parametric forms are conditioned. The greater these values, the smaller is the effect of errors σ_π and σ_{ext} of the optical measurements on the results of the inversions. This enables one to make a best choice of the operating wavelength for optical sensing of the atmosphere.

It should be noted that the above conclusions concerning the convergence of the iteration method as well as the conditioning of inverse problems like (3.44) are, on the whole, valid for the system (3.34). In the latter case, the variations of the polydispersed integrals β on the functional sets Φ and $\Omega_m(\Lambda)$ should be introduced instead of the logarithmic derivatives of optical characteristics with respect to the microstructural parameters. The elements of the set $\Omega_m(\Lambda)$ are the functions $\bar{m}(\lambda)$, where λ is in the interval Λ. Omitting a detailed discussion of these questions, we now consider some peculiarities encountered when solving the inverse problems like (3.44) in the IR range. As before, the sensing of atmospheric hazes with R_2 = 1μm will be dis-

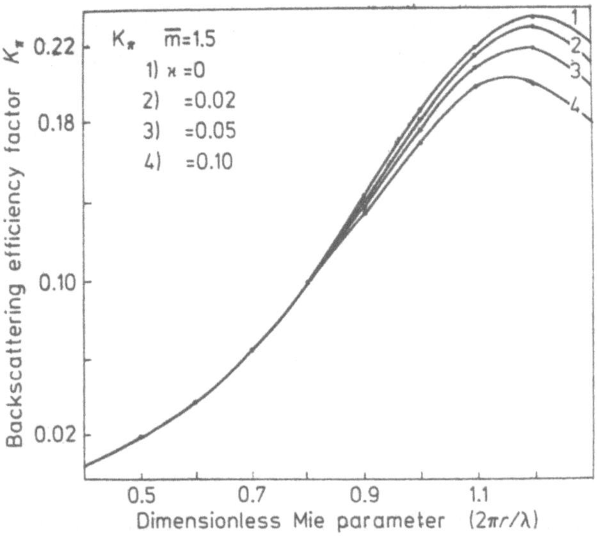

Fig.3.15. Backscattering efficiency factor K_π in the vicinity of small values of the diffraction parameter x for different refractive indices

cussed. The calculational analysis of the variational coefficients Q for the near IR has shown that the characteristic $\beta_\pi(\lambda)$ is weakly sensitive to variations in the optical constants $\bar{m}$ and $\varkappa$ of the aerosol substance. This is explained by the fact that the kernel $K_\pi(\bar{m},\varkappa,x)$ in the integral (3.33) depends weakly on the variables $\bar{m}$ and $\varkappa$ in this spectral region. As an example, the monodispersed efficiency factors $K_\pi(\bar{m},\varkappa,x)$ for $x \leq 1.2$, $\bar{m} = 1.5$ and various $\varkappa$ are presented in Fig.3.15. As is easily seen, the variations of the factor $K_\pi(x)$ with respect to x, when $x \lesssim 1.2$, are significantly weaker than those in the region $x \leq 20$ (Fig.3.6). However, the situation in the case of $\beta_{ext}(\lambda)$ is quite opposite, so that the variations of this characteristic with respect to $\bar{m}$ and $\varkappa$ increase in the IR range. This causes some peculiarities in the solution of inverse problems like (3.33). It is obvious that one should reverse the order of inverting the equations (3.34), since now the operator $K^{-1}_{\pi\alpha}$ is more stable with respect to uncertainties in the optical constants as compared with the operator $K^{-1}_{ext,\alpha}$.

As to the method of multifrequency sensing with a monostatic lidar, it is necessary to emphasize that IR wavelengths are preferable if the refractive index of the aerosol substance is unknown. In this case, the errors of the inversion of optical data due to an arbitrary choice of values of the optical constants of the aerosol substance will be smaller than those in the visible. By this we should like to emphasize, once more, the importance of the fact that various spectral intervals possess different information potentials, with respect to the microphysical parameters of dispersed media which require different methods for inverting optical data. Let us now return to the visual range and consider one of the simplest methods for the joint inversion of the characteristics β_π and β_{ext}.

3.4.5 The Method of Combined Discrepancy, an Example of Experimental Data Handling

The idea of the method under consideration is to combine the functionals (3.38) and (3.39) into one functional of the form

$$T_\alpha(\bar{m},\mathbf{s}) = p_1\rho_\pi^2 + p_2\rho_{ext}^2 + \alpha\Omega^2(\mathbf{s}) \tag{3.48}$$

assuming that $\bar{m}(\lambda)$ = const over the entire sensing interval Λ. The quantities p_1 and p_2 are weighting coefficients. The variational problem is solved with respect to $\bar{m}$ and s simultaneously. As numerical analysis has shown, the minimization of $T_\alpha(\bar{m},\mathbf{s})$ with respect to s only cannot be made satisfactorily when $|\partial\rho_\pi/\partial\bar{m}| \gg |\partial\rho_{ext}/\partial\bar{m}|$. In other words, it appears impossible to find, in the vector space of possible solutions Φ_m, such a vector $\mathbf{s}^*$ that would obey the condition

$$\min_{\mathbf{s}\in\Phi_m} \{T_\alpha(\bar{m},\mathbf{s})\} \leq \delta(\sigma) \quad , \tag{3.49}$$

if the value $\bar{m}$ used is too far from $\bar{m}_0$. Recall that the vectors of the space Φ_m have possible components $s_\ell(\ell = 1,2,\dots,m)$. The condition (3.49) can be fulfilled only if a simultaneous minimization of $T_\alpha(\bar{m},\mathbf{s})$ on the sets of $\mathbf{s}$ and $\bar{m}$ values is performed. This conclusion is valid for the case in which atmospheric hazes are sounded in the visible and is not general. We now consider a numerical example in order to illustrate the efficiency of the variational problem for $T_\alpha(\bar{m},\mathbf{s})$ when determining $\bar{m}$ and $\mathbf{s}$ in investigations of the atmospheric hazes.

Figure 3.16 shows the behavior of the discrepancy ρ_π with respect to the parameters $\bar{m}$ and α (recall that when the quadratic form (3.48) is minimized, the single vector $\mathbf{s}$ corresponds to every value of α). The reference data in the numerical experiment are: spectral interval Λ = [0.353μm; 0.7μm], $\bar{m}_0$ = 1.5, the model size-distribution function $s_0(r)$ is the bimodal one from Fig.3.11, and the perturbations of the reference values of β_π and β_{ext} are oscillatory with an amplitude not exceeding 5% of their maximum values.

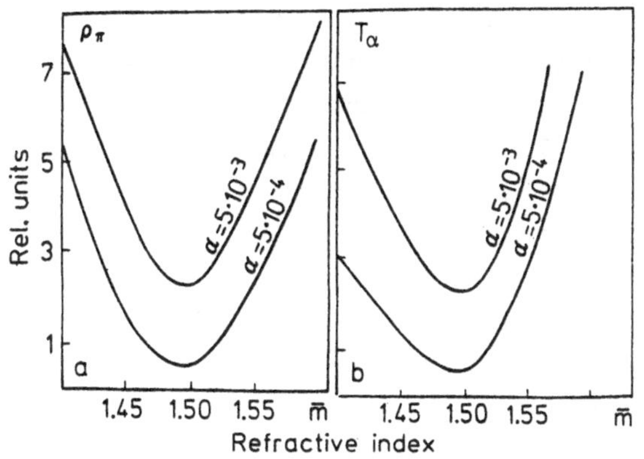

◄ **Fig.3.16.** An example of typical dependences of the discrepancy ρ_π and smoothing functional T_α on m illustrating the inversion of (2.48) on $\bar{m}$ and s(r)

Fig.3.17. The results of estimation of $\bar{m}$ by minimizing T with respect to $\bar{m}$ and s(r) based on the experimental results (a)

▼

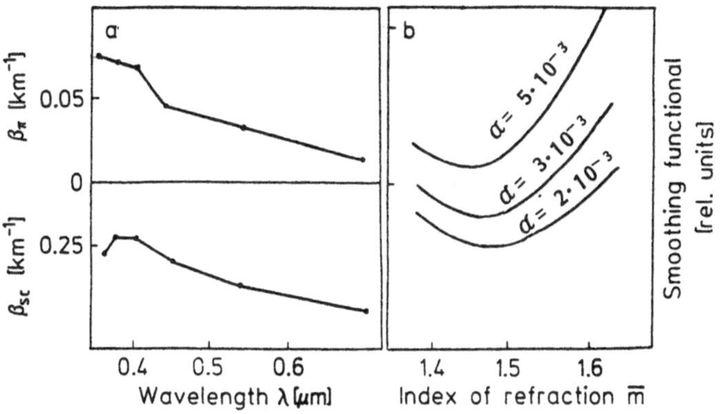

As can be seen from the results of the calculations, the discrepancies $\rho_\pi(\bar{m})$ have a minimum with respect to $\bar{m}$, in the vicinity of $\bar{m}_0$ for all values of α which are comparable with ε^2. This property of the functions $\rho_\pi(\bar{m})$ can be transferred completely to the parametric family $T_\alpha(\bar{m})$ (Fig. 3.16b). The behavior of $T_\alpha(\bar{m})$ in the vicinity of $\bar{m}_0$ does not differ significantly from that of $\rho_\pi(m)$, due to the weak dependence of the optical discrepancies ρ_{ext} on $\bar{m}$. The sharpness of the maxima of the above curves determines the precision of the estimation of the parameter $\bar{m}$. Unfortunately, as the results of the calculations presented in Fig.3.16 show, this precision is not very high. The extrema of the above peaked functions appear to be very wide. As a consequence, the variational problem (3.48) provides an assessment of the optical constants of aerosol substance only to very rough approximation. In the case of experimental data, the minima of $T_\alpha(\bar{m})$ are even more "diffused", as is clearly demonstrated by the examples presented in Fig.3.17.

The reference data on β_π and β_{sc} (Fig.3.17a) come from nephelometric measurements in atmospheric hazes (Sect.3.2). The width of the maximum of the functional T_α, in this case, is caused most probably by incorrectness in the initial assumptions on which (3.48) is based. At the same time, one should keep in mind that the method of combined inversion of β_π and β_{sc} is more useful for investigating the microstructures of scattering ensembles than inversion schemes which are based on an a priori choice of the optical constants of the aerosol substance. In any case, the simultaneous assessment of the refractive index $\bar{m}$ enables one to determine more reliably the microstructure of the aerosol ensemble sounded even if the sensitivity of an optical method is low. Let us consider, in this connection, the results of microstructure investigations based on the combined inversion of β_π and β_{sc} and minimization of $T_\alpha(\bar{m},s)$.

The reference data are presented in Table 3.1. The upper line presents the measured values of β_{sc} while the values of β_π are presented in the lower line. The experiment was carried out in the semi-desert area near Kirbaltaby of the Alma-Ata region during October 1973 [3.23]. The atmospheric temperature and humidity varied, during this period, from 10° to 15°C and from 62% to 75%, respectively. Since the errors of the optical measurements did not exceed 20% on the average the use of the variational problem (3.48) for inverting these data was justified. The use of more general methods, in this case, could not be regarded as advisable. The estimated value of the refractive index of the aerosol substance is $\bar{m} \cong 1.46 \mp 0.04$ ($\varkappa \leq 0.003$). The inverse results on the microstructure parameters are presented in Table 3.2. These data are, on the whole, in good agreement with those from the literature for atmospheric hazes in the visual range $S_M \geq 10$ km. The analysis of the histograms presented in Table 3.2 reveals the temporal variations of separate haze fractions during nighttime. The microstructure of the aerosol ensemble studied is characterized by two histograms, i.e., by the distribution of geometrical cross-sections S of particles of separate fractions (left column) and by the distribution of the particles' number density (right column). As should be expected, the variations are stronger for the larger particles in the ensemble [3.24,25].

Let us now consider some methodological aspects concerning the inversion technique based on (3.48). Recall, first of all, that the variational problem for $T_\alpha(\bar{m},s)$ can be regarded, in a certain sense, as a problem of finding the best approximation of the measured optical characteristic $\beta(\lambda)$ with the functions $\beta_\alpha(\lambda) = \int_R K(r,\lambda)s_\alpha(r)dr$, where $s_\alpha(r)$ is from the family of

Data and local time (October 1978)	0.353	0.38	0.405	0.446	0.542	0.7
	Wavelengths λ [μm]					
15.20^h35^m	0.293	0.343	0.298	0.238	0.157	$0.689\ 10^{-1}$
	$0.984\ 10^{-1}$	0.101	$0.632\ 10^{-1}$	$0.560\ 10^{-1}$	$0.408\ 10^{-1}$	$0.255\ 10^{-1}$
15.20^h57^m	0.255	0.266	0.234	0.183	0.124	$0.656\ 10^{-1}$
	$0.991\ 10^{-1}$	$0.831\ 10^{-1}$	$0.703\ 10^{-1}$	$0.479\ 10^{-1}$	$0.347\ 10^{-1}$	$0.173\ 10^{-1}$
15.21^h07^m	0.217	0.242	0.241	0.198	0.134	$0.600\ 10^{-1}$
	$0.774\ 10^{-1}$	$0.734\ 10^{-1}$	$0.681\ 10^{-1}$	$0.461\ 10^{-1}$	$0.312\ 10^{-1}$	$0.134\ 10^{-1}$
16.00^h04^m	0.219	0.253	0.229	0.176	0.110	$0.502\ 10^{-1}$
	$0.620\ 10^{-1}$	$0.519\ 10^{-1}$	$0.395\ 10^{-1}$	$0.237\ 10^{-1}$	$0.219\ 10^{-1}$	$0.118\ 10^{-1}$
16.00^h15^m	0.192	0.227	0.135	0.108	$0.722\ 10^{-1}$	$0.328\ 10^{-1}$
	$0.531\ 10^{-1}$	$0.547\ 10^{-1}$	$0.374\ 10^{-1}$	$0.239\ 10^{-1}$	$0.182\ 10^{-1}$	$0.106\ 10^{-1}$
16.00^h25^m	0.199	0.215	0.220	0.163	0.103	$0.513\ 10^{-1}$
	$0.648\ 10^{-1}$	$0.633\ 10^{-1}$	$0.484\ 10^{-1}$	$0.291\ 10^{-1}$	$0.200\ 10^{-1}$	$0.158\ 10^{-1}$

Table 3.1. Values of aerosol coefficients β_{sc} (upper line) and backscattering coefficients β_π (lower lines) measured at different wavelengths [3.23]

Size interval [μm]	$\Delta(S)/\Delta(r)$ [cm^{-3} μm]						$\Delta(N)$ [cm^{-3}]					
	15.29^h	35^m	15.20^h	57^m	15.21^h	07^m	16.00^m	04^m	16.00^h	15^m	16.00^h	25^m
0.125-0.175	420	312	340	247	348	253	330	242	280	204	289	210
0.175-0.225	390	158	310	123	317	128	305	123	248	100	265	107
0.225-0.275	323	83	249	64	257	66	252	65	192	50	221	57
0.275-0.327	229	41	174	31	176	32	179	32	120	21	160	29
0.325-0.375	128	17	94	12	86	11	98	13	48	6.3	93	12
9.375-0.45	34	5	22	3	1	0.14	22	3	14	0.2	29	4.1
0.45-0.55	-	-	-	-	-	-	-	-	-	-	-	-
0.55-0.65	-	-	-	-	-	-	-	-	-	-	-	-
0.65-0.75	11	0.7	31	2	9	0.6	-	-	22	1.4	9	0.6
0.75-0.85	39	2	46	2.3	40	2	-	-	5	0.25	21	1

Table 3.2. Aerosol microstructure parameters reconstructed from the optical data in Table 3.1

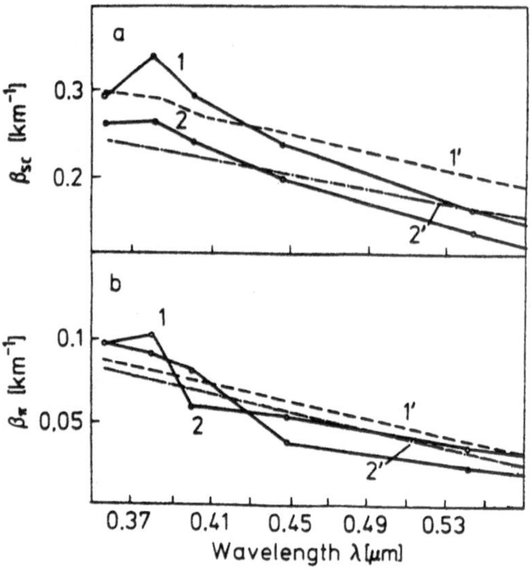

Fig.3.18. Spectral behaviors of the measured optical characteristics (Curves 1 and 2; for parameters see first two lines of Table 3.1) and corresponding approximations $\beta_{sc,\alpha}(\lambda)$ and $\beta_{\pi,\alpha}(\lambda)$ (Curves 1' and 2', respectively)

possible regularized inverse solutions that correspond to different values of the parameter α. It is obvious that if all the initial assumptions are valid and the measurement accuracy is the same over the sensing interval Λ, then the errors of the approximation $\Delta_i(\beta) = |\beta(\lambda_i) - \beta_\alpha(\lambda_i)|$ should not differ from each other significantly. Moreover, when the error σ of optical measurements vanishes, the following condition must be valid for all α values:

$$|\beta(\lambda) - \beta_\alpha(\lambda)| \leq \delta(\sigma) \quad . \tag{3.50}$$

The non-fulfillment of this condition can be caused by the fact that the initial functional equations do not entirely fit the given optical experiment [3.26,27]. Let us consider again the previous example. Figure 3.18 shows the reference characteristics $\beta_{sc}(\lambda)$ (the first two lines from Table 3.1) and the corresponding approximating functions $\beta_{sc}(\lambda)$. It is seen from this figure that the curves $\beta_{sc\alpha}(\lambda)$ are very smooth. It is also seen that the deviations $\Delta_i(\beta) = |\beta_{sc}(\lambda_i) - \beta_{sc\alpha}(\lambda_i)|$ are not uniform over the interval Λ, and in the region of small λ they exceed the errors of the optical measurements. This last remark is especially valid for the approximation of $\beta_\pi(\lambda)$ (Fig.3.18b). Since the backscattering coefficient of aerosols varies most strongly when the refractive index of the aerosol substance varies with respect to the wavelength, one can suppose that, in this case, the initial assumption on the constancy of the refractive index over the above spectral interval is wrong. In this respect, the assumption of the selective absorption by aerosols observed in the ultraviolet might be more valid. It is possible that iron oxides can cause such an absorption since these oxides may be contained in the aerosol substance [3.23]. Therefore an approach taking into account the spectral behavior of the refractive index $\bar{m}(\lambda)$ is better justified for inverting these optical measurements. In principle, the method of inversion based on the variational problem (3.48) allows the numerical solution of such problems. Therefore, the approach to the inversion of optical data in which the spectral behavior of the refractive index $\bar{m}(\lambda)$ is estimated together with the size spectrum, is more appropriate here. The inversion method based on solution of the variational problem (3.48) enables one, in principle, to solve numerically the problems that are analogous to the example given here. Unfortunately, the data presented in Table 3.1 contain significant errors which makes the formulation of more informative inverse problems impossible.

3.4.6 Further Consideration of Inverse Problems in the Method of Multi-Frequency Sensing

Let us consider one more formulation of the inverse problem dealing with the determination of optical constants of the aerosol substance as well as microstructures of aerosol ensembles with the use of a sensing scheme for measuring $\beta_\pi(\lambda)$ and $\beta_{ext}(\lambda)$.

For investigations of aerosols in the ground layer,the following form for the representation of optical characteristics may be more suitable:

$$\beta(\lambda) = \int_{R_1}^{R_2} K(\bar{m}_1,r,\lambda)s_1(r)dr + \int_{R_3}^{R_4} K(\bar{m}_2,r,\lambda)s_2(r)dr \quad . \tag{3.51}$$

Here the microstructure of the aerosols is represented by two fractions having the size-distributions s_1 and s_2. It is assumed that these fractions have different indices of refraction (i.e., different chemical composition). Such models can be found elsewhere in the literature [3.28,29]. In those papers the modal radii of the fractions r_{n_1} and r_{n_2} are 0.2 and 1μm, respectively, (the standard deviation is 0.3 for both components). There is also a second model for which one suggested a bimodal distribution of volumes $v(r)$ with the first mode $r_{v_1} = 0.2$μm and the second one being somewhere in the region from 4 to 5μm, depending on the atmospheric transmittance. In both these models,the refractive indices are assumed to be equal for both fractions,which is physically not justified.

The use of such models in inverse problems may be considered well justified if the size intervals are wide and the aerosols are from different sources. The vector form of the corresponding inverse problem is written as follows

$$K_{\pi}(\bar{m}_1)s_1 + K_{\pi}(\bar{m}_2)s_2 = \beta_{\pi}$$

$$K_{ext}(\bar{m}_1)s_1 + K_{ext}(\bar{m}_2)s_2 = \beta_{ext} \quad . \tag{3.52}$$

This system of equations can be solved with the use of the iteration method described above. It should be noted that the inverse problems in the form (3.52) require the optical sensing to be performed in wide spectral intervals.

3.5 Investigations of Aerosols Microstructures in the Boundary Layer Using Multi-Frequency Lidars

3.5.1 Methods for Determining the Local Variations of the Aerosol Microstructures

The aerosols of the boundary layer of the atmosphere are characterized by the major spatial inhomogeneities of the microstructure and their variability in time. This, naturally, makes it difficult to interpret the data of multi-frequency sensing since, in this case, the initial assumptions cannot be made reliably at all points of the sensing path. For more clarity, let us consider the initial functional relationships. In order to take into account all the spatial inhomogeneities of the microphysical parameters of the aerosol, (3.33) should be written as follows

$$\beta_{\pi}(\lambda_i,z) = \int_{R_1(z)}^{R_2(z)} K_{\pi}(\bar{m},z,r,\lambda)s(r)dr$$

$$\beta_{ext}(\lambda_i,z) = \int_{R_1(z)}^{R_2(z)} K_{ext}(\bar{m},z,r,\lambda)s(r)dr \tag{3.53}$$

$$i = 1,2,\dots,n \quad .$$

Even if one assumes that the shapes of particles are close to spherical everywhere along the sensing path and the refractive index of the aerosol substance $\bar{m}(z) = \text{const}(z_1 \le z \le z_2)$, (3.53) remain indefinite with respect to the size-distribution function $s(r)$ since the dependences $R_1(z)$ and $R_2(z)$ on z are unknown.

In the case of atmospheric hazes (background aerosols), the values of R_1 are, as a rule, very small. Thus, for example, in [3.10] the value $R_1 = 0.02\mu m$ was suggested to fit the aerosol model to be used in problems on optical sensing of the lower atmosphere. In other studies [3.4], $R_1 \lesssim 0.05\mu m$ was chosen. In any case, the value $R_1 \lesssim 0.1\mu m$ is quite reasonable for atmospheric hazes. It then follows that a reliable determination of the size spectrum in the size region [0.02μm, 0.05μm] requires, first, very high precision of optical measurements and, secondly, very short wavelengths of sensing radiation.

In the following discussion, we shall mainly deal with assessments of the information potentialities of multi-frequency lidars with ruby and neodymium lasers, including their second harmonics, as sources [3.32]. As numerical analysis has shown, the inversion methods allow one to determine the fractions of aerosol particles with sizes not less than 0.2 or 0.3μm if the lower boundary of a sensing interval is $\lambda_{min} \cong 0.53\mu m$ and the measurement accuracy $\varepsilon \simeq 10\%$. In this region of sizes, the reconstruction errors ε_s in the value of $s(r)$ may be expected to be no greater than 2ε. The use of the second harmonic of a ruby laser provides λ_{min} which allows the least particles with radii r_{min} as small as about 0.1 to 0.15μm to be revealed.

The value of r_{min} can be determined from the following relationship

$$\int_{R_1}^{r_{min}} K(r,\lambda_{min})s(r)dr \le \eta\sigma \quad , \tag{3.54}$$

where R_1 is the actual left boundary of a size spectrum $s(r)$, σ is the standard deviation of optical measurements of $\beta(\lambda)$ and η is the confidence coefficient. When $\lambda_{min} \simeq 0.347\mu m$, the value r_{min} for the hazes of the atmospheric boundary layer may be assumed to satisfy the condition

$$\int_{R_1}^{r_{min}} q(r)dr = \alpha_1(r_{min}) \le \eta\varepsilon \quad , \tag{3.55}$$

where $q(r)$ is the normalized size-distribution function $S^{-1} \int_{R_1}^{r} s(r')dr'$. This means that particles with sizes in the region $[R_1, r_{min}]$ make not only a small optical contribution to the total scattering coefficient, but have also a small geometrical cross-section as compared with the integral S over all particles per unit volume. In this case, we mean the aerosol of the ground layer, for which the values of R_2 and of the total cross-section S sometimes reach 3 or 4μm, and 0.3 - 0.5 km^{-1}, respectively. For this case one can assume, as a first approximation, that in (3.53) $R_1(z) \lesssim r_{min}$ for all z and $r_{min} \in [0.1\mu m, 0.3\mu m]$ depending on λ_{min} and ε.

Now let us consider the estimations of $R_2(z)$ in the system (3.53). Unfortunately, the qualitative assessments as used above cannot be regarded as satisfactory. It is characteristic for the boundary layer of the atmosphere that R_2 varies significantly along the sensing path and depends, in addition, on the dimensions of a scattering volume. It is obvious that the smaller the scattering volume, the less reliable are the a priori assessments of R_2. The value of R_2 can be estimated if the data on the spectral transmittance of the atmosphere are inverted, especially when the sensing paths are long enough. However, if the sensing is performed with short light pulses, this is no longer possible. It is also difficult to use the techniques described in Sect.3.3, since those require a large number of spectral measurements.

We have thus seen that the inversion of data of multi-frequency lidar sensings is a typical example of problems which should be solved regardless of the deficit in experimental information. This makes it necessary to develop some specialized techniques for estimating the proper values of R_2 in the course of the inversion of sensing data. One such technique is based on numerical experiments on the inversion of data of multi-frequency sensing of atmospheric hazes in the spectral interval [$0.347\mu m$, $1.06\mu m$].

For simplicity, the equation $K_\pi s = \beta_\pi$ is assumed to be solvable with any of the methods described above as a variational approach. A regularized solution $s_\alpha(r)$ which corresponds to every value of the regularization parameter is, in turn, an approximation to the actual size-distribution function $s_0(r)$. If the operator K of the initial equation is exact, i.e., if all of the assumptions on the integral representation of $\beta(\lambda)$ are valid, then the optical discrepancy $\rho(K s_\alpha, \beta) = \|K s_\alpha - \beta\|_{L_2}$ must decrease monotonically as α decreases. In this case, the value α_0 can be found, such that the inequality

$$\rho(K s_\alpha, \beta) \leq \delta(\sigma) \tag{3.56}$$

will be correct for all $\alpha \leq \alpha_0$. However, if the operator K does not fit a given experiment on the optical sensing of a dispersed medium, then the condition (3.56) cannot, as a rule, be satisfied by varying α within reasonable limits (Sect.2.4). For the ensemble of particles whose shapes do not deviate significantly, on the average, from spheres, the disagreement between the operator K and the optical data may be caused either by errors in assessing the refractive index of the aerosol substance or by an improper choice of the boundaries R of a solution (which then do not agree with the boundaries of an actual size-spectrum). Uncertainties in the boundaries of the size-interval R are a characteristic feature of the integral equations of the theory of optical sensing of aerosols. This becomes a principal difficulty in laser sensing of the atmosphere since the desire to realize a very high spatial resolution inherent in the pulsed sensing schemes excludes the possibility of choosing the boundaries R based on these or other models.

In studies of the local variations of microstructures of aerosols using methods of multi-frequency lidar sensing, the above difficulty can be avoided if one includes the estimation of boundaries of a size spectrum in the inversion scheme. One possible version of such an interpretation of the optical data is considered below. Assume that the refractive index of the aerosol substance can be assessed from supplementary measurements of the spectral transmittance of the atmosphere using any of the methods presented in Sect. 3.4. In this case, the discrepancy ρ of the initial equation may be considered to be dependent not only on α but on the boundaries of the size interval R as well. In order to emphasize this, we shall write $\rho_\alpha(R_2)$, assuming R_1 to be known accurately enough.

As numerical analysis has shown, the discrepancies $\rho_\alpha(R_2)$ for integral equations like (3.53), with kernels taken from the Mie theory, are continuous functions of R_2 for all admissible values of α. That means that the minimization of the discrepancy $\rho_\alpha(R_2)$ with respect to R_2 has a certain practical sense. Recall, in this connection, that the functions s(r), obeying the condition (3.56), are generally called quasi-solutions of the equation $Ks = \beta$ for a given error σ in the optical measurements [3.18]. The concept of quasi-solution can be widened, when inverting polydispersed optical characteristics under the condition of an unknown right boundary R_2 of the size spectrum sought, if one requires that the condition

$$\rho_\alpha(R_2,s) \leq \delta(\sigma) \qquad (3.57)$$

to be fulfilled with respect to α and R_2. In so doing, one includes in the concept of inverse solution not only the size-distribution itself but also its boundaries. The condition (3.57) is natural in those cases for which (3.56) is not fulfilled for any reasonable α. Let us present examples of the experimental data inversion in order to illustrate the usefulness of the technique suggested.

Figure 3.19 presents the family of optical discrepancies $\rho_{\pi\alpha}(R_2)$ for several values of the refractive index of the aerosol substance. The data presented in this figure have been obtained at the test field of the Institute of Atmospheric Optics (USSR, Tomsk, July 1975) [3.18]. The standard deviations of the measurements did not exceed 5%. The measurements were performed at the wavelengths: 0.53; 0.69; 1.06μm. The sensings were made at night in the atmospheric layer at heights ranging from 200 m to 800 m.

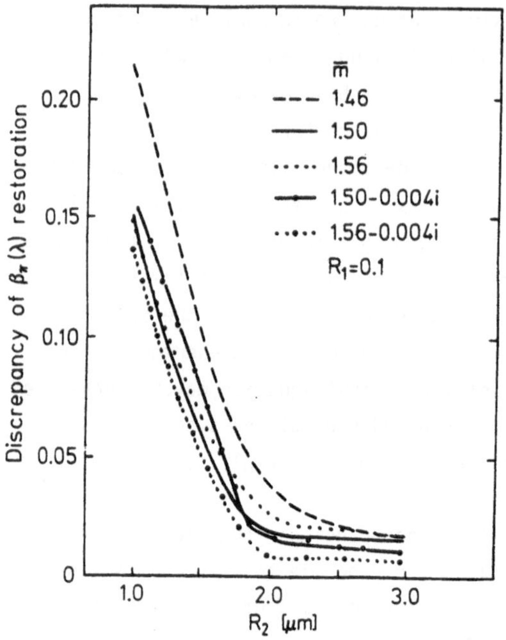

Fig.3.19. Typical dependence of optical discrepancy $\rho_{\pi\alpha}$, see (3.58), of the inversion of three-frequency lidar sensing data on the right effective boundary of atmospheric-haze size spectra R_2 [μm]

The visual range was monitored with a photometer at $\lambda = 0.63\mu m$ along a horizontal path of length 920 m. The aerosol formations sounded could be classified as mists. In such a case,the maximum size of the particles had to be 2μm, which is confirmed by the data presented in Fig.3.19.

It is clearly seen from this figure that the inequality $\rho_{\pi\alpha} \leq (0.02\text{-}0.03\ \mathrm{km}^{-1})$ is not fulfilled for all the calculated values of the refractive index if R is less than 2μm. The discrepancy $\rho_{\pi\alpha}(R_2)$ which in this case is calculated according to the expression

$$\rho_{\pi\alpha}(R_2) = \left\{(1/n)\sum_{i=1}^{n}\left[\beta_\pi(\lambda_i) - \int_{R_1}^{R_2} K_\pi(\bar{m},r,\lambda_i)s_\alpha(r)dr\right]^2\right\}^{1/2} \tag{3.58}$$

reaches admissible values only for $2 \leq R_2 \leq 3\mu m$. It is typical that an increase of R_2 above 3μm leads to an increase of $\rho_{\pi\alpha}$. The value of the regularization parameter α used in this calculation had been chosen in accordance with the estimation $c_1\sigma^2 \leq \alpha \leq c_2\sigma^2$, and the regularized solutions were sought with the use of the inverse operator $K_{\pi\alpha}^{-1} = (G\,K_\pi^*K_\pi + \alpha I)^{-1}G\,K_\pi^*$ [scheme of (2.67)]. In this example,the reference vector β_π with components $\beta_\pi(\lambda_i)$ $(i = 1,2,3)$ relates to a fixed volume on the sensing path. Some interesting features are revealed when interpreting the data obtained for differ-

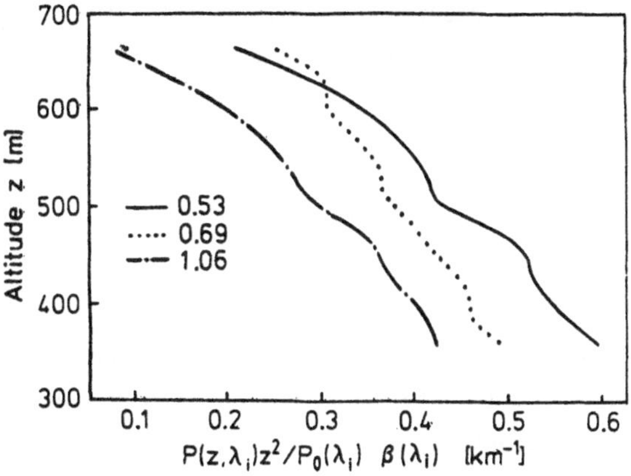

Fig. 3.20

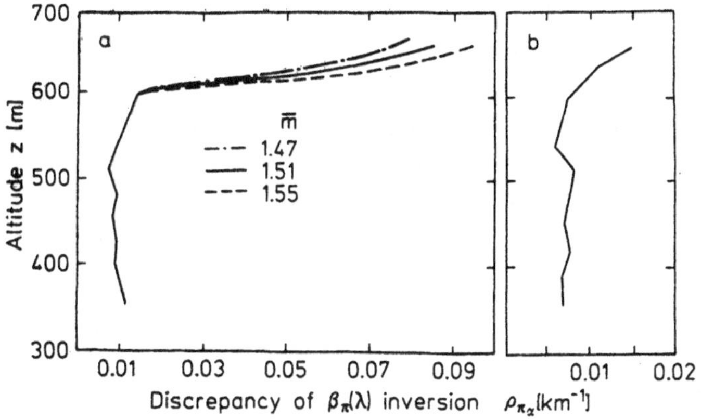

Fig. 3.21

Fig. 3.20. Profiles $\bar{S}(z,\lambda)$ obtained experimentally from the data of three-frequency lidar sensing data

Fig. 3.21. The altitude behavior of the optical discrepancy $\rho_{\pi\alpha}$ in the data on inverting the reference profiles $\bar{S}(z,\lambda)$ (Fig. 3.20), for $R_2(z)$ = const. = 3μm (Curves a) and for varying $R_2(z)$ (Curve b), see (3.57)

ent altitudes, which are related to the construction of quasi-solutions according to (3.57).

Figure 3.20 presents the profiles

$$\bar{S}(z,\lambda_i) = P(z,\lambda_i)z^2/P_0(\lambda_i)B(\lambda_i)$$

where $P(z,\lambda_i)$ are the amplitudes of the lidar returns measured at wavelength λ_i and z is the height varying within the limits mentioned above. For interpretation, the sensing path was divided into layers of 30 m thickness (Δz = 30 m). First, the size-distribution function s(r) and the value of R_2

were assessed according to the technique described above. The estimated value of R_2, in this case, was about 3μm. The value of the index of refraction of the aerosol substance ($\bar{m}^* \simeq 1.51$) was fitted to the data on spectral transmittance of the atmosphere at $\lambda = 0.63\mu m$, as has been discussed already. Further, it was assumed that R_2 is constant within the height interval from 360-660 m.

The values of $\rho_{\pi\alpha}$ calculated for the height increment $\Delta z = 30$ m and for a given R_2 value are shown in Fig.3.21a. As is seen from the figure, $\rho_{\pi\alpha}$ increases dramatically in the layer at 600 m height. An attempt to decrease it by varying the refractive index from 1.47 to 1.54 failed, which distinctly showed that the assumption on the constancy of R_2 at this height is not valid. By demanding the condition $\rho_{\pi\alpha}(R_2) < 0.02$ km^{-1} to be fulfilled everywhere along the path, one can determine the profile $R_2(z)$ and corresponding size-distribution functions for every layer.

The final results of the inversion are presented in Fig.3.22 in the form of the density profiles n(r). The profile $R_2(z)$, indicates that R_2 was about 3μm for heights below 480 m, and then decreased to 1.4μm at 660 m. The curves n(r) in Fig.3.22 demonstrate the variations with respect to height of the size spectrum of the haze sounded. The variations are strongest for the fraction of large particles ($r \geq 1\mu m$), as expected.

3.5.2 Efficiency of Multi-Frequency Sensing in Studying the "Details" of Aerosol Microstructure

The characteristic feature of the size spectra of aerosol ensembles are the bimodal size-distribution functions which describe them. The question arises, in this connection, as to how well the curves presented approximate the actual ones, as well as the question whether it is even possible to reconstruct the bimodal size-spectra from sensing data at three frequencies. Since these questions are very important for the assessment of the information content of three-frequency lidars, they have to be discussed in more detail.

Figure 3.23 presents an example of a two-mode size spectrum $s_0(r)$ (solid curve) reconstructed from the data on $\beta_\pi(\lambda_i)$ ($i = 1,2,\dots,n$). The first solution (dashed curve) corresponds to $n = 4$ (the wavelengths $\lambda = 0.347$; 0.53; 0.69 and 1.06μm). The initial uncertainty in the characteristic $\beta_\pi(\lambda_i)$ inverted does not exceed 3 or 5% (the quadrature error). The approximate solution $s_\alpha(r)$ corresponds to the so-called optimum value of the regularization parameter α, which is determined from the condition $\min_\alpha \|s_0 - s_\alpha\|$. This value of α deter-

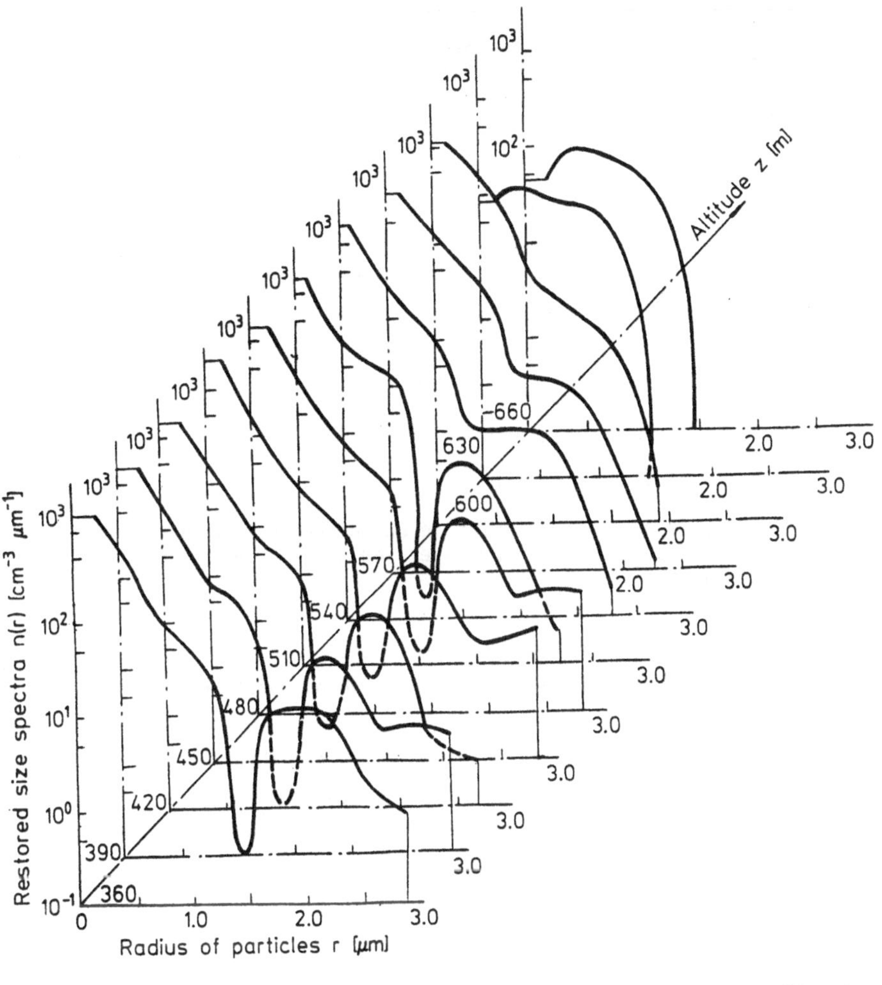

Fig.3.22. Aerosol size spectra obtained by inverting the profiles $\bar{S}(z,\lambda)$ (Fig.3.20); altitude increment Δz = 30 m

mines, for a given dimensionality of the vectors β and $\mathbf{s}$ and error σ, the maximum admissible error of the reconstruction of the initial size-distribution function $s_0(r)$ from optical measurements.

Obviously, the norm $\|s_0 - s_\alpha\|$ cannot be made infinitely small, even if $\sigma \to 0$. In this sense, this norm determines the methodological uncertainty of the optical method which is caused by the small amount of reference information. The amount of information can be increased only if the number of read-outs of an optical characteristic $\beta(\lambda)$ under the inversion is increased (see the dashed and dotted curve in Fig.3.23). As our numerical analysis has shown, the sensing should be carried out at least at nine wavelengths

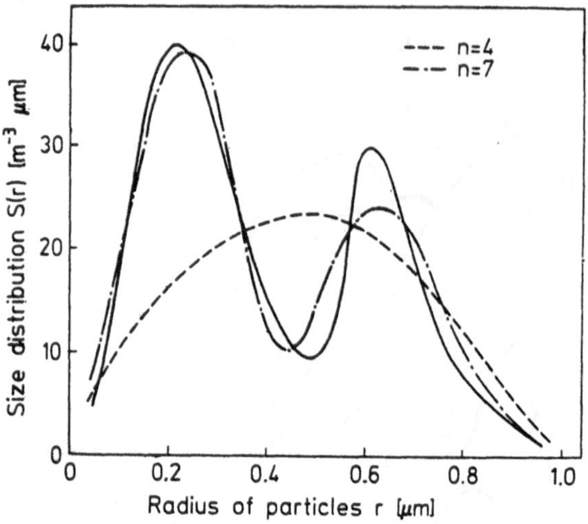

Fig.3.23. Results of numerical simulation on the reconstruction of the model double-peaked size-spectrum (——) by inverting $\beta_\pi(\lambda)$ obtained using different numbers of wavelengths

chosen within the most informative spectral regions in order to achieve reliable resolution of the bimodal size spectra when inverting the characteristic $\beta_\pi(\lambda)$.

Thus, when the inversion is carried out with only three measurements of $\beta_\pi(\lambda)$, there is no reliable basis for distinguishing the bimodal structure of an actual size spectrum. As a matter of fact, this result is obvious and does not require any special validation. Nevertheless, the numerical analysis shows that situations may occur for which measurements at three wavelengths might be sufficient for discovering the bimodality of the size spectrum sought. Such an example is presented below (Fig.3.24).

The information content of sensing within a particular spectral interval depends on the sensitivity of $\beta(\lambda)$ on the function $s(r)$. As one may see from numerical analysis,the backscattering coefficient $\beta_\pi(\lambda)$ of aerosols [or, equivalently the polydispersed factor $\bar{K}_\pi(\lambda)$] is, as a rule, a monotonically decreasing function in the spectral interval [0.53μm, 1.06μm], regardless of the type of size-distribution if $R_2 \leq 1\mu m$. This is valid for aerosol formations such as the haze of H-type [3.2] (Fig.3.24a) as well as for the bimodal size-distribution functions used in the present work as a model (Fig. 3.23). For these size-distribution functions,the interval [0.53μm, 1.06μm] is the region of large λ (in the sense of Sect.3.3) and the spectral behavior of $\beta_\pi(\lambda)$ is determined by the behavior of $K_\pi(x)$ in the region of small x. However, as the number density of large particles in the initial size-

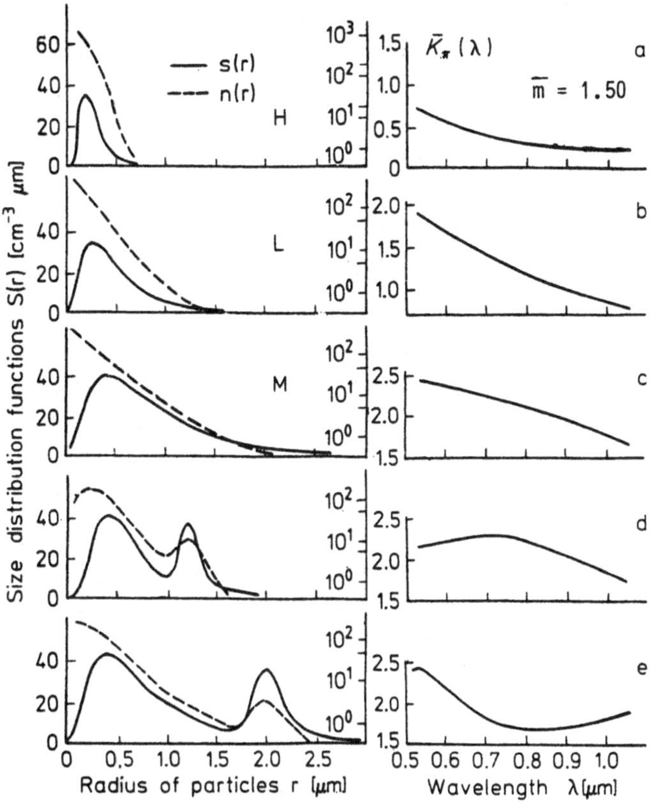

Fig.3.24. Spectral behaviors of the polydispersed backscattering efficiency factor $\bar{K}_\pi$ (also β_π) for different model size-distribution functions

distribution increases, its effect on the spectral behavior $\beta_\pi(\lambda)$ steadily increases within the sensing interval Λ mentioned above. Appropriate examples are presented in Fig.3.24.

The increase of the number density of large particles in unimodal size-distributions such as for the hazes of L and M types results in a change in the slope of the curve $\bar{K}_\pi(\lambda)$ (Fig.3.24b and c). In the case of a bimodal size-distribution s(r) which has a second mode located to the left of the point $r = 1\mu m$, the efficiency factor $\bar{K}_\pi(\lambda)$ becomes convex (from below) (Fig.3.24d), so that its analytical behavior in the interval [0.53μm, 1.06μm] is qualitatively different from that of the previous examples. Also, at least when the second mode is located far enough from the first one (e.g., Fig.3.24e, where $r_{s_2} = 2\mu m$), the function $\bar{K}_\pi(\lambda)$ becomes a concave function of λ.

This circumstance is decisive for checking whether the reference size-distribution is bimodal or unimodal from the data of three-frequency sensing in the interval [0.53μm, 1.06μm]. If the function $\beta_\pi(\lambda)$ is concave on this interval (this can be reliably checked along the whole sensing path by measuring it at λ = 0.53; 0.69; 1.06μm), then it is not possible to choose an appropriate quasi-solution from the class of unimodal size-distributions, since condition (3.56) cannot be fulfilled. Any unimodal size-distribution yields an optical characteristic $\beta_\pi(\lambda)$ which decreases monotonically in the interval Λ. On the other hand, the characteristics $\beta_\pi(\lambda)$ may be convex when one mode of the corresponding size-distribution is markedly displaced to the region of large sizes, although this is not very probable for atmospheric hazes. This was illustrated for inversion results based on experimental data in Fig.3.22.

In conclusion, consider one more example on three-frequency sensing of aerosols in the boundary atmospheric layer, namely the investigation of the influence of the air humidity on the parameters of aerosol microstructures carried out with the use of the inversion of the sensing data. For the inversion, a series of successive profiles $\bar{S}(z,\lambda_i)$ was used together with the data on air humidity measured at the same times. Using the inversion results, the profiles of the number density of particles $N(r \geq a)$ were inferred, where a is some fixed size. When a was smaller than 0.6-0.8μm, the effect of humidity on $N(z,r \geq a)$ was not clearly observed. However, the situation changed drastically when the only number density of large particles was assessed. For example, Fig.3.25 shows three profiles $N(z,r \geq 1\mu m)$. These profiles correspond to the successive increase of air humidity (73; 86; 98%). This example illustrates the well-known fact that the air humidity affects large particles most strongly. It is important, in this case, to emphasize that the method of multi-frequency sensing enables one not only to study qualitatively the physical processes under natural conditions but also to assess quantitatively the corresponding characteristics of the processes.

3.6 Investigation of the Microstructure of Low Stratospheric Aerosols Using Ground Based Lidars

3.6.1 General Remarks

The lidar sensing of stratospheric aerosols has some peculiarities which must of course be taken into account when interpreting optical data. First, the attenuation of a light pulse propagating through the atmosphere at

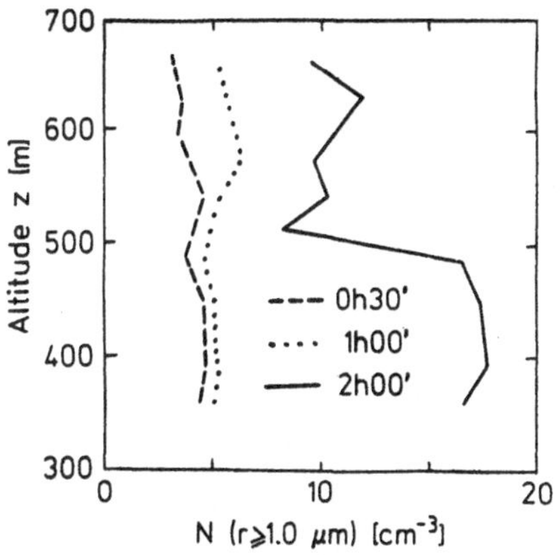

Fig. 3.25. Examples of height distributions of the number density of particles with radii greater than 1μm, obtained using a three-frequency lidar at different air humidity values

altitudes above 10 km can be neglected since the transmission function $T^{\frac{1}{2}}(z)$ is close to unity provided, of course, that no selective absorption is observed at the wavelength used. The transmission function $T^{\frac{1}{2}}(z)$ for altitudes $z \leq 10$ km, should be determined experimentally beforehand. Usually the method of slant paths, which is based on the assumption of horizontal homogeneity of the atmosphere, is used to determine T(z). The validity of this assumption is better for smaller angles between the directions of sensing. On the other hand, the method is more sensitive to errors in the measurement of lidar signals when this angle is small. Therefore, this method can be regarded only as a technique for qualitative assessment of the atmospheric transmission. It is preferable to measure the transmission of the whole atmosphere using either stellar or solar spectral photometers. In this case, the determination of the microstructures of stratospheric aerosols from the data of multifrequency lidar sensings can be reduced to the problem on inverting the set of data $\{\beta_\pi(z,\lambda_i)\}$ ($z \geq 10$ km, $i = 1,2,\ldots,n$).

The second peculiarity of lidar sensing of stratospheric aerosols is that the contribution of molecular Rayleigh scattering to the total backscattering is, as a rule, comparable to or greater than the contribution due to aerosol (Mie) scattering. This means that the ratio $\eta = \beta_\pi^{(a)}(z)[(\beta_\pi^{(R)} + \beta_\pi^{(a)}]^{-1}$ is close to unity at all altitudes above 10 km and for all relevant values of λ. As a result, one should estimate the profiles $\beta_\pi^{(R)}(z)$ in order to infer the information on $\beta_\pi^{(a)}(z,\lambda)$. Obviously, errors in the determination of the former profiles as well as errors in T(z) restrict the accuracy with which the profile $\beta_\pi^{(a)}(z)$ can be inferred from lidar measurements. These errors are evidently systematic since they cannot be decreased by improving the accuracy of the lidar recording system.

Consider now the peculiarities directly inherent in the interpretation of the backscattering coefficient as applied to sensing the stratospheric aerosols at two or three wavelengths. These problems are currently of great practical interest.

3.6.2 Determination of the Boundaries of Aerosol Size Spectrum

The information content of lidar measurements made at two or three wavelengths is obviously insufficient for making a useful inversion based on the use of general and rigorous methods. Therefore, it is expedient to examine more closely the method of model assessments for solving this particular problem. The basic concepts of this method have already been discussed in Sect.2.6. Recall that if the size-distribution function $s_0(r)$ differs strongly in its analytical form from that of the model used in (2.80a) and (2.80b), then this method can at best provide an estimation of some integral characteristics of the microstructures of the aerosol sensed. Let us assume that the size-distribution function $s_0(r)$ differs insignificantly in most respects from the model so that the methodological errors caused by this difference between the forms of $s_0(r)$ and $s_M(r)$ can be neglected to a first approximation. Irrespective of the simplicity of the problem stated, there are a lot of questions that arise in the practice of handling the experimental data and concerning the efficiency of lidar schemes for the optical investigations of aerosols.

The first question concerns the choice of wavelengths for multi-frequency lidar operation, since they must be sufficiently informative for investigating several types of atmospheric aerosol formations. As was shown above, the efficiency of a particular spectral interval of sensing $\Lambda = [\lambda_{min}, \lambda_{max}]$ which contains the wavelengths λ_i $(i = 1,2,\ldots,n)$ at which the optical system operates depends, to a great extent, on the positions of its boundaries with respect to the boundaries of the size spectrum of the aerosol particles $R = [R_1, R_2]$.

It goes without saying that two or three wavelengths are not sufficient for the reconstruction of the size spectrum of particles within accuracy limits comparable to the errors of optical measurements over the whole size interval R. This is especially true in the case of broad particle size spectra or when $R_1 \ll R_2$. In this case, it is expedient to select some subinterval R from the interval R, in which the errors of the reconstruction of the aerosol size-distribution function are comparable with those of optical measurements. The errors of the inversion at the remaining points of the

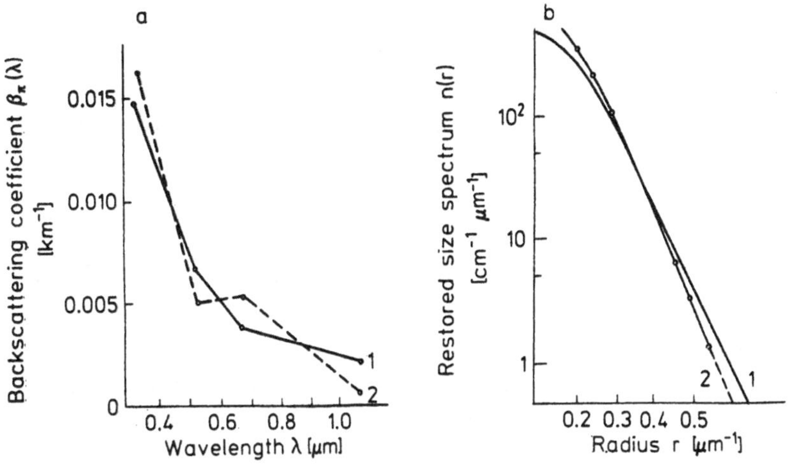

Fig.3.26. An example of the reconstruction of aerosol particles size-spectrum from $\beta_\pi(\lambda)$ by the method of model assessment

interval are then considered to be large as a result of insufficient information content available for the solution of the problem. From the standpoint of the reliability of inversion results, it is very important to know the boundaries of the interval R. For more clarity, let us consider some numerical examples.

Figure 3.26 presents the size spectrum $n_0(r)$ reconstructed with the use of measured characteristic $\beta_\pi(\lambda_i)$, $(i = 1,2,\ldots,4)$ (Curve 1 in Fig.3.26a). The inversion was performed with the method of model assessment, i.e., by estimating the parameters r_s and s beforehand, and while α and γ were set a priori. The data inverted were perturbed by fluctuations of about 10% from $\max\beta_\pi(\lambda)$ (Curve 2, Fig.3.26a) in the values of $\beta_\pi(\lambda)$.

The inversion results are presented in Fig.3.26b. The effective interval $\tilde{R}$, where the deviations of the inverse solution (Curve 2) from the exact one (Curve 1) are small,is limited by the radii 0.25 and 0.5μm. The boundaries of the reference interval R are approximately 0.1 and 0.6μm. These values for R_1 and R_2 satisfy the condition $\int_0^{0.1} q(r)dr = \int_{0.6}^{\infty} q(r)dr \lesssim 0.3$. In such a situation it is common to consider the size regions [0.1 m, 0.25μm] and [0.5μm, 0.6μm] to be represented weakly in the solution with the above reference data. An improvement in the accuracy of the measurements, or an increase in the number of wavelengths allow one to broaden the interval $\tilde{R}$.

According to the data on the microstructure parameters of aerosols at altitudes from 20 to 25 km investigated in [3.5], the size-distribution functions s(r) are mainly defined for sizes smaller than 0.6 or 0.8μm. Hence one can believe that $R_2 \simeq 0.6\mu m$. The efficiency (or significance(of a_r at this boundary for the size spectrum is estimated by the value of the quantiple of the integral size-distribution q(r). In particular, the condition $\alpha_2(0.6) = 1 - q(0.6) \leq 0.1$ can be considered fulfilled, with large probability, for aerosols of the low stratosphere. This means that if the uncertainty of optical measurements does not exceed 10% or 15%, then using such wavelengths as $0.69\mu m$ and especially its second harmonic ($0.347\mu m$), one can easily determine the values of s(r) in the vicinity of $0.6\mu m$ with an acceptable accuracy using inversion methods. The optical contribution of the fraction of particles having sizes about $0.6 - 0.8\mu m$ will be distinguished by the error mentioned above, if this fraction is significant in the size distribution s(r). Thus, for stratospheric aerosols sounded in the spectral region [$0.345\mu m$, $1.06\mu m$], one can assume that $\tilde{R}_2 \simeq R_2 \simeq 0.6\mu m$. Now let us discuss the estimation of the efficiency of lidar methods for determining the aerosol microstructure in the region of small sizes, i.e., the estimation of $\tilde{R}_1$.

Although the fraction of particles with the sizes less than $0.2 - 0.3\mu m$ in the total geometrical cross-section S is relatively small in the case of aerosols of the boundary layer, this is not so for stratospheric aerosols. The absence of large particles leads to a situation in which the value of the smaller size $\tilde{R}_1$ determined from the expression

$$\int_{R_2}^{R_1} K(r,\lambda_{min}) s(r) dr \leq \eta\sigma$$

is not already small, in the sense that now $\alpha_1(\tilde{R}_1) = q(\tilde{R}_1)$ significantly exceeds the value of the relative error ε of the optical measurements. These errors in the interval [R_1, $\tilde{R}_1$] are so large that a reliable estimation of such an important parameter as S cannot be assured. This is also true for assessments of r_s. It is necessary to choose the wavelengths for the lidar operation properly. It also becomes clear that in such a situation the estimations of the whole geometrical cross-section are not reliable; one can only determine a fractional cross section, namely that from the interval [$\tilde{R}_1$, $\tilde{R}_2$].

Thus, for stratospheric aerosols sensed with a lidar, it is advisable to determine the value $S(r \geq \tilde{R}_1)$ (geometrical cross-section of particles whose sizes exceed the value $\tilde{R}_1$) since it is the most reliable result of the optical data inversion. The value $S(r \geq R_1)$ is calculated using the estimates S^* and r_s^* that are found with the use of the model size-distribution function $S_M(r)$. In this respect, it would be interesting to assess the expected

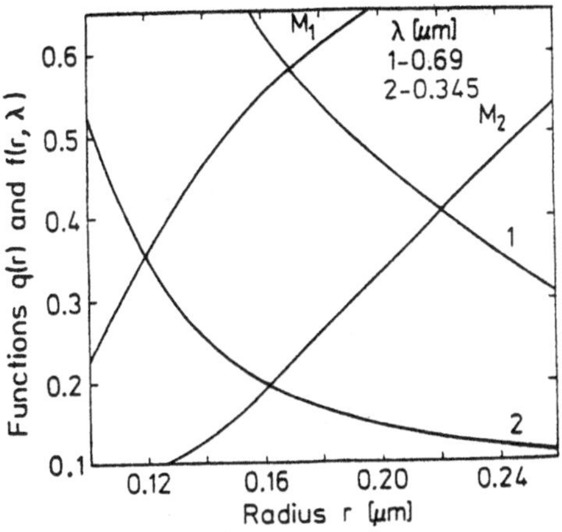

Fig.3.27. Numerical solutions of (3.59) with reference to R_1 for two models of the size-spectrum g(r) (M_1 is the atmospheric haze H, and M_2 is the Junge size-distribution function). Curves 1 and 2 represent the function $f(r,\lambda)$ [right-hand side of (3.59)] at two wavelength: $\lambda_1 = 0.69$ μm and $\lambda_2 = 0.345$ μm

values of $\tilde{R}_1$ for the case when the aerosols of the lower stratosphere are sensed at the wavelengths of 0.69μm and its second harmonic of 0.347μm. An approximate equation for estimating $\tilde{R}_1$ can be constructed from the preceding inequality as follows

$$q(\tilde{R}_1) = \varepsilon_\pi \bar{K}_\pi(\lambda)/\tilde{K}_\pi(\tilde{R}_1,\lambda) \quad , \tag{3.59}$$

where $\bar{K}_\pi(\lambda)$ is the polydispersed efficiency factor for the system of particles; $\tilde{K}_\pi(\tilde{R}_1,\lambda)$ is the smoothed efficiency factor of backscattering (Sect. 3.3). The solution $\tilde{R}_1^*$ of this equation can be found graphically as the point of intersection of the curves $q(\tilde{R}_1)$ and $f(\tilde{R}_1,\lambda) = \varepsilon_\pi \bar{K}_\pi(\lambda)/\tilde{K}_\pi(\tilde{R}_1,\lambda)$ at a given λ. In [3.30,31], questions of the optimum choice of the wavelengths for sensing various aerosols were discussed in detail. It was shown that the value $\tilde{R}_1^*$ found in this manner gives an upper limit for $\tilde{R}_1$, i.e., $\tilde{R}_1 \lesssim \tilde{R}_1^*$. An example of the graphical solution of (3.59) is given in Fig. 3.27. Two model size-distribution functions q(r) (curves M_1 and M_2) were used in the calculations. The first model corresponds to the power law (2.74) for $r_1 = 0.05$μm, $r_2 = 1$μm and $N = 100$ km^{-3}, while the second is the haze H [3.2] ($S = 0.096$ km^{-1}, $r_s = 0.2$μm). The first model is characterized by a large fraction of small particles, viz. about 65% of the total geometrical cross-section of the whole model is formed by particles with sizes less than 0.2μm. In the second model, the situation is opposite, i.e., about 65% of the total geometrical cross-section is contributed by particles having sizes larger than 0.2μm.

In order to solve (3.59), the values ε_π and the refractive index of the aerosol substance were assumed to be equal to 0.05 and 1.5-0i, respectively. As is seen from Fig.3.27, the roots of the equation at $\lambda = 0.69$ and 0.347μm are, for the first model, 0.17 and 0.12μm, respectively. The value $\tilde{R}_1^* = 0.17\mu m$ is too large for this polydispersed system, which is composed mainly of small particles. The value $q(\tilde{R}_1)$ is about 0.6, i.e., more than 60% of the particles (according to the measure S) are out of the range of sensitivity of the optical measurements. The second value of the smallest radius sensitive to radiation with $\lambda = 0.347\mu m$ is more reasonable, in this respect, since $q_1(0.12) \simeq 0.35$. It is obvious, for this reason, that when the lower boundary of the sensing interval is $\lambda = 0.347\mu m$, the corresponding multi-frequency lidar is more useful for studying the microstructures of aerosol ensembles whose size-distribution functions are close to the first model.

These remarks are valid to an even greater degree for the second model. Indeed, the second root of (3.59) is approximately equal to 0.16μm and since small particles with radii less than 0.16μm contribute only about 20% of the total geometrical cross-section of that model, the value $S(r \geq 0.16\mu m)$ may be regarded as a consistent estimate of the parameter s_0 of the model. Assuming that the above models are the extreme models of the set of possible size-distributions $q(r)$, for the given type of aerosols, one can assess the interval of expected values of $\tilde{R}_1$. In particular, when $\lambda = 0.347\mu m$ and $\varepsilon_\pi = 0.05$, the values of $\tilde{R}_1$ are contained in the interval [0.12μm, 0.16μm]. As ε_π increases, the boundaries of the interval move to the region of large sizes (examples are presented in [3.30]). For $\lambda_{min} = 0.69\mu m$ one has $0.16 \leq \tilde{R}_1 \leq 0.25\mu m$.

Note in conclusion that the values $\tilde{R}_1$ and $\tilde{R}_2$ introduced into the theory of lidar sensing of the atmosphere as the minimum and the maximum sizes of the optically detectable particles play an important role not only for assessing the reliability of the interpretation but also for the choice of optical parameters for the optical schemes. In particular, (3.59) allows one to solve a corresponding inverse problem, i.e., to estimate λ_{min} and ε_{min} for a given $\tilde{R}_1$ and a set of possible size-distributions $q(r)$. For the estimation of λ_{max}, one can use the relationships of Sect.3.3. The above analysis also shows that a purely formal approach to the inversion of optical data cannot be effective per se. The use of any calculational scheme requires a prior analysis of the inverse problems, especially if there is a "deficit" of experimental information. The optical sensing of stratospheric aerosols using ground-based lidars is one such problem.

3.6.3 An Example of Analyzing the Data of Two-Frequency Lidar Sensing of Low Stratospheric Aerosols

Let us consider the previous example of data inversion presented in Fig. 3.26 from the standpoint of the behavior of the value $S(r \geq a)$. The corresponding data for the analysis are given in Table 3.3. These data illustrate the efficiency of the assessments $S(r \geq a)$ and $N(r \geq a)$. As one sees from the table, the choice of $a = 0.1\mu m$ for R_1 cannot be regarded as good for the reference data: $\lambda = 0.347\mu m$ and $\varepsilon_{min} \geq 0.1$. The estimate of the value $\tilde{R}_1$, according to (3.54) is close to $0.2\mu m$, and this is confirmed by the results of the numerical inversion presented in Table 3.3. In this case, the value $s(r \geq 0.2)$ is estimated with minimum possible error.

Note that the estimation of the number density of particles is possible only for $r \geq R_1$, since the value $N(r \geq a)$ increases sharply as a decreases so that even small deviations $\Delta(S)$ yield significant errors in the number density. This is illustrated in Table 3.3 by the values $N(r \geq 0.1)$.

Table 3.3. Errors in the determination of the microstructure characteristics $S(r \geq a)$ and $N(r \geq a)$ by inverting $\beta_\pi(\lambda)$ (Fig.3.26) for various values of a

a [μm]	0.1	0.15	0.2	0.4
$\varepsilon_S = \Delta(S)/S[\%]$	20	15	10	20
$\varepsilon_N = \Delta(N)/N[\%]$	40	15	15	20

It was assumed above that the actual size-distribution function $s_0(r)$ is unimodal, and hence the solution of the inverse problem of light scattering in terms of convex unimodal functions can, in principle, be free of methodological inaccuracies even if the number of optical measurements is insufficient. It is clear that in the case of two or three frequency sensing, a solution is possible only in the class of elementary size-distributions (Sects.2.6 and 3.5); therefore, it is interesting to consider, within the framework of the method of model assessments, the errors due to improper choice of the reference model. Such an example is presented in Fig.3.28. The left part of the figure presents the results of the reconstruction of n(r) (Curve 1) using the model (2.75) while the working model is the gamma size-distribution (Curve 1'). The opposite case, when the reference model is the gamma size-distribution (Curve 1) and the power law function is used as a working model, is presented in the right part of Fig.3.28. The figure indicates that methodological errors are unimportant within the interval $(\tilde{R}_1, \tilde{R}_2)$ if $S(r \geq \tilde{R}_1)$ and $N(r \geq \tilde{R}_1)$ are estimated properly, but that $n_0(r)$

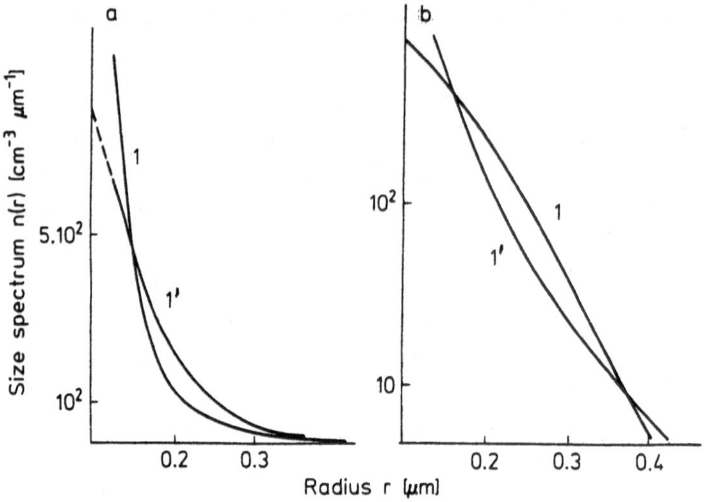

Fig. 3.28. An example illustrating the methodological errors in the method of model assessments due to a poor choice of model size-distribution function in (2.80)

and $n_M(r)$ differ significantly from each other if they are directly compared. The comparison of the size-distribution functions considered is based on integral relationships of the type (2.77) and (2.78) and shows that the deviation of $S^*(r \geq \tilde{R}_1)$ from $S_0(r \geq \tilde{R}_1)$ does not exceed 10% in both cases. This is also true for the mean radii $\bar{r}^*$ and $\bar{r}_0$. We now proceed to some examples of the interpretation of experimental data.

We discuss as an example the inversion of the data on lidar sensing of low stratospheric aerosols at altitudes about 14, 25-27 and 75 km, performed with a two-frequency lidar at λ = 0.53 and 0.69μm. The lidar parameters are given in Table 3.4. The lidar signals in this experiment have been processed with the use of a 33-channel photon counting system. Thirty channels of the system are used to monitor the lidar signals while the other three are used to control the laser radiation power, to count the number of sensing pulses, and to measure the background radiation.

The technique for measuring the profiles $\beta_\pi(z,\lambda)$ of the backscattering coefficient of the aerosols was as follows: light pulses at the above wavelengths were successively fired into the atmosphere. The backscattered signals were recorded from 30 altitude intervals, each having the same length Δz which was regulated by the duration of a time gate. The lower altitude, from which the signals were recorded, was determined by a time delay between the laser firing and the start of the recorder. Since the signals which were

Table 3.4. Parameters of the two-frequency lidar used for sensing low stratospheric aerosols

Lidar parameters	λ = 0.530	λ = 0.6943 [μm]
Laser pulse energy	0.03	0.8
Beam divergence rad	2.9 10^{-4}	4.5 10^{-4}
Receiving area [m^2]	0.17	0.17
Receiver field of view [rad]	1.5 10^{-3}	1.5 10^{-3}
Maximum pulse repetition frequency [Hz]	100	10
Filter halfwidth [nm]	1.7	2.4
Overall quantum efficiency of the receiver	0.86	0.36
Duration of a time gate [μs]	2.5 ÷ 40	
Duration of signal channel and background channel [μs]	2.5 - 10^4	
Maximum count rate [MHz]	25	
Maximum number of sensing pulses	10^4	
Storage frequency [kHz]	2	
Time delay between sending laser pulse and the recorder start [μs]	10 ÷ 10^3	

backscattered from the layers above 10 km were too weak, the signal storage from N realizations was used in order to obtain reliable estimates of the mean values. At the same time, the total energy of the laser pulses was measured with the corresponding channel of the recorder.

The value $T^{\frac{1}{2}}(z)$ up to z = 14.25 km was evaluated using the method of slant paths. For this purpose, a layer Δz = 1.5 km at the altitudes z = 30 or 33 km was selected, and backscattered signals from it were recorded at zenith angles 0° and 32°. Assuming horizontal homogeneity of the scattering properties of the atmosphere at these altitudes, one can obtain the atmospheric transmittance by comparing the signals.

In order to evaluate the optical losses in the transmitter and receiver optics and also to estimate the quantum efficiency of the photocathode of the PMT at the operating wavelengths, a calibration was usually performed using a reflecting screen with a known reflectivity. By comparing the amplitudes of the transmitted pulse and that reflected from the screen, the overall quantum efficiency k_λ of the lidar was evaluated. Then, using the measured values of the mean lidar returns $P(z,\lambda_i)$ and known T_i and k_λ, the profiles

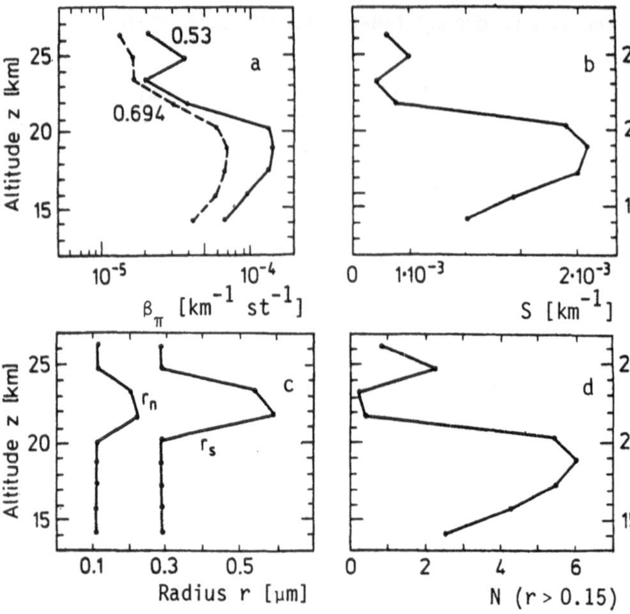

Fig. 3.29. Results illustrating the interpretation of sensing data for low stratospheric aerosols using a two-frequency lidar

of the backscattering coefficient $\beta_\pi = \beta_\pi^{(a)} + \beta_\pi^{(R)}$ were determined. The profiles $\beta_\pi^{(R)}$ were calculated according to the barometric formula with the use of data on the ground layer temperature and pressure as well as of data on the standard atmosphere. The lidar measurements were carried out in June 1975 near Tomsk, USSR, under clear skies with a meteorological visual range of about 25 km. The data acquisition and processing system provided the evaluation of signals $P(z,\lambda_i)$ with an error less than 2% over the whole path. Typical profiles of the aerosol backscattering coefficients measured at $\lambda = 0.53$ and $\lambda = 0.694\mu m$ on June 15, 1975 are shown in Fig.3.29a. The parameters S and r_s (Fig.3.29c) were calculated using (2.80a) and (2.80b) with a gamma size-distribution used as a model ($\alpha = 2$; $\gamma = 0.7$). The remaining parameters of the aerosol microstructures were found using the determined values of S and r_s. The value $\bar{m} = 1.43 - 0i$ of the refractive index of the aerosol substance was assumed since this is widely used in the literature for analyzing the results of lidar measurements of low stratospheric aerosols.

The possible errors due to uncertainties in the choice of the value of the refractive index on the accuracy of the reconstruction are illustrated by the data presented in Table 3.5. The second value of the refractive index used is $\bar{m} = 1.46 - 0i$. As is seen from the table, the value $S(r \geq \tilde{R}_1)$ is

Table 3.5. Results of the evaluation of the microstructure parameter $S(r \geq a)$ for the low stratospheric aerosol obtained by inverting $\beta_\pi(\lambda)$ (Fig.3.29a) for two values of the refractive index

Z [km]	$S(r \geq 0.15)$ [km^{-1}]			$S(r \geq 0.2)$			$S(r \geq 0.25)$		
	$\bar{m} = 1.43$	$\bar{m} = 1.46$	ε_s[%]	$\bar{m} = 1.43$	$\bar{m} = 1.46$	ε_s[%]	$\bar{m} = 1.43$	$\bar{m} = 1.46$	ε_s[%]
14.25	$0.88\ 10^{-3}$	$1.11\ 10^{-3}$	21	$0.82\ 10^{-3}$	$0.87\ 10^{-3}$	6	$0.75\ 10^{-3}$	$0.63\ 10^{-3}$	16
15.75	1.38	1.66	16	1.27	1.29	2	1.12	0.94	16
17.25	1.77	2.13	16	1.63	1.66	2	1.44	1.21	16
18.75	1.98	2.34	15	1.82	1.83	1	1.61	1.35	16
20.25	1.77	2.15	17	1.63	1.67	3	1.44	1.22	15
21.75	0.32	0.29	10	0.32	0.28	11	0.31	0.27	12
23.25	0.18	0.16	12	0.18	0.16	13	0.17	0.15	15
24.75	0.48	0.58	17	0.44	0.45	2	0.39	0.33	15

Table 3.6. Parameters of the lower stratospheric aerosol microstructure obtained by inverting the data of two-frequency lidar measurements (Fig.3.29a)

z [km]	$N(r \geq 0.2)$ [cm^{-3}]	r_n [μm]	$\frac{N(r > 0.15)}{N(r > 0.25)}$
14.25	1.9	0.12	1.7
15.75	3.1	0.11	2.0
17.25	4.0	0.11	2.0
18.75	4.4	0.11	2.0
20.25	4.0	0.11	2.0
21.75	0.32	0.22	1.3
23.25	0.21	0.20	1.3
24.75	1.1	0.11	2.0

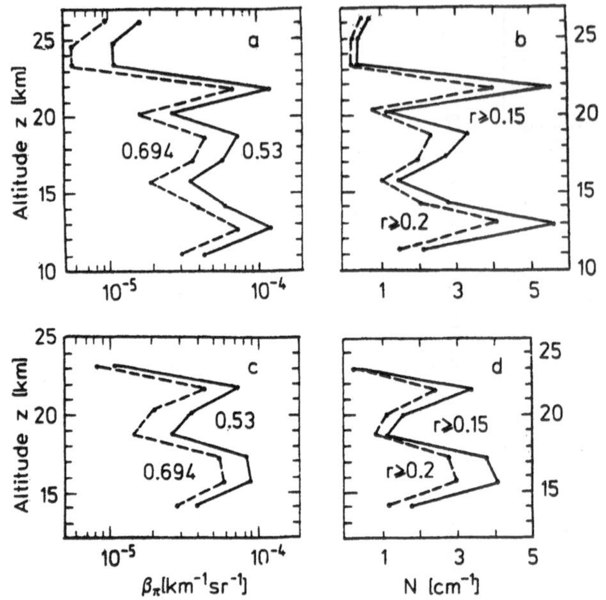

Fig.3.30. Height distributions of number density of the low stratospheric aerosol particles reconstructed from the data of a two-frequency lidar sensing, obtained in a thirty-minute time interval

reasonably stable with respect to the variations of the refractive index within the limits $\Delta\bar{m} = 0.03$. In particular, this is true for the value of $s(r \geq \acute{R}_1)$, where $R_1 = 0.2\mu m$ and $\lambda = 0.53\mu m$. It is for this reason that the most reliable estimate of the number density of aerosol particles is the value $N(r \geq 0.2)$ given in Table 3.6.

The values of the aerosol parameters presented in Tables 3.5 and 3.6 as well as in Fig.3.29 are typical for the lidar measurements made in June, 1975 at the test field of the Institute of Atmospheric Optics, Tomsk, USSR. These results represent the basic behavior of the height distributions of the aerosol microstructure parameters in the low stratosphere observed in the direct measurements during 1975 [3.5]. We note that the reproducibility of the modal radius r_n of the size-distribution $n(r)$ is quite good.

The analysis of the results revealed noticeable spatial-temporal variations in the characteristics of light scattering by aerosols in the low stratosphere. This is clearly demonstrated by the data presented in Fig. 3.30a,c, which were obtained from measurements of β_π carried out on June 14, 1975 at 23^h00^m and 23^h30^m local time near Tomsk. Using these data, the profiles of $N(r \geq 0.15)$ were obtained (Fig.3.30b,d). Note that the mean value of the number density of aerosol particles in the layer at a height of 17-19 km was about 3.5-4 cm^{-3}, as estimated by inverting the data obtained on June 15 and 16, 1975, and was in a good agreement with the data obtained in [3.5] using a photoelectric counter. For more details, the reader is referred to [3.32].

The above results indicate that multi-frequency lidar sensing may be an effective method for investigating remotely the microstructure of low stratospheric aerosols provided that proper techniques for inverting the optical data are used.

3.7 Determination of the Microstructure of Cloud Aerosols Using Lidar Sensing

One of the most interesting tasks of atmospheric physics is the study of the initiating processes and the dynamics of cloud formation. The pulsed laser systems for multi-frequency sensing of the atmosphere allow, in principle, a determination of microstructure of clouds to be made in the natural, nondisturbed state of the clouds, as opposed, e.g., to on-board flying laboratories. At the same time, the use of lidars for investigating clouds requires an appropriate mathematical apparatus for inverting optical data to be developed. The general approach to the solution of such inverse problems has been discussed in Sect.2.3, based on the use of operators of the light scattering theory by polydispersed systems.

This section is devoted to the discussion of qualitative aspects of the theory of lidar sensing of clouds, based mainly on the results of numerical experiments. We observe that the inverse problems of multi-frequency sensing of clouds are generally simpler, as compared, e.g., with those for sensing hazes, at least if no systems of transfer equations for optically dense media have to be solved.

Indeed, the water droplets in clouds are spherical, and hence the use of the Mie theory, as a working hypothesis, is well justified. The index of refraction for water droplets is also known quite accurately for almost any spectral interval used for optical sensing. Furthermore, in most cases, the size-distributions of particles in the water droplet clouds are unimodal and well approximated by the gamma size-distribution function. This makes it possible, in many cases, to determine the microstructures of clouds by two or three appropriate optical measurements.

Let us consider now some numerical experiments that illustrate the possibilities of the use of multifrequency lidars to study the microstructures of clouds when the reference size-distribution function $s_0(r)$ is the model C_1 from [3.2]. The basic parameters of this model are: number density of particles $N = 100\ cm^{-3}$, water content $Q = 6.255 \cdot 10^{-8} g \cdot cm^{-3}$, total geometrical cross-section $S = 7.819\ km^{-1}$, modal radius $r_s = 5.33 \mu m$ ($\alpha = 6$; $\gamma = 1$). It is characteristic for the model size-distribution chosen that it is sharply localized near its maximum r_s. It can be assumed, as a first approximation,that there are no particles having sizes less than $2\mu m$ and greater than $8\mu m$ in the ensemble. It is expedient to choose the sensing interval so that the wavelengths within it are comparable to the sizes of particles of the medium sensed. It is obvious that such wavelengths should be in the IR region. In the numerical experiments discussed here the chosen wavelengths are: 2.36; 3.51; 5.3 and $10.6 \mu m$.

First, we consider the influence of the measurement errors on the accuracy of the determination of the aerosol microstructure parameters. The determination of these parameters is assumed to be made by inverting the aerosol backscattering coefficient measured at the above-mentioned wavelengths. As usual, systematic and statistical errors will be taken into account. The systematic errors are modelled by shifting the measured characteristic to some value that is determined by $\Delta\beta_\pi(\lambda_i) = \varepsilon\beta_\pi(\lambda_i)$, $(i = 1,2,3,\ldots)$, where the constant ε is determined by the expected systematic errors for the given optical experiment. The statistical errors are simulated by oscillatory perturbations of the reference data. The values of these perturbations as well as their signs can be chosen according to the expression $\Delta\beta_\pi(\lambda_i) = (-1)^i \varepsilon\beta_\pi(\lambda_i)$ $(i = 1,2,\ldots,n)$.

The results of the inversion of $\beta_\pi(\lambda_i)$,which corresponds to the cloud model C_1, at the above four values of λ and for various perturbations are given in Table 3.7.

The analysis of the results obtained shows that the errors in the microstructure parameters are, as a rule, comparable to the errors of the optical measurements. It is characteristic that the modal radius r_s depends weakly on the systematic shifts of the optical characteristic $\beta_\pi(\lambda)$. This can be explained by the fact that the modal radius of a size-distribution (also the mean radius and median) is mainly determined by the form of $\beta_\pi(\lambda)$ but not by its absolute value. Taking this into account, we may assume that multiple scattering will have less effect on the inverted data for r_s than it does on such parameters as water content, particle number density and so on.

Table 3.7. Microstructure parameters of the model cloud C_1 reconstructed from the spectral characteristic $\beta_\pi(\lambda)$ for different perturbations of the reference data

Parameters	Systematic perturbation		Oscillating perturbations			
	$\varepsilon = 0.1$	$\varepsilon = 0.1$	$\varepsilon = 0.05$	$\varepsilon = 0.1$	$\varepsilon = -0.05$	$\varepsilon = -0.1$
r_s [m]	5.3	5.3	6.7	6.1	5.1	4.9
S [km^{-1}]	7.04	8.61	7.76	7.69	7.92	7.99
N [cm^{-3}]	90	110	88	86	112	123
Q [g cm^{-3}]	$5.6\ 10^{-8}$	$6.9\ 10^{-8}$	$6.6\ 10^{-8}$	$7\ 10^{-8}$	$6\ 10^{-8}$	$5.8\ 10^{-8}$

The value of the parameter α used in this calculation was assumed to be known. In practice, its value is chosen a priori within some given accuracy limits. In this case, an estimate of the effect of uncertainties in the a priori choice of α on the inversion accuracy was made. It appeared that variations of α within the limits ∓ 1,with respect to the reference value $\alpha_0 = 6$, have almost no effect on the accuracy of the parameters of microstructures presented in the table. In any case, if the accuracy of optical measurements is about 5 or 10%, these errors can be neglected. If the inverted data are more accurate, the choice of the reference value α should be made more accurately. It should be noted that for a smaller number of sensing wavelengths (say, for two wavelengths),the errors of the determination of the parameters shown in Table 3.7 become larger. In this case, the modal radius and the number density become more sensitive to errors in the measurements. The estimations of water content of clouds are more stable in the case of two-wavelengths sensing.

One of the interesting peculiarities of inverse problems of the type considered is the relative stability of estimations of N with respect to various perturbations, as opposed to such estimations for the case of atmospheric hazes, since in this case the sizes of the particles of the aerosol ensemble do not differ very much from each other. In addition, the use of the number density of particles instead of the size-distribution s(r) [according to the expression $n(r) = s(r)\ (\pi r^2)^{-1}$] for $r \geq 2\mu m$ does not result in a sharp increase of the errors in the region of small particles, as occurs for the atmospheric hazes. It follows from this fact that estimates of the number density of aerosol particles in clouds from lidar sensing data are as reliable as those of the total geometrical cross-section.

In the above experiment, the values of $\beta_\pi(\lambda)$ at the wavelengths of lidar operation are considered as known. In practice, it is also necessary to know the transmittance T(z) inside a cloud as well as the contribution of multiple light scattering to the lidar signals P(z). More sophisticated numerical experiments have been carried out, in this connection, with the use of the Monte-Carlo technique that enabled one to simulate the transfer of the laser radiation in clouds [3.33]. The modelled cloud layer is assumed to have plane-parallel boundaries and the size-distribution of particles is assumed to be the gamma size-distribution function ($r_s = 5.33\mu m$; $\alpha = 6$). The number density of particles (water content as well) varies with the height inside the cloud, as is shown in Fig.3.31a. The lower boundary of the cloud layer is located at a height $z_0 = 200$ m above the earth's surface. The variable z increases from the bottom of the cloud. The range increment Δz used for calculating the profiles P(z) is equal to 10 m. Figure 3.31 presents histograms of the value $\bar{S}(z,\lambda) = P(z,\lambda)z^2P_0^{-1}$, for $\lambda = 3.51\mu m$ and various angular apertures of the receiver, $\varphi_0 = 3'$; 20' and 1° (Curves 2, 3 and 4, respectively). Curve 1 characterizes the contribution of single scattering to the total backscattering. The behaviors of $\bar{S}(z,\lambda)$ at $\lambda = 2.36$ and $\lambda = 5.3\mu m$ as a function of altitude were, on the whole, identical to that shown in Fig.3.31b.

Let us consider now the question of extracting the profiles of any of the optical characteristics β_π or β_{ext} from $\bar{S}(z,\lambda)$. For large values of τ (at least,greater than unity) the lidar equation in the single scattering approximation gives more information on β_{ext} than on β_π (Sect.3.3). Therefore, all other conditions being equal, the estimates of β_{ext} are more reliable than those of β_π. Let us consider briefly a technique for determining $\beta_{ext}(z,\lambda)$ using the profile $\bar{S}(z,\lambda)$, having in mind the above numerical experiment.

According to the assumptions on the microstructure of the aerosol sensed ($r_s(z)$ = const) the value of the lidar ratio $b = \beta_\pi/\beta_{ext}$ is independent of the altitude. This leads to a situation in which any of the ratios $\int_{z_j}^{z_{j+1}} \bar{S}(z,\lambda)dz / \int_{z_{j+1}}^{z_m} \bar{S}(z,\lambda)dz$, ($j = 1,2,\dots,m$) is independent of β_π and to a first approximation is equal to $1 - \exp(-2\Delta z_j\bar{\beta}_{ext\ j})$ where $\bar{\beta}_{ext\ j}$ is the mean value of $\beta_{ext}(z)$ over the path interval $z_j \leq z \leq z$. This circumstance allows one to determine, at least approximately, the profile $\beta_{ext}(z)$ from the given experimental data. The results are presented in Fig.3.31c. Using the deviation of the reference profile (Curve 1) from the Curves 2,3,4 which correspond to the curves of Fig.3.31b, one can estimate the errors of this technique for determining $\beta_{ext}(z)$.

Starting from altitudes z of about 160 or 180 m (correspondingly $\tau \geq 3$) the solutions are unstable. The results of the inversion of the profiles $\beta_{ext}(z,\lambda_i)$ ($i = 1,2,3$) obtained in this way are presented in Fig.3.32. The

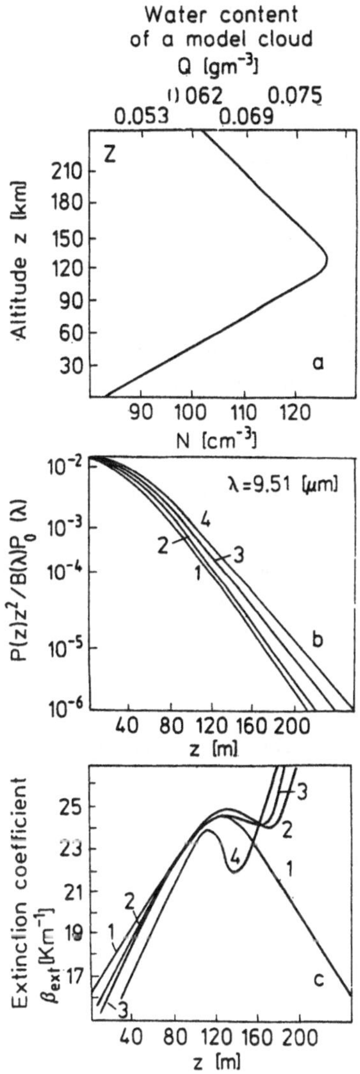

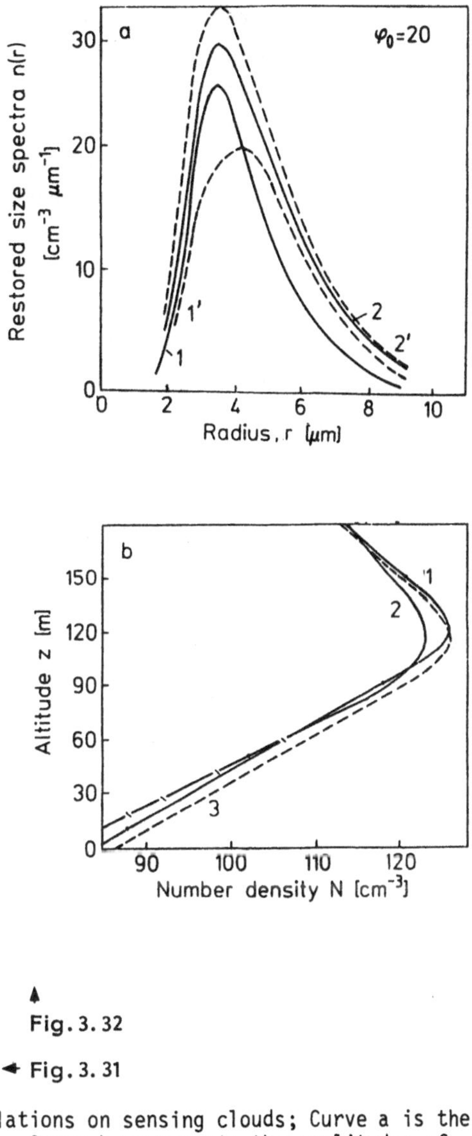

▲ Fig. 3.32

◄ Fig. 3.31

Fig. 3.31. Results of numerical simulations on sensing clouds; Curve a is the reference size-distribution function; Curve b represents the amplitudes of the lidar returns for different receiving areas; Curve c is the reconstructed altitude dependence of the extinction coefficient β_{ext}

Fig. 3.32. Results of numerical simulations on the reconstruction of cloud microstructure at different depths within the cloud (Curve 1 at a depth of 10 m and Curve 2 at a depth of 140 m) size spectra of the aerosol ensemble were determined for the layer $\Delta z_1 = 10$ m near the edge of the cloud (Curve 1') and for the layer $\Delta z_2 = 10$ m located inside the cloud at an altitude z = 140 m from the cloud base (Curve 2'). Underestimation of the size spectrum of the first layer, as compared with the reference one, is caused by systematic underestimation of the optical characteristic β_{ext} which is determined using the above method.

As might be expected, the errors of the determination of the size spectrum of aerosol particles are not significantly revealed in the estimations of the profile of the number density of particles which is presented in Fig. 3.32b, where Curve 1 is the profile $\beta(z)$ reconstructed from the component of the total backscatter due to the single scattering, while Curve 2 takes into account the contribution of multiple scattering for $\varphi_0 = 20'$. The dashed Curve 3 in Fig.3.32b represents the reference function $N_0(z)$. These give an idea of the capabilities of multifrequency lidar sensing of cloudy formations to remotely determine their microstructures.

The principal source of error in the determination of the microstructure parameters in the case considered above is the error in the determination of the optical characteristics from the lidar equation in the single scattering approximation.

It should be emphasized that the use of this equation in the theory of optical sensing when τ is large is not justified at all, especially if one recalls that its solutions are characterized by an explicit instability. This circumstance makes the interpretation of the optical data difficult regardless of whether the background due to multiple scattering is known or not. At the same time,it appears to be possible to use another equation instead of (3.18), which would be characterized, on the one hand, by better stability of its solutions with respect to perturbations of the reference data and, on the other hand, is close to the former in some sense. One such modification of the lidar equations (3.1) has been suggested in [3.34]. This equation is written as

$$P(z)z^2/P_0B = \bar{S}(z) = \beta_\pi(z)T^\xi(z) \quad , \tag{3.60}$$

where ξ is a small parameter. For $\xi = 1$,(3.60) is identical to (3.1). In the case of a uniform path (i.e., b(z) = const),(3.60) is easily solved with respect to T(z) as is also the reference equation (3.1). The solution can be written as follows:

$$T_\xi(z) = \left(1 - 2\xi b^{-1} \int_{z_1}^{z} \bar{S}(z')dz'\right)^{-1/\xi} \quad . \tag{3.61}$$

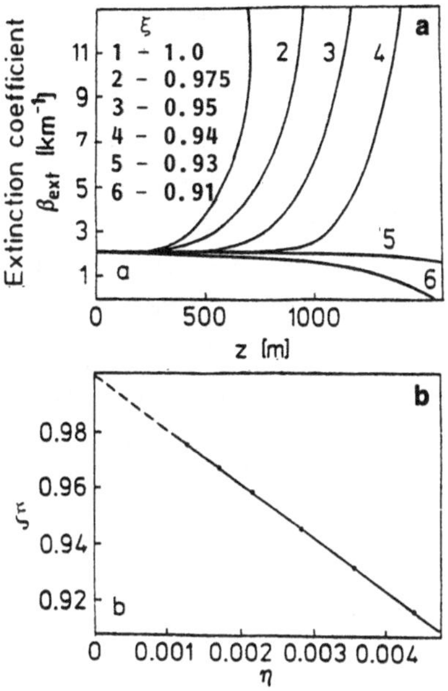

Fig.3.33. (a) An example of the numerical solution of the modified lidar equation (3.60) for clouds with different values (Curves 1-4) of the parameter ξ; (b) the dependence of the optimal ξ values on the coefficient η which is determined by the geometrical parameters of the lidar sensing scheme

For $\xi \to 1$, the function $T_\xi(z) \to T(z)$ and is determined by (3.13). Solutions of (3.60) are stable even at large values of τ, as can be illustrated by the convergence of the iteration scheme for finding the profile $\beta_{ext}(z)$. By analogy with (3.16) one can write

$$\beta_{ext,k}^{(m)} = \bar{S}_k(T_{k-1}b_k)^{-1} \exp\left(2\xi\tau_k^{(m-1)}\right) \quad . \tag{3.62}$$

The scheme (3.62) converges if $|\xi\tau| < 1$. Choosing an infinitesimal value of ξ, one can formally provide the convergence of (3.62 for any τ. It is obvious that in this case the solutions may be far from the actual profile $\beta_{ext}(z)$ and some additional studies are required for the proper choice of the parameter ξ.

Certain recommendations as to the choice of ξ, depending on the optical depth of the layer sensed, can be given based on numerical experiments. An example is given in Fig.3.33a, illustrating the determination of ξ inside an extended cloudy layer with the use of lidar sensing. The reference value of β_{ext} used was 2 km^{-1} and was constant inside the layer. The altitude z is counted, in Fig.3.33a, from the base of a cloud located at z_0 = 100 m.

The amplitudes of the lidar signals (at λ = 0.694μm and the angular aperture of the receiver φ_0 = 1°) were calculated using a Monte-Carlo method. The solutions of (3.60) for various z (also τ) give an idea as to the values of ξ required for the given numerical experiment. For $\xi = 1$ the solution of (3.1) diverges rapidly when z approaches z = 500 m (or τ approaches unity). As a consequence, it becomes impossible to interpret the signals received from the depth of the cloud layer. Successive decreases of ξ enable one to penetrate into the cloud in determining the extinction coefficient β_{ext}. In the above example, the layer has been assumed homogeneous, but in the general case it can be inhomogeneous. It has been stated, based on analogous calculations, that to a first approximation the choice of $\beta_{ext}\xi$ may be related to the value $\eta = z_0\bar{\beta}_{ext}$ tg φ_0 where $\bar{\beta}_{ext}$ is the mean value of $\beta_{ext}(z)$ averaged over the layer sensed. The dependence of ξ on η is close to linear (Fig.3.33b) when $z_0 \geq 200$ m. These are some of the main peculiarities of the interpretation of lidar sensing data in optically dense media, if one proceeds from the lidar equation written in the single scattering approximation. The development of a rigorous theory for inverting the lidar data, taking into account the contribution due to multiple scattering, is beyond the scope of the present book.

4. Methods for Inverting Polarization Data. The Theory of Bistatic Lidars

Optical investigations of aerosol systems utilize not only methods of optical sensing within spectral intervals but also methods based on measurements, with further inversion, of characteristics which describe the spatial distribution of the intensity of light scattered by a volume of the illuminated medium. In the most general form, the corresponding inverse problems of light scattering lead to the inversion of scattering phase matrices for polydispersed systems of particles. The determination of the matrix elements can be based on measurements with laser nephelometers and bistatic lidars. The polarization nephelometers allow one to make in situ measurements, while the use of bistatic lidars makes possible remote studies of the atmosphere.

The methods for solving the inverse problem of laser sensing of the atmosphere (for the case of the above optical schemes) are presented below assuming that the particles of aerosol polydispersed systems are spherical. Note that only in this case it is possible to develop a rigorous theory of optical sensing.

There are as yet no acceptable analytic models that would allow the determination of the scattering phase matrices for polydispersed systems of nonspherical particles. Naturally, such a situation makes the interpretation of the data impossible. As a result, only qualitative methods for the interpretation of data are available in the cases when the effect of nonsphericity of particles is clearly observed from the polarization characteristics of lidar returns. Possible ways to solve such problems are considered in this chapter.

The basic problem of the theory of bistatic lidars, as for any instruments determining remotely the aerosol microstructure, is the study of their information potentialities and an adequate mathematical formulation of the corresponding inverse problems. These problems are then realized in the form of regularizing algorithms and calculation schemes. Such investigations suggest the amalgamation of monostatic and bistatic lidars into a simple complex optical instrument allowing the determination of parameters of both aerosol and molecular components of the atmosphere. In the following we give an analysis of the information potentialities of such "monobistatic" lidar complexes and calculation schemes for the inversion of sensing data which are suggested by numerical simulations.

As previously, the material presented in this chapter concerns mainly the sensing of atmospheric hazes based on the single-scattering theory.

4.1 General Characterization of Integral Equations Related to the Inversion of Elements of Polydispersed Scattering Phase Matrices for the Case of Sperical Particles

The previous "ill-posed" inverse problems dealt with the spectral characteristics of light scattered by dispersed media. In addition, measurements of the angular characteristics of scattered radiation propagating through the media under investigation can serve as a basis for optically sensing the atmospheric aerosol even when laser systems are used. In such a formulation of the problem, one has to deal with inverse problems of scattering phase matrices for polydispersed ensembles of particles. Let us consider a general description of the scattering phase matrix for electromagnetic waves scattered by polydispersed systems of spherical particles. The description will be adequate for the analysis of the fundamental properties of the integral equations, as applied to the remote determination of the aerosol microstructure.

The processes related to light scattering are usually described in terms of the Stokes vector parameter and the corresponding scattering phase matrices. In this approach, the elementary process of wave scattering can be described by the matrix transformation $\mathbf{I}^{(s)} = k^{-2}S\,\mathbf{I}^{(0)}$, where the components of vectors $\mathbf{I}^{(s)}$ and $\mathbf{I}^{(0)}$ are the Stokes parameters {I, Q, U, V} for the scattered and incident fluxes, respectively, and S is the scattering phase matrix. A description of the properties of these matrix elements depending on the parameters of the scattering medium can be found in, e.g., [4.1-4]. If the particles of a scattering ensemble are randomly oriented in space and are not spherical, the corresponding scattering phase matrix for a single particle is

$$S = \{S_{ij}\} = \begin{vmatrix} S_{11} & S_{12} & 0 & 0 \\ S_{12} & S_{22} & 0 & 0 \\ 0 & 0 & S_{33} & S_{34} \\ 0 & 0 & -S_{34} & S_{44} \end{vmatrix} \quad . \tag{4.1}$$

This matrix corresponds to isotropic light scattering and has six independent elements. For spherical particles, S has four independent elements. In particular, the following relations are valid

$$\begin{aligned} S_{11}(x,\theta) &= S_{22}(x,\theta) = [i_2(x,\theta) + i_1(x,\theta)]/2 \\ S_{12}(x,\theta) &= S_{21}(x,\theta) = [i_2(x,\theta) - i_1(x,\theta)]/2 \\ S_{33}(x,\theta) &= S_{33}(x,\theta) = i_3(x,\theta) \\ S_{43}(x,\theta) &= S_{34}(x,\theta) = i_4(x,\theta) \quad . \end{aligned} \tag{4.2}$$

The functions $i_j(x,\theta)$ in (4.2) are defined in the Mie theory [4.5], which enables one to calculate the functional relations between the optical parameters of the scattering medium and the parameters of their microstructures. When the particles are not spherical, such a calculation becomes too complicated. For an example, the reader is referred to [4.3] which is devoted to the study of the scattering properties of spheroidal particles.

Let the shape of the particles in the media investigated be spherical. When formulating the inverse problem for light scattering, the corresponding integral equations have to be constructed.

In the case of light scattering by a system of independent spherical particles,the dimensionless intensities i_j ($j = 1,2,3,4$) can be assumed to be additive [4.5]. If the polydispersed ensemble of particles is characterized by the size-distribution function $n(r)$ defined on the size interval $R = [R_1, R_2]$, then the intensity parameter of the whole ensemble can be written in the form of the polydispersed integral $\int_R i_j(r,\lambda,\theta)n(r)dr$. It turns out to be convenient to use the matrix $k^{-2}S$ instead of S, denoting it as D and relating it to the polydispersed ensemble of particles.

Thus, taking into account the additive properties of $i_j(x,\theta)$, one can write the matrix element D_{11} as

$$D_{11}(\theta) = k^{-2} \int_R [i_1(\theta,r) + i_2(\theta,r)]n(r)dr/2 \quad . \tag{4.3}$$

The other elements of the matrix D can be written in the same way. It is more convenient to use the distribution $s(r) = \pi r^2 n(r)$, instead of the number density. In this case (4.3) will take the following form with $x = 2\pi r/\lambda$ being the Mie parameter

$$D_{11}(\theta) = \int_R \{[i_1(x,\theta) + i_2(x,\theta)]s(r)(2\pi x^2)^{-1}\}dr \qquad (4.4)$$

Below, (4.4) will be written in general form as $D_{11}(\theta) = \int_R Q_{11}(x,\theta) s(r)dr = (Q_{11}s)(\theta)$, where $Q_{11}(x,\theta) = Q_{11}(r,\lambda,\theta)$. It is defined on $[R\times\Theta]$ while λ can be interpreted as some parameter of this kernel. The possibility to vary λ over the interval Λ can be of certain practical interest, and will be frequently used in our further discussion.

In this connection, it is necessary to construct some new integral relations to utilize this possibility. In some problems of light scattering, it is useful to work with a system of functions $P_j(x,\theta)$ $(j = 1,2,3,4)$, such that the functions P_1 and P_2 obey the normalizing relation $\int[(P_1+P_2)/8\pi]d\omega=1$ for all values of x in addition to the intensity parameters i_j $(j = 1,2,3,4)$. The functions $P_j(x,\theta)$ are introduced according to the relations given in [4.1]

$$k^{-2}i_j(x,\theta) = \beta_{sc}(x)P_j(x,\theta)/4\pi \quad , \qquad j = 1,2,3,4 \quad . \qquad (4.5)$$

For ensembles of independent scatterers, the single particle functions $i_j(x,\theta)$ and $\beta_{sc}(x)$ are used additively. Hence, denoting

$$\beta_{sc}(\lambda) = \int_R \beta_{sc}(r,\lambda)n(r)dr = \int_R K_{sc}(r,\lambda)s(r)dr$$

one finds that

$$D_{11}(\lambda,\theta) = \int_R K_{sc}(r,\lambda)[P_1(x,\theta) + P_2(x,\theta)]s(r)dr/8\pi \quad . \qquad (4.6)$$

Assuming that $D_{11}^{(\lambda)}(\theta)$ for fixed λ is measured on the interval $[\theta_1, \theta_2]$, one obtains the integral equation $D_{11}^{(\lambda)} = Q_{11}^{(\lambda)}s$ with respect to s(r). In the same way, one can construct the integral equation $Q_{11}^{(\theta)}s = D_{11}^{(\theta)}$ if the characteristic $D_{11}^{(\theta)}(\lambda)$ is measured within the spectral interval Λ for fixed scattering angle Θ.

Since $\beta_{sc}(\lambda) = \int D_{11}(\lambda,\theta)d\omega$, the characteristic $D_{11}(\Theta)$ is the scattering coefficient along the direction θ, or in other words the "directional coefficient" of light scattering. In addition to D_{11}, it is possible, according to (4.2), to write down three other elements, denoting them as D_{ij}. In turn, corresponding integral equations can be written for the D_{ij} as $Q_{ij}s = D_{ij}$. If the entire scattering phase matrix is measured, in the course of an optical experiment, then the inverse problem is formulated for all the elements simultaneously.

On the one hand, this enables one to choose any of the techniques applied to the inversion of given data, but on the other hand, several calculation schemes and algorithms must be developed. The description of methods for the solution of integral equations like $Q_{ij}s = D_{ij}$ begins in its simplest version with the inversion of the polydispersed scattering phase function. Optical experiments to measure such scattering phase functions occur frequently in atmospheric research.

4.2 Determination of the Aerosol Microstructures by Inverting the Polydispersed Scattering Phase Functions

4.2.1 Numerical Inversion of Aerosol Scattering Phase Functions

The polydispersed scattering phase function $\mu(\theta)$ is related to the scattering coefficient along the Θ direction and D_{11} is given by

$$D_{11}(\theta) = \beta_{sc}\mu(\theta)/4\pi \quad . \tag{4.7}$$

It is not difficult to write an integral equation of the aerosol size-distribution function s(r), taking into account (4.4,7)

$$\mu(\theta) = (4\pi/\beta_{sc}) \int_R \{[i_1(x,\theta) + i_2(x,\theta)]/2\pi x^2\}s(r)dr \quad . \tag{4.8}$$

Equation (4.8) does not really differ from (4.4). Therefore, the inversion of $\mu(\theta)$ is identical to the inversion of the first element of the polydispersed scattering phase matrix D.

The regularizing algorithm for the numerical solution of (4.8) is based on the use of (2.42). In this connection,consider briefly the application of this scheme to invert the elements of a polydispersed scattering phase matrix.

It should be noted, first of all, that besides being a rapidly oscillating function in the region $[R \times \theta]$, the kernel $Q(r,\theta)$ of (4.8) has a much wider range than that of the efficiency factors $K_{sc}(r,\lambda)$ and $K_{\pi}(r,\lambda)$. For example, the function $i_1(r,\theta)$ for $r \in [0.1\mu m, 1\mu m]$ $0 \leq \theta \leq 180^{\circ}$ and $\bar{m} = 1.56 - 0i$ varies over about three orders of magnitude at $\lambda = 0.55\mu m$. As a consequence, some of the optical measurements data $\{\mu(\theta_i)\}$ $(i = 1,2,\ldots,n)$ are considerably larger in magnitude than the others. Therefore, linear systems for the algebraization of corresponding integral equations should be constructed carefully.

In this connection, the discrepancy of the smoothing functional constructed is written as $\rho^2(Qs,\mu) = \sum_{i=1}^{n} p_i[(Qs)(\theta_i) - \mu(\theta_i)]^2$, where the p_i are weighting factors. The use of such factors allows the lines of linear systems to be "planned". As a result, the distorting influence of measurement errors on the accuracy of the inverse solution is reduced [4.6].

In addition, the quadrature formulas from [4.7,8] developed for polydispersed integrals with kernels from the Mie theory are used when constructing matrix analogs Q_m of the initial integral operators Q. These formulas are most effective when applied to oscillating kernels. In the case of equations of type (4.8) the above formulas provide an admissible dimensionality of the solution **s**. As numerical analysis shows, for an index of refraction $\bar{m}\leq 1.5$ five points from the interval $[R_1, R_2]$ ($R_2 \leq 1\mu m$) are sufficient for the quadrature error to be not greater than 5%. This fact is very important because, as will be shown below, the inverse problem for (4.8) is poorly conditioned.

This quadrature processes for polydispersed integrals also enable one to formulate boundary conditions at the points R_1 and R_2 for the functions s(r) being sought in the course of the solution of an inverse problem. The kernels $Q(r,\theta)$ are characterized by the relation $Q(R_1\theta) \ll Q(R_2,\theta)$ for $R_1 \ll R_2$ for almost all θ values (excluding perhaps angles from 90° to 120°) and this fact is of great importance for improving the inversion of $\mu(\theta)$. This is seen clearly in the case of small particles in the vicinity of the left boundary R_1. Introducing the condition $s(R_1) = 0$ into the inverse problem,one can obtain a solution whose behavior is sufficiently regular in the vicinity of the point R_1. Without this condition the regularized inverse solution s_α is quite irregular for small particle sizes because the inverse problem (4.8) is practically noninformative with respect to such particles. Any attempts to smooth the solution s_α in this region by increasing the regularizing parameter α lead immediately to an undesirable distortion of the solution s_α in the other regions of the size spectrum and is thus not justified. As a result of the numerical experiments, one may conclude that the use of special quadrature formulas and the normalization of matrix lines allows one to construct quite efficient algorithms for the inversion of polydispersed scattering phase functions. It was shown in [4.9] that the same conclusion can be drawn for any element of the polydispersed scattering phase matrix. In the following, we give a short summary of the results of numerical investigations on the inversion of $\mu(\theta)$, based on the Mie theory.

4.2.2 Examples of the Numerical Inversion of Scattering Phase Functions

Figure 4.1 gives an example of an inversion of $\mu(\theta)$ corresponding to an initial aerosol size distribution similar to (2.50) for $\bar{m}_0 = 1.56 - 0i$, $\lambda = 0.55\mu m$. The solution was sought in the interval [0.06 - 0.5μm] and the regularization parameter α was assumed to be equal to α_q according to [4.10]. The "measured" vector $\boldsymbol{\mu}$ was distorted by oscillating deviations in each component which were of the order of 5% of that component. Curve 1 represents the exact size-distribution $s_0(r)$ for the given size interval, while Curve 2 is the corresponding inverse solution $s_\alpha(r)$. Curve 3 is the regularized inverse solution of the type (4.8), obtained for the same initial data as Curve 2, but assuming $\bar{m} = 1.50-0i$. The last example illustrates the effect of an error in the index of refraction on the accuracy of the inverse solution. On the whole, the numerical investigations showed that for $0.05 \leq \varepsilon_\mu \leq 0.1$ the initial error in the real part of the index of refraction should not be larger than $\pm 0.02\ m_0$. Here ε_μ is the error in $\mu(\theta)$. This conclusion (as well as the numerical results presented) is valid for scattering angles in the backward hemisphere, i.e., for angles from 90° to 180°. For scattering angles in the forward direction, the errors can be a little larger.

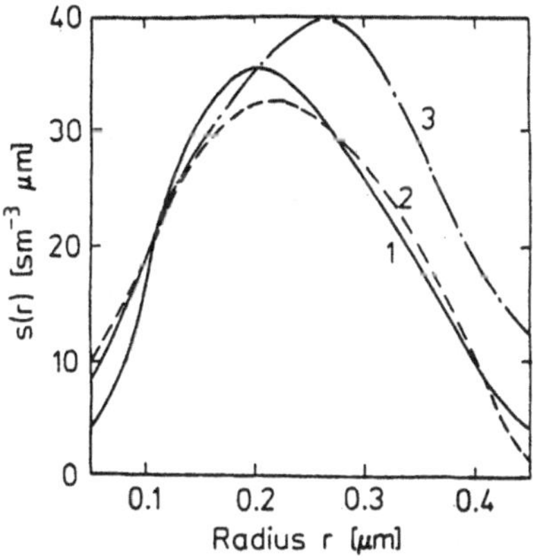

Fig.4.1. Size-distribution functions $s_\alpha(r)$, obtained by inverting the polydispersed scattering phase function $\mu(\theta)$; 1: exact size-distribution function $s_0(r)$, 2: $s_\alpha(r)$ when $\varepsilon_\mu = 0.05$ and $\bar{m} = \bar{m}_0$, 3: size distribution function $s_\alpha(r)$ for $\bar{m} = \bar{m}_0 - 0.06$

We are now able to reach some conclusions and recommendations for the use of the inversion of scattering phase functions $\mu(\theta)$ measured with the use of a nephelometer. This instrument is known to give good angular resolution over the entire region of scattering angles from 0° to 180°, thus providing a large number of points at which the function $\mu(\theta)$ is measured. Even for a 5° interval between measurements, one has more than thirty values of $\mu(\theta)$ measured in this angular region. Thus, large dimensionality of the measured

Table 4.1. Ratios of maximum to minimum eigenvalues of the matrices Q^*Q which occur in the inversion of polydispersed scattering phase functions of atmospheric haze for scattering angles in the forward and backward directions

R_2	n	m	μ_{max}/μ_{min} $5^\circ \leq \theta \leq 90^\circ$	$90^\circ \leq \theta \leq 175^\circ$
0.6	18	5	$1.9\ 10^2$	$1\ \ 10^2$
0.6	18	10	10^7	$1\ \ 10^6$
3	18	5	$6.3\ 10^2$	$1.3\ 10^2$

vector $\boldsymbol{\mu}$ is a characteristic feature of nephelometric measurements, in contrast, for example, to spectral optical measurements in the atmosphere. Since not all of these measurements can be considered independent, it appears necessary to choose the dimensionality of the solution vector $\mathbf{s}$ optimally. It is also evident that for $n \geq 20$, the choice of the dimension of $\mathbf{s}$ according to the requirement $m = n$ is hardly justified, especially when $\varepsilon_\nu \geq 0.1$. Taking it into account, the necessary condition for this class of inverse problems (i.e., for equations like (4.8) with kernels defined by the Mie theory) should be $m < n$. In order to estimate the dimensionality m, one should estimate the degree to which the inverse problem (4.8) is conditioned in terms of eigenvalues of the operator Q^*Q (its matrix analog is meant here). According to the technique given in Sect.2.5, several values of the ratio (μ_{max}/μ_{min}) are presented in Table 4.1.

Since values of (μ_{max}/μ_{min}) greater than 100 are not very useful in practice, Table 4.1 indicates that m should be not greater than 5 or 7 in inverse problems like (4.8). This is valid for scattering angles in both the forward and backward hemispheres. The unnecessary increase of the dimension m results in worse conditioning of the matrix inverted according to (2.62, 65) and the following equations.

Figure 4.2 shows the dependence of the error ε_s of the equation under study on the errors ε_μ of the initial vector μ obtained for various n and m. The curves presented in Fig.4.2 are typical for the solution of (4.8) and are practically independent of the numerical method used in the class of single and bimode size distributions when $R_2 \leq 1$ m and the real part of the refractive index varies between 1.5 and 1.6 ($\varkappa \leq 0.005$). It should be noted that ε_s characterizes the rms error over the whole size interval. At the same time, numerical analysis shows that the largest deviations of $s_\alpha(r)$ from $s_0(r)$ occur in the region of small particles, primarily near the lower boundary R_1 of the distribution. In the particular case of the above initial

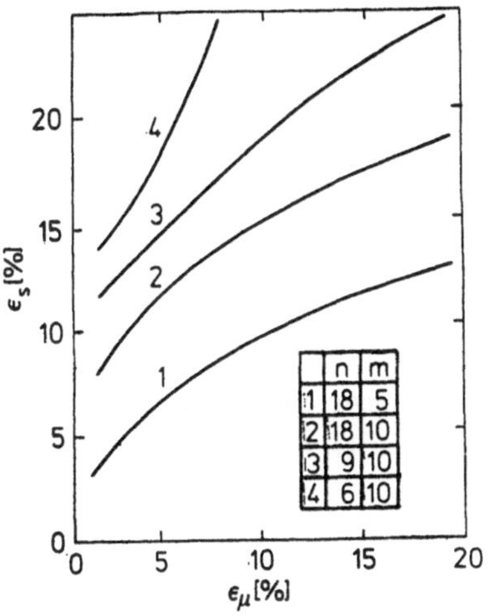

Fig.4.2. Typical dependences of errors in the inversion of polydispersed scattering phase functions $\mu(\theta)$ on the measurement errors ε_μ for different dimensions n and m

data and for $\lambda \leq 0.7\mu m$, this interval occupies the region to the left of the point $0.2\mu m$. Since this size interval contributes weakly to the angular measurements of a polydispersed scattering phase function, a decrease in the ε_μ over the whole interval of scattering angles $(0, 180^\circ)$ results in little improvement in the values of ε_s. This can readily be seen from the curve ε_s in Fig.4.2 for $\varepsilon_\mu \leq 0.05$. This fact explicitly reveals the effect of optical "screening" of small particles by larger particles in the polydispersed ensemble. The nature of this fact, and the methods for its qualitative evaluation, were discussed in [4.11] (see also Sect.3.6). A significant decrease of ε_s with a decrease of ε_μ can be achieved only if the "screening" effect is weakened, e.g., by the use of shorter wavelengths. The calculations presented here were made for $\lambda - 0.55\mu m$. Calculations have shown that the use of radiation of wavelength $\lambda = 0.4\mu m$ can provide a 1.5 times decrease in ε_s. Therefore, the choice of the radiation wavelength for nephelometer operation plays an important role for formulating the inverse problems.

We should also make a few remarks concerning the choice of the regularizing parameter α for the inversion of $\mu(\theta)$. The selection of α according to the equation $\alpha = \alpha_q$ is not as simple as it seems at first sight following the recommendations given in [4.10]. In particular, the function $\|d\, s_\alpha(r)/d \ln \alpha\|$ has, as a rule, several extrema to the left of the point α_0 which is determined by the condition $\rho(Q\, s_\alpha, \mu) \leq \delta(\sigma)$. Therefore the minimum point of the above norm "closest" to α_0 from the left can be chosen as a suitable value of α. In general, such a method is reasonable if $0.05 \leq \varepsilon_\mu \leq 0.15$, while more efficient and sophisticated methods for selecting the proper value of α

are desirable if the initial data are more precise (Chap.2). Lidar and nephelometric measurements made simultaneously enable one to increase the reliability of the data on optical and microphysical parameters of the aerosols under investigation.

4.2.3 Determination of the Lidar Ratio from Nephelometric Measurements Using the Inverse Operator Method

In connection with the development of laser radar methods for an operative monitoring of atmospheric optical properties,it is important to be able to estimate the values of the "lidar ratio" $b = \beta_{\pi}/\beta_{ext}$. If lidar measurements are made in parallel with the nephelometric measurements, the value of b can be easily evaluated by using the above methods for the inversion of $\mu(\theta)$. For example, when sensing the lower troposphere, the concomitant measurements can be performed with airborne nephelometers [4.12]. For this case,one can obtain data on the "lidar ratio" profile by using the inversion method developed above. In another measurement procedure,ground-based nephelometric measurements can be made and a value $b(z_0)$ estimated. The altitude dependence of b can be estimated based on the value of $b(z_0)$ in accordance with the relationship between the lidar ratio and humidity profiles (Sect.5.3). In order to assess the efficiency of the method suggested here, we have processed the scattering phase functions measured in Gelendgic (August 1974) [4.13]. In addition to scattering phase functions, values of b were measured in these experiments with the use of a special photometer designed for measurements at large scattering angles.

Thus, it was possible in this case to compare the value b obtained with the use of the inversion technique for $\mu(\theta)$ with those measured experimentally. The results of this comparison are given in Table 4.2. It is seen that b^* has been estimated accurately enough. The values of the error ε_b presented in Table 4.2 characterize the efficiency of the method for the estimation of b which was numerically checked by using the vast amount of experimental material. The selection of the value of the refraction index was made taking into account the measured values of humidity following the recommendations given in [4.14,15]. Some decrease of ε_b for shorter wavelengths can also be seen from the table. This behavior of ε_b is explained by the fact that the inverse solution of the corresponding inverse problem is more efficient for smaller λ, particularly in the region of small particles, which , in the given experiment, played a significant role in light scattering by the atmosphere. It is interesting to note that the experimental values ε_{μ} were too

Table 4.2. Lidar ratio values obtained by inverting the polydispersed scattering phase functions

λ [μm]	b^*	$b_{measured}$	ε_b [%]
0.7	0.41	0.35	16
0.552	0.36	0.41	12
0.446	0.32	0.36	11

large, being not less than 20%, and also systematic. As a consequence, the expected error in s_α obtained by the inversion of $\mu(\theta)$ could be of the order of 30%-40%. Nevertheless, this fact does not prevent one form obtaining quite acceptable estimates of b^*. This is explained simply by the fact that the value of b is determined by the ratio of two integrals and, hence, the variations of the function $s_\alpha(r)$ (especially variations having the same sign) would affect it weakly (Sect.2.6). There is another peculiarity of this method for estimating b, which can be of practical use. The values of the "lidar ratio" presented in Table 4.2 for fixed wavelengths were determined using the values of $\mu(\theta)$ measured at the same wavelengths. It is evident that if the known function $s_\alpha(r)$ is an acceptable approximation to the function $s_0(r)$, then, using only numerical calculations, one can find an approximate estimate of b for any wavelength including λ_0, at which the measurements of the scattering phase function used for the inversion were made. The corresponding value of the error ε_b will evidently be dependent not only on ε_μ but also on λ and λ_0. For $\lambda_0 \leq \lambda$ the results obtained are better, which is quite evident by taking into account the previous discussion.

The above material concerning the estimation of b from nephelometric measurements is interesting not only by itself, but also as an example illustrating the potentialities of the inverse operator method in its application to the solution of purely optical problems based on the theory of light scattering by polydispersed ensembles of particles.

4.2.4 Examples of the Inversion of Experimentally Measured Scattering Phase Functions

In conclusion,we shall give some examples of the inversion of scattering phase functions $\mu(\Theta)$ made for determining the real aerosol microstructure from nephelometric measurements. The results of such an inversion made for two experimentally measured scattering phase functions are presented in Fig.4.3 and compared with results obtained from direct sampling of the aerosol, using filters. The scattering phase functions $\mu(\Theta)$ inverted were

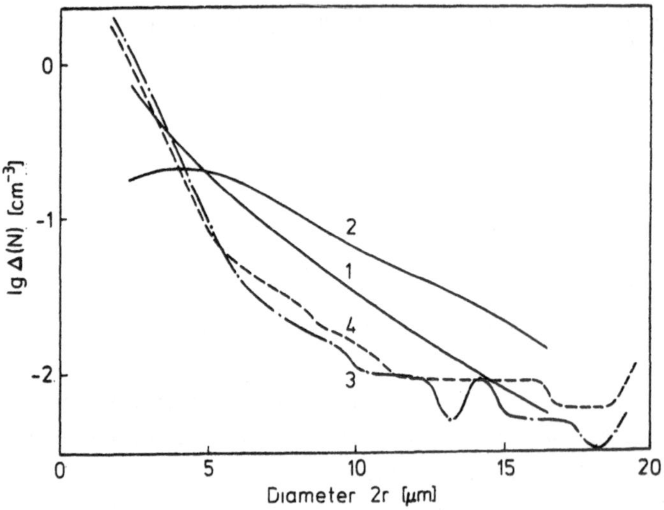

Fig.4.3. Results of the inversion of scattering phase functions (Curves 1, 2) compared with the data obtained by direct sampling of aerosol particles (Curves 3, 4)

measured during the night of August 16-17, 1973 in Gelendgic [4.13] at 21^h25^m and 00^h16^m local time, respectively. The corresponding inverse results are given by Curves 1 and 2 in Fig.4.3. Aerosol sampling was made simultaneously with the optical measurements. Equation (4.8) was solved for the functions s(r) which were used for further estimation of the number density of the ith fraction according to the expression $N_i = s(\xi_i)(r_{i+1} - r_i)/\pi\xi_i^2$, where $r_i < \xi_i < r_{i+1}$. The points $\{r_i\}$ $(i = 1,2,\dots)$ were selected according to the fractions available from the direct measurements of the particle number density using filters. It is well known that direct measurements of the particle number density give information on the so-called dry fraction of aerosols, while the optical methods enable one to obtain information about the aerosol in its natural ("unperturbed") state. It is for these reasons that the absolute coincidence of the inversion results with those of the direct microstructure analysis can hardly be expected. It should be noted that the values of the number density obtained by inversion methods exceed slightly those from the direct sampling. This can be explained, to a certain degree, by the insufficient effectiveness of the impactors used.

Characterizing, on the whole, the inversion of scattering phase functions $\mu(\Theta)$ as a method for estimating aerosol microstructure, it is necessary to keep in mind, first of all, the fact that corresponding inverse problems are conditioned very poorly.

As a consequence, the regularized solutions of (4.8) are, as a rule, overly smooth functions which can be regarded only as a first approximation to the actual size spectrum of the aerosol under study. This remark is valid particularly in the case of broad size spectra, when $R_1 \ll R_2$. Examples of such spectra are shown in Fig.4.3, where Curves 1 and 2 are very smooth due to the above reasons. In the case of narrow size distributions, when the interval R is so narrow that the averaging in the polydispersed integral (4.8) is negligible in its effect on the conditionality of the corresponding inverse problem, the reconstructed size spectra are quite acceptable (Fig.4.1). It should be emphasized that in any case the error is largest for the distribution of small particles. From this point of view, optical measurements in different spectral regions are more efficient for the investigation of aerosol microstructure, especially in the region of small particles.

4.3 The Use of Polarization Measurements for the Determination of Aerosol Microstructure and Index of Refraction

4.3.1 Methods for Determining the Optical Constants of the Aerosol Substance Based on the Use of Light Scattering Operators for Polydispersed Systems

Let us assume that an optical experiment gives information on several elements of the polydispersed scattering matrix. For spherical particles, the matrix D has four independent elements and, therefore, according to the Mie theory, four independent integral equations are available, in principle, for the interpretation of optical data. For convenience, we shall denote these integral equations as follows

$$\int_R Q_i(\theta,r)s(r)dr = \mu_i(\theta), \qquad i = 1,2,3,4 \quad .$$

In order to write down the function s(r) as a solution of the equation $Q_i s = \mu_i$, it is necessary to know the index of refraction of the aerosol substance even if the number of data points for any characteristics $\mu_i(\theta)$ in the interval $[\theta_1, \theta_2]$ is sufficient for determining s(r) within the measurement errors. In this connection, consider the inverse problem for the determination of optical constants of aerosol substance in more detail than previously in Sect.2.6.

The determination of the refractive index of particulate matter in polydispersed ensemble can be considered, to a certain extent, as a separate mathematical problem. If, for example, the aerosol size distribution s(r) is the solution of equation $Q_1 s = \mu_1$, then formally we have $s = Q_{1\alpha}^{-1}\mu_1$ where

$Q_{1\alpha}^{-1} = (Q_1^*Q_1 + \alpha H)^{-1}Q_1^*$. All the operators related to $Q_i(r,\theta)$ are functions of $\bar{m}$. This, in turn, implies that any of the operators $W_{i1} = Q_iQ_{1\alpha}^{-1}$ is also a function of the parameter $\bar{m}$. The operators W_{i1} are analogous to the corresponding operators introduced earlier into the theory of multi-frequency lidars [4.9,16] (see also Sect.2.5).

In the inverse problem considered here, these operators define the transformation $W_{i1}: M_1 \to M_i$ where M_i is the array of measurement data for the optical characteristic μ_i. For any two cases μ_1 and μ_2 one has the relationship $\mu_i = W_{i1}(\bar{m})\mu_1$, which can be regarded as an equation for the unknown parameter $\bar{m}$. Now, we have only to show how to calculate the unknown value of $\bar{m}$. The simplest method to do this can be related to the minimization of the function

$$F_{i1}(\bar{m}) = \|W_{i1}(\bar{m})\mu_1 - \mu_i\|_{L_2} \tag{4.9}$$

with respect to the parameter $\bar{m}$ in the domain of its possible values Ω. If this method is to be of practical use,the value $\bar{m}^*$ giving the minima of $F_{i1}(\bar{m})$ on Ω must satisfy the condition $|\bar{m}^* - \bar{m}_0| \leq \varepsilon$ for $F_{i1}(\bar{m}^*) \leq \delta(\sigma)$, ε and δ being selected in accordance with the value of the error σ in the optical measurements. Since the formalism used for the construction of the operators W_{i1} (also Q_i) is, as before, the Mie theory, the proofs of the above statements cannot be given explicitly due to analytical difficulties. In this connection,we shall give below only some general comments and consider the numerical analysis of the functions (4.9) in their application to the problem of optical sensing of atmospheric hazes.

Since the factors $Q_i(r,\theta)$ $(i = 1,2,3,4)$ in the Mie theory are continuous functions of the complex parameter $\bar{m}$, the operators Q_i (as well as $Q_{1\alpha}^{-1}$) and W_{i1} are also continuous functions of $\bar{m}$. As a result,the functions $F_{i1}(\bar{m})$ are continuous on Ω and hence can, in principle, take on their absolute minimum values. Of course, this does not yet prove that the condition $F_{i1}(\bar{m}^*) \leq \delta(\sigma)$ is fulfilled at the absolute minimum $\bar{m}^*$. It is therefore necessary that the assumption that the scattering particles are spherical be more or less correct and that the real part of the refractive index $\bar{m}_0$ be selected from the interval Ω. The minimization of $F_{i1}(\bar{m})$ as a method for the evaluation of $\bar{m}$ can be considered justified provided all these requirements are satisfied. Now the question of the uniqueness of such an inverse solution arises. When the function $F_{i1}(\bar{m})$ is convex in Ω, the value $\bar{m}^*$ is obviously unique. However, it can be shown that the convexity of the function $F_{i1}(\bar{m})$ requires the norms of operators $W_{i1}(\bar{m})$ to be monotonic for $\bar{m}$ varying within Ω. However, calculations have shown that such behavior can be expected only for narrow intervals Ω. The specific feature of the above method for determining the optical constants of the aerosol substance is that the estimates $\bar{m}^*$ are independent of the aerosol microstructure, since formally none of the operators W_{i1} depends upon the size-distribution. For making numerical calculations when constructing the matrix analogs of W_{i1} (also for Q_i and $Q_{1\alpha}^{-1}$), it is necessary to know only the boundaries R_1 and R_2 of the real size spectrum.

This peculiarity is very important when the amount of experimental data is quite insufficient to make a consistent evaluation of the size spectrum, but are sufficient for the determination of the refractive index of the aerosol substance. Thus, this method of operators allows one to find the optical constants of the aerosol substance by solving the independent inverse problem of light scattering by aerosols.

It was assumed above that the measurements of μ_i and μ_1 were made at a particular wavelength λ. To make a study of $\bar{m}(\lambda)$ as a function of λ, one should make several such measurements within the spectral interval Λ.

4.3.2 Numerical Examples for the Determination of Optical Constants from the Polydispersed Scattering Phase Matrix

Now consider briefly some results of numerical investigations of the analytical behavior of $F_{i1}(\bar{m})$ in the region Ω of parameters whose points are denoted by $\{\bar{m}, \varkappa, \alpha\}$. The initial data for the calculations were taken to be useful in optical sensing of atmospheric hazes. Thus, in particular, the values $\bar{m}_0 = 1.56$, $\varkappa = 0$ were chosen but allowed to vary within the intervals [1.4-1.7] and [0-0.08], respectively. The characteristics μ_i in (4.9) correspond to an aerosol size-distribution similar to (2.50). Note that the choice of the model size-distribution $s_M(r)$ is of no great significance here. Figure 4.4 shows the behavior of the functions $\tilde{F}_{i1} = F_{i1}^{1/2}\|\mu_i\|^{-1}$ in the vicinity of $\bar{m}_0$, with $\varkappa = \varkappa_0$ and $\alpha \cong \alpha_q$. Perturbations of the initial data μ_i were taken to be oscillatory and did not exceed 10%. As is seen from the figure, the minimization of $F_{41}(\bar{m})$ is most effective for the estimation of $\bar{m}$ because the minimum of $F_{41}(\bar{m})$ in the vicinity of $\bar{m}_0$ is the sharpest. Keeping this in mind, one can state that the pair (μ_1, μ_4) or, equivalently, the operator W_{41} is most informative with respect to the real part of the complex index of refraction. Of course, these conclusions are not absolutely valid but are relevant only for the given numerical experiment. The second derivative of a function can be regarded as a measure of the sharpness of the extremum (or measure of localization) of this function at the corresponding

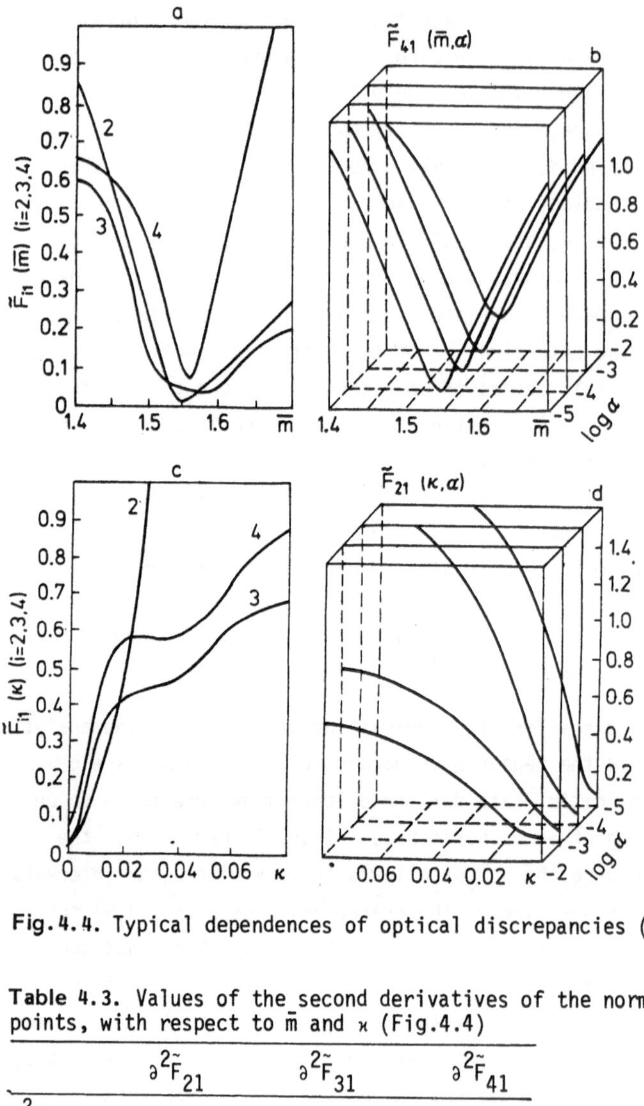

Fig.4.4. Typical dependences of optical discrepancies (4.3) on $\bar{m}$, $\varkappa$ and α

Table 4.3. Values of the second derivatives of the norms F_{i1} at the minimum points, with respect to $\bar{m}$ and $\varkappa$ (Fig.4.4)

	$\partial^2\tilde{F}_{21}$	$\partial^2\tilde{F}_{31}$	$\partial^2\tilde{F}_{41}$
$\partial^2\bar{m}$	91	32	272
$\partial^2\varkappa$	918	2487	2512

point. Several values of the second derivatives of the functions F_{i1} (i = 1,2,3,4) at the point ($\bar{m}_0$, $\varkappa_0$) are given in Table 4.3.

Since the functions F_{i1} (i = 1,2,3,4) depend not only on $\bar{m}$ and $\varkappa$ but also on α, the question of the influence of the value of α selected on the accuracy of the estimations of $\bar{m}$ and $\varkappa$ arises. Figure 4.4b shows that in practice variations of α from 10^{-5} to 10^{-2} do not modify the behavior of the functions F_{i1} near the extremum. As a matter of fact, a weak dependence of $\bar{m}^*$ on α is a result of the structure of the operators W_{i1}. This explains also the weak dependence of $\bar{m}^*$ (or $\varkappa^*$) on the errors in the measurements of the vector μ_1. This follows, in particular, from the fact that the operator W_{i1} is the compressing operator. The efficiency of the estimation of $\bar{m}^*$ depends much more strongly on the measurements errors for the second vector μ_i, which enters into (4.9). The last fact can be used in optical experiments. Thus, for example, $\bar{m}^*$ can be estimated efficiently provided that several accurate measurements of μ_i are made, although high precision is not necessary for measurements of μ_1, for which the number of measured components is important. All the above regarding the determination of $\bar{m}^*$ is also true for $\varkappa^*$, as is illustrated by Fig.4.4c,d. Some brief methodological remarks concerning the use of the scheme (4.9) in practice should now be made. In numerical experiments on the estimation of $\bar{m}$ and $\varkappa$ the minima of the functions F_{i1} ($i = 2,3,4$) in the vicinity of $\bar{m}_0$ and $\varkappa_0$ are very sharp, for operators W_{i1} ($i = 2,3,4$) and characteristics μ_i ($i = 1,2,3,4$) defined numerically according to the Mie theory. This ensures good accuracy of the estimates $\bar{m}_0$ and $\varkappa_0$. Nevertheless, the use of this method to interpret the experimentally measured optical characteristics does not always give good results. This is caused by the fact that the region Ω of $\bar{m}$ values, satisfying the condition $F_{i1}(\bar{m}) \le \delta(\sigma)$, is too broad even for sufficiently small values of δ. This means that the minimum of the norm $\|W_{i1}\tilde{\mu}_1 - \mu_i\|_{L_2}$ is not sharp. Here $\tilde{\mu}_i$ are the experimentally measured characteristics. Apparently the substitution of the vectors $\tilde{\mu}_i$ ($i = 1,2,3,4$) instead of the calculated μ_i ($i = 1,2,3,4$) into (4.9) shifts the above norms into the region in which their dependence on the parameters $\bar{m}$ and $\varkappa$ is weak. Mathematically, such a "bias" can be explained by the discrepancy between the operators W_{i1} and the array of measured characteristics $\tilde{M}_i$. Physically it can be observed when nonspherical particles appear in the ensemble of scattering particles. The last explanation need not be unique. When, for example, the structure of spherical particles has a radial inhomogeneity, then the corresponding operators Q_i should be rewritten. In the above calculations only the functions F_{i1} ($i = 1,2,3,4$) were considered. However, it is obvious that the number of such functions can be greater and, accordingly, the capabilities of the use of the Mie theory to interpret the optical parameters are wider. In the general case, one has a system of functions F_{ij} ($i,j = 1,2,3,4,\ i \neq j$). One can choose the best pair of functions like (4.9) depending on a given experiment. Finally, if the series of measurements $\{\mu_i^{(q)}\}$ ($q = 1,2,\ldots$) has been made

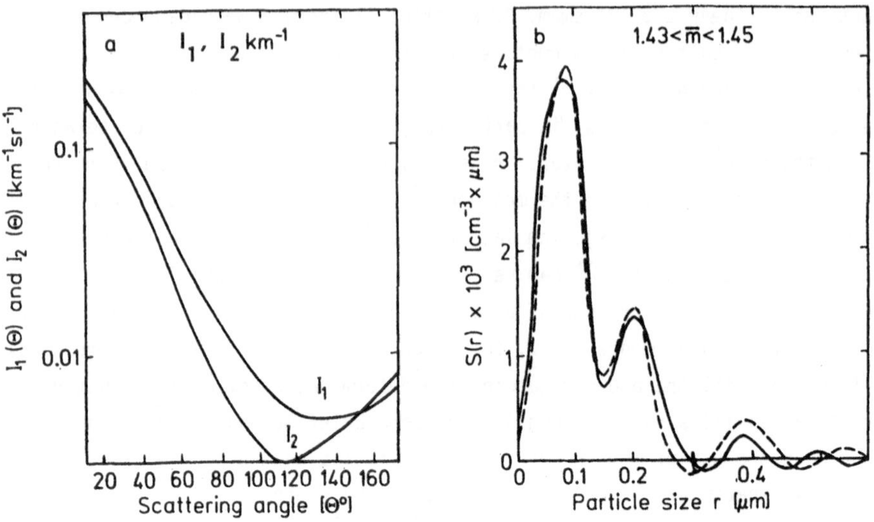

Fig. 4.5. (a) Polarization scattering phase functions $I_1(\Theta)$ and $I_2(\Theta)$. (b) Results of their simultaneous inversion

(as usually occurs in an experimental study) for a dispersed medium, the determination of $\bar{m}$ and $\varkappa$ can be performed by minimizing the averaged discrepancies, i.e., $\langle F_{ij}\rangle = \langle \| W_{ij}\mu_j^{(q)} - \mu_i^{(q)} \| \rangle$. Numerical results of the inversion of polarization scattering phase functions illustrate the applications of the above theory. This inverse problem is of great practical importance because polarization nephelometers are widely used in atmospheric-optics research. In the case of polarization scattering phase functions, which we shall designate as $I_1(\Theta)$ and $I_2(\Theta)$, the kernels of the corresponding integral equations are $x_1^{-2} i_1(x,\theta)$ and $x^{-2} i_2(x,\theta)$ (Sect.4.2). Specific features of the inversion, in this case, can be seen from the example given in Fig. 4.5a. The measuring apparatus and technique are described in [4.17]. Therefore only the conditions under which the experiment has been carried out and some of its capabilities are given below.

The experimental measurements were made at a place 60 m from the sea shore and 40 m above sea level, in Karadag, Crimea, USSR during October, 1975. The effective wavelength of the radiation used was $\lambda = 0.546\mu$m; the diameter of the sensing beam was 95 mm; the angular resolution of the nephelometer was about 40'; the angular region from 5° to 175° was covered. The total inaccuracy of the device did not exceed 10%. Measurements of the scattering diagram with a 5° increment over the above angular region took at most 3 or 4 minutes.

The results of the inversion of $I_1(\theta)$ and $I_2(\theta)$ (Fig.4.5a) are shown in Fig.4.5b. It was assumed that the real part of the complex index of refraction of the aerosol substance $\bar{m}$ was in the interval from 1.35 to 1.5 and the imaginary part $\varkappa \leq 0.002$ (the relative humidity of the air did not exceed 55-57%). An estimation of $\bar{m}$ according to (4.9) gave the interval $\Omega^*_{\bar{m}} = [1.43\text{-}1.45]$. This interval of $\bar{m}$ values is too narrow to be accurately defined a priori. The particle size-distribution function was determined according to the expression $s_\alpha = Q^{-1}_{1\alpha}(\bar{m}^*)I_1$ (solid curve in Fig.4.5b).

A specific feature of the inverse solution obtained is its localization in the size region $r \leq 0.4\mu m$.

As numerical investigations have shown, the "tail" of the distribution for $r > 0.3\mu m$ is caused mainly by measurement errors. The main result of this inversion can be summarized as follows: the investigated aerosol did not contain particles with radii larger than 0.4 - 0.5µm, i.e., it is formed only by the fine fraction. Consider now the estimation of the particle number density. The calculations showed that the number density of particles with radii from 0.08 to 0.12µm was about $6 \times 10^3 cm^{-3}$ and of those with radii from 0.15 to 0.25 was about 1.5×10^3 cm^{-3}. Thus, the values of the aerosol number density obtained are in a good agreement with the data on the number density in the ground layer. The absence of large particles in the inverse solution can be explained as follows.

First of all, the conditions under which the experiment was carried out can be characterized by high transmission with no wind observed. Secondly, the scattering volume was small (less than 5×10^3 cm^3). For fast measurements under these conditions, the probability for large particles to be in the scattering volume should be small [4.18].

Except for the absence of large particles, these results are characterized by a high resolution with respect to details of the aerosol microstructure which, in turn, shows the capabilities of this optical method. By "optical method" we mean the nephelometric measurements themselves and also the numerical procedure for their inversion. To illustrate the fact that the maxima at $r_{s_1} = 0.1$ and $r_{s_2} = 0.2\mu m$ in the inverse solution $s_\alpha(r)$ (Fig.4.5b) are not caused by instability in the inverse problem (or by poor smoothing of the regularized inverse solution $s_\alpha(r)$ for a given measurement error) the control calculations on the reconstruction of multimode distributions, using the polarization scattering phase functions, have been carried out [4.19]. One of the inverse solutions obtained from the number of nephelometric measurements in the above experiment was taken as an exact solution for such calculations. These calculations confirmed the possibility of revealing two maxima in the distribution to the left of 0.4μm, using combined inversion of the polarization scattering phase functions. The measurement errors influence the solution mainly in the region to the right of 0.4μm where the values of the initial size distribution are small. Since particles are localized in a narrow size range in the model distribution, the effect of optical screening can be neglected. As a consequence, the optical resolution with respect to microstructure details in the region of small particles is higher, all other conditions being equal.

It should be noted that the above questions concerning the possibility of revealing "fine structure" in the particle size spectrum when inverting the optical characteristics are the most important and difficult aspects of the theory of the indirect method. This is especially true for those based on the solution of integral equations. We have already considered this problem earlier in Chap.3, when dealing with the study of the analytical properties of optical characteristics. However, only the application to spectral measurements was considered there.

It is hardly possible to make an analogous analytical study of the angular characteristics (such as, e.g., elements of the polydispersed scattering phase matrix), using the Mie theory, except, perhaps for some special cases (Sect. 4.6). It is worthwhile to recall here (Sect.2.4) that in any case the inverse solutions should be not less smooth than the kernels of the equations solved, merely due to the fact that the measured functions do not contain information on the fine structure of a real size-distribution in this case. In this connection, consider directly the kernels of the integral equations for the polarization scattering phase functions. The family of functions $\tilde{Q}_1(r/\theta) = Q_1(r,\theta)\cdot\{\max_r Q_1(r/\theta)\}^{-1}$, $(r \in [0;\ 0.5\mu m])$, produced by the kernel $Q_1(r/\theta)$ for the function $I_1(\theta)$ is shown in Fig.4.6. It is not difficult to see that the curves $Q_1(r/\theta)$ for $\theta = 20^\circ$ and $\theta = 30^\circ$ are too smooth over the size interval $[0;\ 0.4\mu m]$ except, perhaps, the amplitude variation which, in fact, is unimportant for the structure resolution in s(r). The situation is quite different for scattering angles θ between 80° and 140°. The positions of the local minima of $Q_1(r/\theta)$ are distinctly localized in this case and are noticeably shifted with respect to each other even if the difference between the corresponding angles is not very large (see, e.g., the curves for $\theta = 90^\circ$ and $\theta = 100^\circ$). Such a variability of the kernel $Q_1(r,\theta)$ of the initial integral equation provides a reliable possibility to determine the sought solutions from the class of bimode distributions (Sect.3.1).

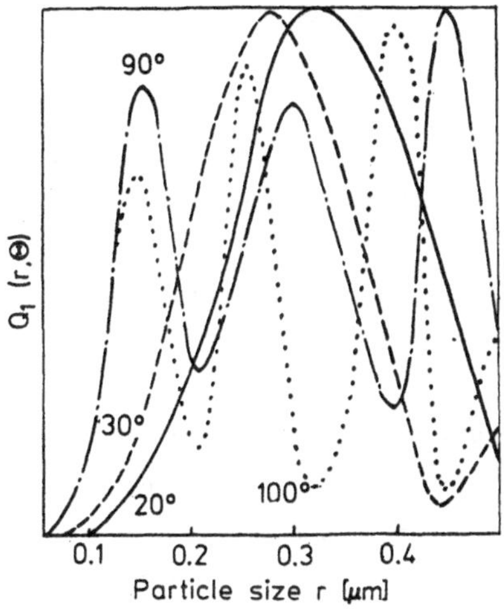

Fig.4.6. The kernel $Q_1(r,\theta)$ of the equation $(Q_1s)(\theta) = I_1(\theta)$ for some scattering angles Θ

Finally, note another aspect of the inversion of several optical parameters, e.g., μ_i and μ_j. It is easy to show that the functions μ_i and μ_j can commute in (4.9), and, as a consequence, the distribution s_α can be determined using two operators, namely, $Q_i(Q_j^*Q_j + \alpha H)^{-1}Q_j^*$ or $Q_j(Q_i^*Q_i + \alpha H)^{-1}Q_i^*$. It is obvious that the corresponding solutions would coincide within the error limits if the index of refraction is determined accurately and if the shapes of the particles are close to spherical. This confirms the reliability of the inversion results. The dashed curve in Fig.4.5b represents the second of these solutions obtained by the combined inversion of $I_1(\theta)$ and $I_2(\theta)$ shown in Fig.4.5a.

Some aspects which are inherent in the solutions of inverse problems of the theory of light scattering by polydispersed ensembles of particles for polarization scattering phase functions $I_1(\Theta)$ and $I_2(\Theta)$ were considered above in detail because the methods suggested here for the inversion can be widely used in optical studies of other polydispersed scattering media.

4.3.3 An Example of Investigation of Microphysical Parameters of Smoke Aerosols Using a Polarization Nephelometer

In addition to the above example of atmospheric haze, we carried out investigations of the microstructure parameters of smoke aimed at the development of remote methods for monitoring atmospheric pollution using optical sensing. Results of the inversion of the scattering phase functions were found to be in a good agreement with the data from direct microstructure analysis.

Let us consider the results of the investigations on optical sensing of wood smoke. The investigations were based on the measurements of the polarization scattering phase functions as in the above example. The smoke was obtained by burning wood in a heating cell inside the nephelometer. A more detailed description of the experiment and the apparatus used,as well as the smoke microstructures,can be found in [4.20]. The task of the investigation was to study the influence of the air humidity on the optical parameters and the microstructure of wood smoke,as well as the dependence of its microstructure upon the mass of wood burnt.

Typical results of the investigations are given graphically. In Fig.4.7 the scattering phase functions, the degree of linear polarization and the asymmetry of the scattering phase functions (a, b and c, respectively) are shown as functions of the scattering angle Θ measured at the air humidity values of 27, 86, 91 and 96% inside the nephelometer cell (Curves 1,2,3,4). Figure 4.8a presents the dependence of the real part of the complex index of refraction on air humidity, obtained by the inversion of optical measurements data (Fig.4.7). The inversion was performed using the methods given above. The expected errors of the inversion are shown in parentheses. The inversion errors were determined by the initial measurement errors which were at least 10% in the experiment discussed here. The size-distribution functions for smoke particles (distribution of volumes) shown in Fig.4.8b, characterize, on the whole, the transformation of the microstructure as the air humidity changed. Omitting a detailed discussion of the inversion results, we should like to note that these results illustrate the possibilities of optical methods to investigate microphysical parameters of real media. The trustworthiness of these results is confirmed, firstly, by direct sampling which showed that the shapes of the smoke particles were close to spherical and, hence, the use of the Mie theory could not produce significant methodological inaccuracies. Secondly, a comparison made between the size-distribution obtained numerically by inverting the optical measurements data and that estimated from particle counter data (particle radii $r \geq 0.2\mu m$ in both cases) showed good agreement.

It is interesting to note that the curves V(r) shown in Fig.4.7b can be satisfactorily approximated by the log-normal size-distribution function. It is a well-known fact that the size-distribution of smoke particles is well described by this law. At the same time, the size distributions presented in Fig.4.7b are more complicated analytically in comparison to the corresponding log-normal models and, as a consequence, bear more information about the smokes, microstructure. This is true particularly for the fraction of large

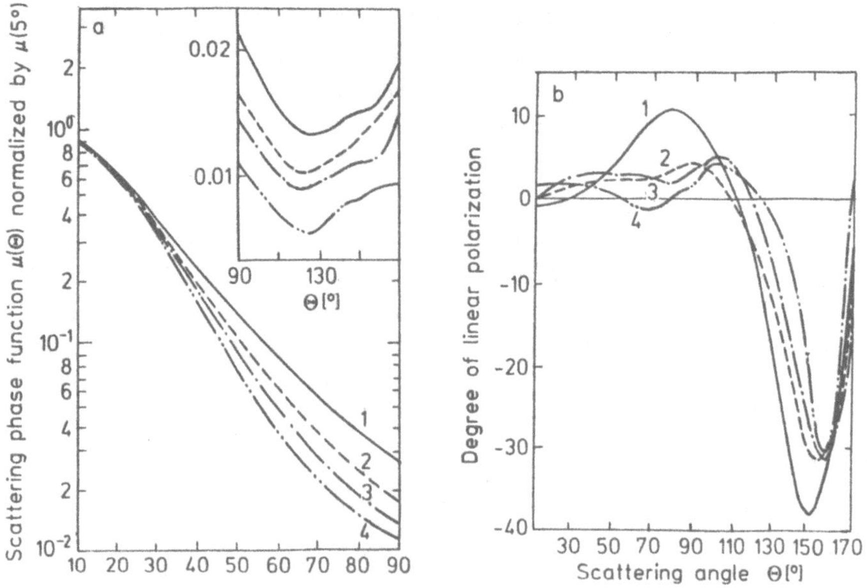

Fig.4.7. Scattering phase functions μ(Θ) and the degree of linear polarization of the radiation scattered by smokes measured with polarization nephelometers at different values of air humidity

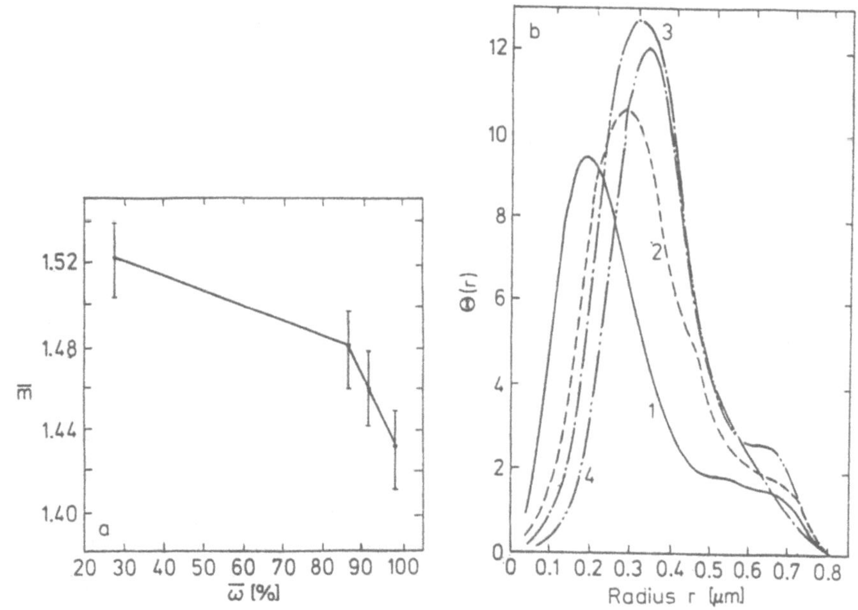

Fig.4.8. The dependences on air humidity of the refractive index and microstructure of smokes obtained by inverting the optical data shown in Fig.4.7

particles, where the influence of humidity is more significant. The above example gives a good illustration of the possibilities of optical methods to investigate smokes using polarization nephelometers and also the efficiency of numerical methods for optical data handling developed above.

The above methods for interpreting the polarization characteristics of light scattering are based on the Mie theory, i.e., they can be regarded, in this respect, as a rigorous theory of the polarization sensing of polydispersed systems of spherical particles. In Sect.4.5, the application of this theory to the remote sensing of aerosols with the use of bistatic lidars will be presented. It is obvious that the applicability of this theory in practice is limited, since the particles of a real atmospheric aerosol are not spheres. This circumstance should be taken into account when interpreting polarization measurements.

4.4 On the Inversion of Polydispersed Scattering Phase Matrices for Nonspherical Particles

4.4.1 General Comments

To a first approximation, the inversion of such optical parameters as β_{ext} and β_{sc} measured in a spectral interval Λ can be made regardless of the shape of the scattering particles. However, the inversion of the data on other optical parameters of aerosols obtained with the use of nephelometers and monostatic and bistatic lidars requires the estimation of the applicability limits of the Mie theory used as an analytical apparatus for interpreting the data. This is especially important when measurements of the polarization characteristics of lidar signals are performed to improve the information capabilities of measuring complexes. The experiments [4.21], indicate that the polarization characteristics of the lidar returns are characterized by significant spatial inhomogeneity and in many cases cannot be interpreted correctly if spherical symmetry is assumed. It is known that the polarization properties of scattered electromagnetic waves depend on the geometry of the scattering body more strongly than the energy dependence. It is sufficient, in particular, to note that the scattering phase matrices for ensembles of spherical and nonspherical particles are essentially different (Sect.4.1). It is necessary to point out, in this connection, that for the inversion of spectral optical measurements one has no direct experimental information on the shape of the particles sensed. The situation changes drastically when the polydispersed scattering phase matrix is used for the inversion.

The structure of this matrix and the relations between its elements indicate the degree to which the use of the Mie theory for data inversion can be regarded as correct. In addition, the array of elements of the scattering phase matrix permits the formulation of more complicated and informative inverse problems. Such problems can include not only the determination of the microstructure, but also of the geometrical characteristics of the scattering bodies. Mathematically such problems could be expressed in terms of multidimensional integral equations or, preferably, in the form of a set of integral equations. However, the current state of the theory of scattering of electromagnetic waves on bodies of arbitrary shape does not yet allow such an approach to this problem. Here the reader is referred to the detailed review concerning the problem of particle nonsphericity in lidar sensing applications given in [4.22].

Analytical results obtained in this work on the solutions of Maxwell's equations for the case of symmetrical particles with smooth surfaces require long computation time. This makes unrealistic the use of corresponding theories for the description of the optical properties of polydispersed ensembles of particles, taking into account their orientations in space.

In addition to Maxwell's equations for bodies having elementary shapes, some other analytical methods can be used to study the optical properties of the dispersed media which require, however, some very specific conditions to be fulfilled. Among these formalisms, two approximations for the optical properties of particles of arbitrary shapes should be mentioned. These are based on the Rayleigh and Rayleigh-Hans theories. Since the use of these theories is strongly restricted in applicability [4.5], the corresponding approximations permit the solution of quite specific problems (direct and inverse) of the theory of light scattering by polydispersed media.

Such an example can be found in [4.23,24], where estimates of the microstructures of polydispersed ensembles of spherical and nonspherical particles are given, based on the inversion of scattering matrix elements in the Rayleigh-Hans approximation. However, the use of these theories to describe the polarization properties of the optical waves propagating through the real media is very limited. The applicability of the Mie theory in this case is much wider [4.22]. Therefore, the Mie theory remains the single formalism applicable to the interpretation of optical measurements. As is shown in [4.7], the possibility of using it as an approximate formalism can be significantly widened if additional information on the geometry of the scattering particles is used. We shall consider some results of numerical studies for the case of an ensemble of randomly oriented nonspherical particles.

4.4.2 An Example Illustrating the Influence of the Particles' Nonsphericity

Calculations were made to illustrate the possibilities of using the Mie theory to estimate the microstructure parameters of an ensemble of nonspherical particles by inverting certain elements of the polydispersed scattering phase matrix. The calculations are based on the experimental material given in [4.25], where the elements S_{11}, S_{22}, S_{12}, S_{33} and S_{34} of the polydispersed scattering phase matrix for an ensemble of randomly oriented nonspherical particles were measured. The particles' shapes, index of refraction and size distribution were known. In particular, the particles of the aerosol studied were discs (practically plane particles) and, hence, were far from spherical. Thus, the following estimates of the influence of the nonspherical shape of the particles on the deviations of the size-distribution functions obtained by the inversion of the matrix elements, based on the Mie theory for real distributions, are of particular interest.

These estimates make it possible to evaluate the maximum errors which can occur in the inversion of optical sensing data. The regularizing algorithms, used for the inversion of optical measurement data, were the same as those used in Sect.4.2.

The first problem to be solved in the numerical study is the inversion of those elements of the scattering phase matrix, which have analogs in the matrix for spherical particles. Some results of that study are presented in Fig.4.9. The so-called exact solution $s_0(r)$ (Curve 1) characterizes the distribution of the mean areas of the particle projections onto the microscope table with respect to the radii of circles having the same areas [4.26]. Curves 2, 3 and 4 describe the solutions $s_\alpha(x)$ of the equation $Qs = D$ when the elements S_{11}, S_{22} and S_{33} are taken as the right-hand side. It is readily seen that in all three cases the norms of the solutions obtained exceed the norm of the initial solution $s_0(r)$ in the given size interval. Since all the functions are positive, their full integrals, characterizing the cross sections of particles contained in a unit volume of polydispersed ensemble, can be used as the norms for making a comparison between them.

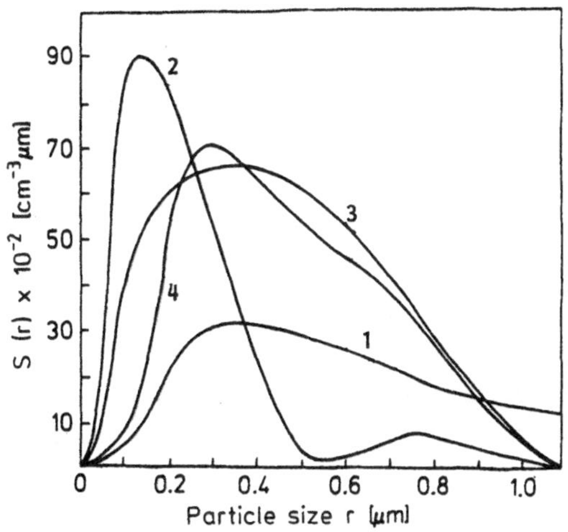

Fig.4.9. Size-distribution functions $s_\alpha(r)$ obtained by inverting the elements S_{11}, S_{22} and S_{33} of the scattering phase matrix (Curves 2,3 and 4, respectively) for an ensemble of spherical particles (Mie scattering efficiency factors); Curve 1 represents the reference size distribution function

From the fact that such norms exceed the norms of the real distributions, the conclusion can be drawn that the efficiency factors of light scattering by nonspherical particles are larger, on the average over the ensemble, than the corresponding factors for optically equivalent spherical particles with a size distribution $s_\alpha(r)$ (Fig.4.9). By optical equivalence we mean that the deviations of the measured elements of the scattering phase matrix from those calculated according to the Mie theory for the ensemble of particles with the size distributions $s_\alpha(r)$ are not larger than the measurement errors. In the calculations, scattering angles from 20° to 160° with 5° increment were used for the inversion. Considering the experimental measurements of the elements of the scattering phase matrices for monodispersed ensembles of nonspherical particles randomly oriented in the scattering volume, one finds the above conclusion concerning the efficiency factors to be confirmed in general.

In such experiments [4.27] the differential factors $Q(r,\theta)$ are 2 to 4 times as large as the corresponding efficiency factors from the Mie theory, especially for forward angles. The opposite situation can occur for backward angles. The result is that the inversion of optical measurements, made for angles in the backward hemisphere, can give underestimated norms for the solutions. As a consequence, a systematic underestimation of the aerosol number density obtained, e.g., from laser sensing data, is possible.

Another very important aspect of this problem is that the inverse size distribution allows the construction of a first approximation of the optical model for the ensemble of nonspherical particles using the Mie theory, independent of how close that distribution is to a real one. In other words, these size-distributions enable one to answer the question whether it is possible to select such a polydispersed ensemble of spherical particles or not, and what the corresponding size distribution should be in order to obtain the measured values of the elements of the scattering phase matrix for the ensemble of nonspherical particles from calculations according to the Mie theory with an acceptable accuracy. Such a problem of applied optics is very important for optical investigations of dispersed media.

At the same time, it is practically impossible to solve this problem using the method of selection [4.28]. The minimization of the regularizing functional on which the inversion methods are based implies also minimization of the discrepancy $\rho(Qs\ ,D)$ of the optical data on a certain class of possible distributions. If, when solving the problem, one finds the relation $\rho(Q_{ij}s\ D_{ij}) \leq \delta(\sigma)$ (σ being the experimental error) to be valid for an element S_{ij}, then the corresponding function $s_{\alpha}(r)$ provides the answer to the above question. Relevant examples are presented in Fig.4.10, where optical models for S_{11} and S_{22} (Curves 1 and 2, respectively) are constructed according to the Mie theory for the aerosol size-distributions $s_{\alpha}(r)$ (Curves 2 and 3 in Fig.4.9).

Such an approximation of the elements can be regarded as quite satisfactory, because the value of $\rho\|D\|^{-1}$ is of the order of the experimental errors. The discrepancies between the calculated values (shown in Fig.4.10) of these elements and those measured experimentally are acceptable. It should be noted that when we are talking about the optical model, we imply some particular element of the scattering phase matrix. A model size-distribution s_{α}, allowing the simultaneous approximation of several elements, having analogs in the Mie theory, is hardly possible. An example of calculations for the matrix element S_{12}, made according to the Mie theory (curve with breaks) using size distributions inferred by the inversion of S_{11} and S_{22} (Curves 2 and 3 in Fig.4.9), is given in Fig.4.11. This result can be considered negative. In other words, the inversion of one element of the measured scattering phase matrix provides no possibility to estimate (in the given examples) the other elements by using only the Mie theory. An attempt to invert the elements S_{12} and S_{34} has also failed. The minimization of the smoothing functional gave negative functions which in this case had no physical meaning. All this indicates the limitations of the Mie theory in the interpretation of optical measurements when scattering from "strongly nonspherical" particles is involved.

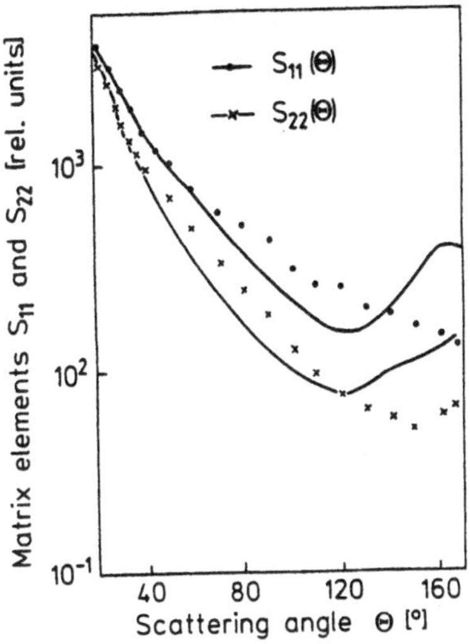

Fig.4.10. Measured values of S_{11} and S_{22} (dashed curves) for a system of nonspherical particles and corresponding analogs from the Mie theory (solid curves)

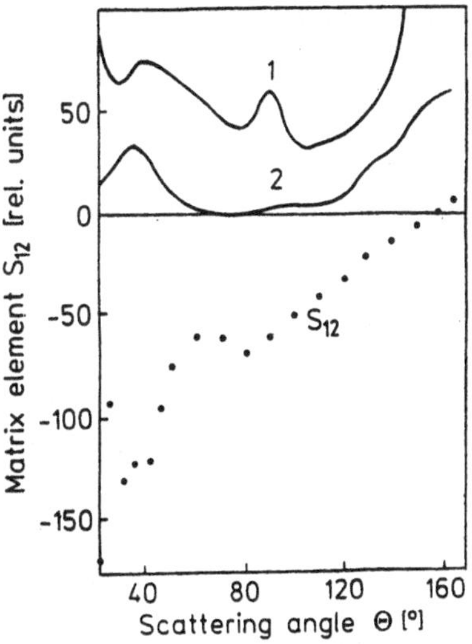

Fig.4.11. Elements S_{12} of the scattering phase matrix for nonspherical particles [4.25] and their analogs from the Mie theory calculated from the results of inversion of S_{11} and S_{22} (Curves 1 and 2, respectively)

Morphological analysis indicates that the particles of atmospheric hazes are, on the whole, oval and, hence, not as asymmetric as the particles considered above. Of course, a possible exception would be crystalline particles, which do occur in the atmosphere.

4.4.3 Parametric Modification of the Integral Equations for Light Scattering by Systems of Nonspherical Particles

To conclude the discussion of the numerical investigations of the inversion of the scattering phase matrices of a polydispersed ensemble of nonspherical particles, consider one possibility to interpret the optical data more correctly. In the simplest cases, when the size spectrum does not differ very much from an unimodal one, the corresponding methodological errors can be reduced. This possibility is related to the averaging formula (2.45).

Using the theorem on the average value of an integral, one can obtain from (2.45) that

$$\bar{K}(\bar{\ell},\lambda) = K(\xi\bar{\ell},\lambda) \quad ,$$

where ξ is some positive number. It is clear that the value of ξ is determined by the weighting functions p and q, i.e., by the geometrical shape of the body, and it also depends upon λ. If it is assumed that the measure of particle symmetry varies weakly, on the average over the ensemble, then one obtains the following integral representation of the spectral optical characteristics

$$\bar{\beta}(\lambda) = \int_L K(\xi\bar{l},\lambda)dS(\bar{l}), \quad \lambda \in \Lambda \quad , \tag{4.10}$$

in which the value ξ is used everywhere in the spectral interval Λ. In the next discussion the value of ξ will be regarded as an unknown parameter. In any particular case, the choice of ξ can be related, in some way, to the geometrical shape of particles of the medium under investigation. An example will be given below.

When formulating the inverse problem for the elements of the polydispersed scattering phase matrices, the following form of (4.10) will be used

$$\int_R Q(\xi r,\theta)s(r)dr = D(\theta) \quad . \tag{4.11}$$

Solving (4.11), one obtains not one function $s_\alpha(r)$ but a parametric family of regularized solutions $s_\alpha(r,\xi)$. The existence of such a family of inverse solutions allows one to make a more efficient interpretation of the optical measurements by choosing the proper value of the parameter ξ. It is for this reason that the above approach to the inversion of the characteristics of light scattering by polydispersed ensembles of particles may be called the method of parametric modification of the initial integral equation [4.28]. A numerical investigation of the effectiveness of this approach has been carried out in [4.4]. This approach to the inversion of elements of the polydispersed scattering phase matrix was suggested in [4.19]. Figure 4.12 presents an example illustrating the capabilities of this method, the solutions $s_\alpha(r\xi)$ obtained by inverting the element S_{22}. The values of the parameter ξ for the Curves 2, 3 and 4 are 1, 0.7 and 0.45, respectively. The actual size distribution $s_0(r)$ is displayed by the Curve 1.

As is seen from Fig.4.12, the real size distribution $s_0(r)$ is best approximated when $\xi = 0.45$. A further decrease of this parameter resulted in an increased discrepancy between $s_\alpha(r,\xi)$ and $s_0(r)$. Analogous results occur also for the case when S_{11} and S_{33} are inverted. The solution of (4.11) in these cases provides the smallest errors in the reconstructed $s_0(r)$ as compared with those for the inversion of S_{22}. Nevertheless, it should be noted that the capabilities of the above approach are limited. Information about the geometrical shapes of the scattering particles is necessary in order to

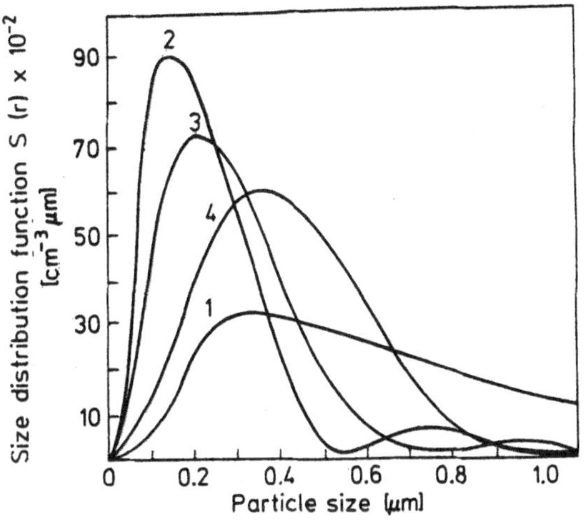

Fig.4.12. Size distribution functions $s_\alpha(r,\xi)$ obtained by solving (3.11) with respect to the element S_{22} (ξ = 1; 0.7; 0.45 - Curves 2, 3 and 4, respectively

find more informative solutions of the inverse problems. Thus, the numerical investigations made allow one to reach the following conclusions.

First of all, quite an acceptable estimation of the size-distribution functions can be obtained by inverting the elements of the scattering phase matrix for an ensemble of nonspherical particles randomly oriented in the scattering volume. In this case, the size-distribution of the particle cross-sections is meant, and not the number density (Sect.2.2). The best results are obtained when the elements S_{11} and S_{22} are inverted. The inversion of S_{22} gives somewhat worse results. The parametric modification (4.11) of the initial integral equations can provide some improvement in the accuracy of estimations of the reference size distributions obtained within the framework of the Mie theory. Finally, it may be noted that the inversion of nondiagonal elements of the scattering phase matrices by solving corresponding integral equations with kernels from the Mie theory is worthless.

Of course, we are talking about those elements of the scattering phase matrices which have analogs in the matrices for systems of spherical particles.

It is interesting to note that all the above conclusions, which were drawn from numerical studies, are confirmed by experimental results. Thus, in [4.2] it was shown that the element S_{22} depends on the geometrical shapes of particles more strongly than the others. It was suggested to use the ratio S_{22}/S_{11} as a qualitative measure of the effect of the particle nonsphericity [4.2]. Thus, in the case of oval particles this ratio has a value of about 0.84, while for cubic particles it is 0.42.

In the example considered above, the ratio S_{22}/S_{11} is about 0.32. Thus, the Mie theory can be considered applicable to the solution of inverse problems for the elements of a scattering phase matrix if the ratio S_{22}/S_{11} is not less than about 0.8. This value of S_{22}/S_{11} can serve as a criterion of the applicability of the Mie theory only if the sizes of the particles do not exceed, for the most part, the value 1μm, and the refractive index of their substance is 1.5 to 1.6 ($\varkappa \leq 0.005$). Evidently, other criteria may exist in other cases. Such a situation may occur, e.g., if a monostatic lidar is used for sensing aerosols. Let us consider, in this connection, some aspects of the interpretation of lidar data obtained for the real atmosphere.

4.4.4 *On the Interpretation of Depolarization of Monostatic Lidar Signals*

Let us consider the results of a complicated investigation of the atmosphere presented in [4.12]. This paper presents the experimental data on optical and microphysical parameters of aerosols in the lower troposphere. According to the data of the microstructure analysis made with airborne impactors followed by the investigations using electron microscopy, the following classification of shapes of the tropospheric aerosol particles could be given. About 50% of the total number of particles have oval shapes (mainly of silica origin) not far from spherical; about 30% are amorphous; their origin is droplets of sulfuric acid; 15% are porous particles and about 5% have crystalline shapes.

For the first group of particles, it was noted that 50 of every 450 particles are ellipsoids with the ratio of the semimajor to semiminor axis being less than or equal to 2. All the other particles are close to spherical. It should also be noted that the porous particles observed were mainly convex bodies. Thus, one can state that the majority of the aerosol particles of the real atmosphere are convex and, as a consequence, the methods developed in Chap.2 are valid for the inversion of lidar data. It is also important to emphasize that aerosol particles are, on the whole, not very asymmetric. The ratio of the maximum diameter of a particle l_{max} to its minimum value l_{min} does not exceed, on the average, the value 2 or 3. Therefore, the interpretation of the elements of polydispersed scattering phase matrices (or, more correctly, of the polarization ratios) measured in a bistatic scheme made within the framework of the Mie theory, could be regarded as acceptable. Thus, the estimate of the refractive index of aerosol substance, obtained from optical measurements, appeared to be in a good agreement with that obtained from the chemical analysis. It is also important to note that

REAGAN et al. [4.12] in practically all cases, managed to construct the calculational analogs for the measured polarization ratios by varying the parameters of the model size-distribution functions. Analyzing the results of this experiment, including their interpretation, one can state that the problem of particles nonsphericity is not as important (within the accuracy of 15 or 20%) for the inverse problems of light scattering by polydispersed ensembles of particles as, for example, the problem of a proper choice of the index of refraction of the aerosol substance. In other words, the use of the Mie theory as a working apparatus for the inversion of optical measurements is justified in practice. This is valid, in particular, for the case of bistatic lidar sensing, when the angles of sensing do not exceed 160°. A further increase of this angle may increase the difficulty of interpretation of the experimental results, as may be seen in [4.26]. It is interesting, in this connection, to survey some experiments on sensing the atmospheric aerosols with a monostatic lidar.

Assume that the laser radiation of such a lidar is linearly polarized and that the polarization of a lidar return is measured. Usually two components of the backscattered signal, $\beta_{\pi\perp}$ and $\beta_{\pi||}$, are measured and the directions of their polarizations are determined relative to the polarization plane of the incident radiation. It is known that in the case of spherical particles, the polarization of the strictly backscattered radiation is the same as that of the incident radiation and, hence, the component $\beta_{\pi\perp}$ should be equal to zero. Under atmospheric conditions, such a situation should be observed in fog, rain and so on, unless the optical depth is so large that the effects of multiple scattering are insignificant, as is confirmed by experiments [4.29]. However, in some cases the ratio $\beta_{\pi\perp}/\beta_{\pi||}$ can attain large values. Thus, e.g., the smoke of the automobile exhaust produces a ratio $\beta_{\pi\perp}/\beta_{\pi||}$ of about 0.1 ± 0.05; for soil dust it is about 0.61 ± 0.16 and so on [4.30]. All these examples explicitly show the importance of the particle nonsphericity for the intepretation of sensing data. Unfortunately, there are as yet no proper analytical models which enable one to interpret the ratio $\beta_{\pi\perp}/\beta_{\pi||}$ and to infer, at least as a first approximation, information on the geometrical shapes of real atmospheric particles. Therefore, investigations in this direction that would widen the information capabilities of the monostatic lidars are needed.

A final remark on the interpretation of the depolarization ratio $\beta_{\pi\perp}/\beta_{\pi||}$ is related to the sensing of clouds, where the depolarization of lidar returns is caused mainly by the effects of multiple scattering. It is often observed experimentally that the maxima of depolarized $\beta_{\pi\perp}$ and polarized $\beta_{\pi||}$

components of the lidar signals are spatially separated when dense aerosol formations are sensed. As is shown in [4.29], the value of this separation is proportional to the optical depth of the formation. It is characteristic that this effect is not observed in crystalline or mixed clouds. This effect gives another possibility to interpret qualitatively lidar measurements in the atmosphere.

4.5 Inverse Problems of the Theory of Bistatic Lidar Sensing

4.5.1 Lidar Equation for Polarization Sensing and Inverse Problems of Light Scattering by Aerosols

The above investigations on the inversion of the elements of polydispersed scattering phase matrices made within the framework of the Mie theory form the basis for the development of the theory of bistatic lidar. It is to be used for the remote determination of microphysical parameters of the scattering determination of microphysical parameters of the scattering components of the atmosphere.

In the case of a bistatic scheme, the basic functional relation (lidar equation) can be written, taking into account the polarization of the lidar signals, in vector form as follows:

$$\mathbf{I}^{(s)} = Z_1^2 B T_1 T_2 D \mathbf{I}^{(0)} \quad , \tag{4.12}$$

where B is the apparatus constant, and T_1 and T_2 are the transmission coefficients for the sensing paths Z_1 and Z_2, respectively (Fig.4.13). The optical characteristics entering into (4.12) depend on θ and h. The scattering phase matrix will be assumed to have the form $D = D^{(a)} + D^{(R)}$. Assuming, as before (Sect.4.1), that $D_{ij}(\theta) = (Q_{ij}s)(\theta)$ and considering spherical particles, one can write, in a general form, four integral equations for the sought function s(r)

$$(\bar{Q}_i s)(\theta) = q_i(\theta) \quad , \quad i = 1,2,3,4 \tag{4.13}$$

where

$$\bar{Q}_i = \sum_{j=1}^{4} I_j^{(0)} Q_{ij} \quad , \quad q_i(\theta) = I_i^{(s)} (BT_1T_2z_1^2)^{-1} - \sum_{j=1}^{4} D_{ij}^{(R)} T_j^{(0)} \quad . \tag{4.14}$$

In some particular cases, (4.13) may take a simpler form. If, for example, the incident radiation is linearly polarized at an angle $\pi/4$ with respect to the scattering plane, then the operators $\bar{Q}_i$ and Q_{i2} are equal since $\mathbf{I}^{(0)}$

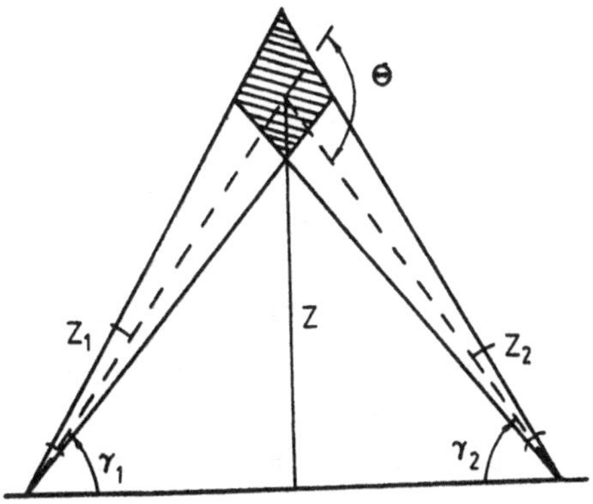

Fig.4.13. Geometrical scheme of the bistatic lidar

has components (0,1,0,0). In this case,the inverse problem formulated above in (4.13) is reduced to the inversion of elements of the polydispersed scattering phase matrix. The kernels of the corresponding integral equations are linear combinations of the functions $Q_{ij}(r,\theta)$. For (4.13) to be well defined it is necessary, according to (4.14), to know the values T_1, T_2 and $D^{(R)}$. This, in turns, means that photometric measurements should accompany the optical sensing using the bistatic scheme when the system (4.13) is used to interpret the results. In addition, the profile of $\beta_{sc}^{(R)}$ should also be determined. If the determination of all the above optical parameters is provided in the course of an experiment and the apparatus constant B is known then one has a well-defined inverse problem for the determination of the microstructure and the index of refraction of the aerosol substance. The numerical methods for solving such inverse problems were presented in Sect. 4.3 and therefore the following discussion will deal mainly with the numerical analysis of the equations $(\bar{Q}_i s)(\theta) = q_i(\theta)$ as applied to the optical sensing of atmospheric hazes.

The range of sensing angles used in the numerical experiments was chosen in the backward hemisphere $(90^\circ, 180^\circ)$. The variations of the index of refraction are assumed to be ±0.05 near the value $\bar{m}_0 = 1.55$. The upper bound of the size spectrum of particles is given by $R_2 \le 1\mu m$, and the set of single-mode size distributions is chosen as a class of solutions Φ. Under these conditions, when the index of refraction $\bar{m}$ is known exactly, the magnification of the error ε_q entering the right-hand side of (4.13) is about 2 or 3 depending on the index i. In other words, the relation $\varepsilon_s = \|s - s_0\|/\|s_0\| \lesssim (2 + 3)\varepsilon_q$ is valid for this case, where ε_s is the error of the solution obtained provided the proper choice of α is made. This relation is also valid for the statistical error in $q_i(\theta)$ if $0.05 \le \varepsilon_q \le 0.1$. The presence of variations in $\bar{m}$ increases the value of the error magnification coefficient to 3 or 4. These numerical investigations have enabled one to estimate the precision of the measuring device required, as well as the appropriate information content of the measurements. The function $q_i(\theta)$ should be measured with an accuracy no worse than 5% in order that the function s(r) be determined from the class of one-mode size-distribution functions with an acceptable accuracy. The total number of independent measurements made using a bistatic lidar should be about 16 to 20. For example, enough measurements can be obtained when two elements of the polydispersed scattering phase matrix (or $I_1(\theta)$, and I_2, see Sect.4.3) are measured at five angles and two wavelengths λ_1 and λ_2 and so on. If, in addition, the determination of the index of refraction is desired, then the increase of the required data can be achieved by measuring the other elements of the polydispersed scattering phase matrix.

4.5.2 Polarization Relations and Methods for Their Inversion

The above variation for solving the inverse problem of bistatic lidar sensing is mathematically the simplest. The corresponding integral equations (4.13) are equivalent to the equations $Q_{ij}s = D_{ij}$. Both of these systems are the "ill-posed" inverse problems with the right-hand sides given only approximately. The kernels of these equations can be considered as known provided the assumption that the particles are spherical is valid. In the following discussion, we shall consider some other formulations of the inverse problems of bistatic lidar sensing that are more complicated for calculations but simpler and more effective for instrumental applications. The latter consideration can be very important. The fact that measurements of T_1, T_2 and $D^{(R)}$ (equivalently $\beta_{sc}^{(R)}$) are necessary for (4.13) is an essential disadvantage that makes the use of (4.13) difficult in practice. In addition, the dependence of the accuracy of the determination of aerosol microstructure parameter from the bistatic lidar measurements on the errors of the accompanying optical measurements does not permit the full realization of the information capabilities of the bistatic schemes as an independent optical method. Taking the above remarks into account, let us consider a different approach to the interpretation of the initial relation (4.12). Let us introduce the vectors $\mathbf{c}^{(s)}$ and $\mathbf{c}^{(0)}$ with components $c_i = I_i I_1^{-1}$ instead of the vectors $\mathbf{I}^{(s)}$ and $\mathbf{I}^{(0)}$. Then one has

$$c_i^{(s)}(\theta) = \left(\sum_{j=1}^{4} D_{ij}^{(a)} c_j^{(0)} + \sum_{j=1}^{4} D_{ij}^{(R)} c_j^{(0)}\right) \times \left(\sum_{j=1}^{4} D_{1j}^{(a)} c_j^{(0)} + \sum_{j=1}^{4} D_{1j}^{(R)} c_j^{(0)}\right)^{-1} \quad (4.15)$$

$$i = 2,3,4 \; .$$

The components of this vector are already independent of T_1, T_2 and B, in contrast to the vector $\mathbf{I}^{(s)}$. Assuming now that the bistatic schemes allow one to measure some components of the vector $\mathbf{c}^{(s)}$, it is not difficult to construct a new system of relations analogous to (4.13), viz.,

$$(\bar{Q}_i s)(\theta) = \beta_{sc}^{(R)} \bar{q}_i(\theta) \qquad i = 2,3,4$$

$$\bar{Q}_i = \sum_{j=1}^{4} c_j^{(0)}\left(c_i^{(s)} Q_{1j} - Q_{ij}\right) \;,$$

$$\bar{q}_i = (1/4\pi) \sum_{j=1}^{4} c_j^{(0)}\left(f_{ij}^{(R)} - c_i^{(s)} f_{1j}^{(R)}\right) \; . \quad (4.16)$$

The values $f_{ij}^{(R)}$ are the elements of a normalized scattering phase matrix of the molecular atmosphere. Equations (4.16) are determined with respect to the size distribution s(r) sought, provided the value $\beta_{sc}^{(R)}$ is known as a function of h. It is interesting to compare (4.13) and (4.16). Such a comparison shows, first of all, that the corresponding integral operators Q_i of these equations are defined in different ways. The feature, by which (4.16) differs essentially from (4.13), is that its kernel $Q_i(r,\theta)$ depends on the measured vector $\mathbf{c}^{(s)}$ and is thus an integral equation with the kernel set approximately. This fact makes the inversion of (4.16) a more complicated calculational task. Let us give, in this connection, some explanatory remarks.

Recall that the equation Qs = D is an equation with the kernel set approximately if the index of refraction $\bar{m}$ is known approximately with a certain inaccuracy. However, the value $\sup_{\|\varphi\| \leq 1} \|Q(\bar{m})\varphi - Q(\bar{m}+\Delta\bar{m})\varphi\| / \|\varphi\|$ as a function of $\Delta\bar{m}$ is, as a rule, regular. Moreover, to a first approximation, it can be considered to be a monotonic function, especially in problems of sensing of atmospheric hazes or for the interpretation of corresponding data using the Mie theory. For this reason, it is not difficult to make the inversion of the equation Qs = D even if there are errors in $\bar{m}$.

In particular, the method of generalized discrepancy can be used for constructing the regularizing algorithm [4.7]. Inaccuracies in the kernel $Q_i(r,\theta)$ of (4.16) are caused by errors in the measurements and, hence, cannot in general be considered as systematic. The statistical errors in the components of $\mathbf{c}^{(s)}$ yield the irregular behavior of the kernels of (4.16). The construction of regularizing algorithms for (4.16) requires, for this reason, a certain care and some prior numerical studies. Let us give a short overview of the results obtained by analyzing numerically the properties of the solutions $s_\alpha(r)$ of (4.16). These solutions are obtained using the method of smoothing functionals. The coefficient of error magnification, i.e., the ratio $\varepsilon_s/\varepsilon_q$ increases now up to 5 or 7, even if the initial data are as those used in the example of numerical solution of (4.13) considered above. Note that the ratio $\beta_{sc}^{(R)}/\beta_{sc}^{(a)}$ is, as earlier, about 0.2. The errors in the inverse solution increase as this ratio decreases and vice versa. It is also important to note that the calculations have been made for the wavelength $\lambda = 0.69\mu m$.

In the case of shorter wavelengths, the inverse solutions s_α of (4.16) show more stability with respect to experimental errors. For example, if $\lambda = 0.53\mu m$ the ratio $\varepsilon_s\varepsilon_q^{-1}$ decreases to 4 or 5. This is caused by the fact that the optical screening of small particles in the ensemble is weaker at shorter wavelengths. This gives an estimate of the accuracy attainable with the scheme (4.16).

4.5.3 Determination of the Normalized Size-Distribution Functions from Sensing Data Using Bistatic Lidars

The low efficiency of the scheme (4.16) is caused mainly by poor conditioning of the inverse problem. Some additional conditions should be imposed on the problem considered so that the ratio $\varepsilon_s\varepsilon_q^{-1}$ has an acceptable value (about 2 or 3). In this connection, the following analytical scheme has been suggested and investigated in [4.9]

$$\bar{Q}\varphi = \psi q_i , \quad \int_R \varphi \, dr = 1 , \quad \varphi \geq 0 , \quad r \in R , \tag{4.17}$$

where $\psi = \beta_{sc}^{(R)}S^{-1}$ serves as a parameter for the inverse problem, and the operators Q_i ($i = 2,3,4$) are defined in the same manner as in (4.16). The use of the supplementary condition $\int_R \varphi \, dr = 1$ enables one to improve the conditioning of the inverse problem (4.17) and, as a consequence, to reduce the maximum values of the ratio $\varepsilon_s\varepsilon_q^{-1}$. Let us consider the accuracy characteristics of (4.17) assuming that the constant ψ is known. As before, the

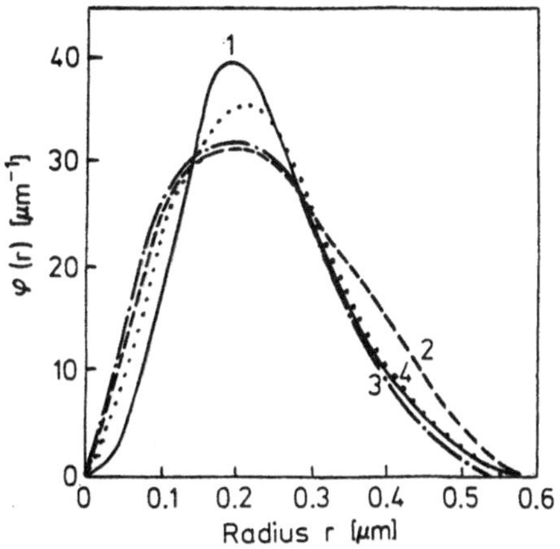

Fig.4.14. An example of the numerical solution of (3.17) for $i = 2,3,4$: the corresponding size-distribution functions are given by Curves 2,3,4; Curve 1 is the reference size-distribution

model size distribution H (Fig.4.14, Curve 1) has been used in the calculations. Curves 2, 3 and 4 in this figure represent the solutions of (4.17) for the indices $i = 2,3,4$, respectively. The uncertainties in the components of the vector $\mathbf{c}^{(s)}$ have been assumed to be about 10%. The value of the constant ψ used is 0.475 (the model values of $\beta_{sc}^{(R)}$ are taken according to [4.31]). The value of the refractive index was taken as $\bar{m} = 1.5 - 0i$. The sensing angles are in the backward hemisphere within the interval from 90° to 170° (the angular increment is 10°). The numerical analysis has shown that the ratio $\varepsilon_s\varepsilon_q^{-1}$ is, on the average, between 2 and 3 provided that α is chosen properly and ε_q is between 0.05 and 0.1. The dependence of the errors of the reconstruction of the normalized function $\varphi(r)$ on the errors in the reference data, i.e., $q_i(\theta)$, is shown more clearly in Fig.4.15. Curves 1,2 and 3 correspond to the inversion of the components $c_2^{(s)}$, $c_3^{(s)}$ and $c_4^{(s)}$. It is characteristic that in order to reduce the errors of the reference data it is necessary, as earlier (Sect.4.1), to use shorter wavelengths, since only in this case one can improve the accuracy of the inversions significantly. In addition to the above assessments, it is important to estimate the effect of uncertainties in the parameter ψ on the accuracy with which $\varphi(r)$ is determined. Corresponding numerical analysis has been carried out for the component $c_2(\theta)$. The analysis shows that uncertainties of about $\pm 0.2\,\psi$ produce errors in the value of φ which do not exceed 6% for $\varepsilon_q \leq 0.1$. This shows that the error $\Delta(\psi)$ perturbs the solution $\varphi_\alpha(r)$ of (4.17) more significantly than the experimental errors when $0.05 \leq \varepsilon_q \leq 0.1$. It may be emphasized here that the error $\Delta\psi$ is systematic at every Z and hence does not have such grave consequences for the solution of the inverse problem.

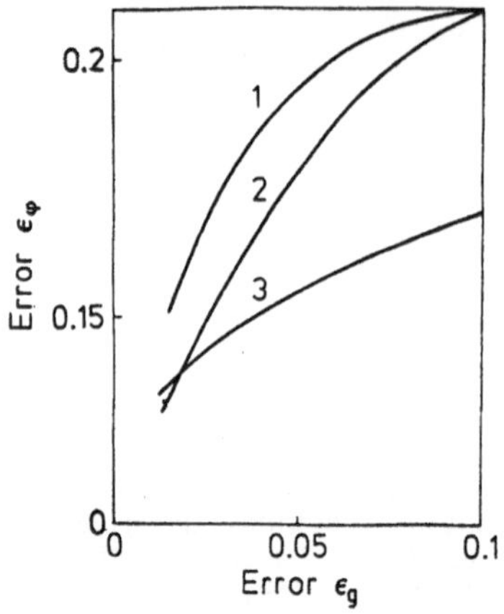

Fig.4.15. Dependence of the errors (ε_φ) in the inversion of equations like (3.17) on the errors (ε_q) in the reference data (Curves 1,2,4 correspond to the indices i = 2,3,4, respectively)

4.5.4 On the Theory of Joint Sensing Using Mono- and Bistatic Lidars. The Separation of the Contributions of Scattering Components

The choice of a proper value for the parameter ψ in (4.17) is not trivial; therefore, the scheme (4.17) is more interesting from the theoretical point of view than for use in practice. Let us consider the development of two practical variants for solving the inverse problems of bistatic lidar sensing, which are based on the use of (4.17). They differ from each other only by the manner in which quantitative information on the relation between the components of light scattering by the atmosphere is introduced in the inverse problem.

Let us introduce the additional function $\tilde{\varphi} = \psi^{-1}\varphi$, which would satisfy the integral equation $\bar{Q}_i\tilde{\varphi} = q_i$. The function $\tilde{\varphi}$ obeys the condition $\int_R \tilde{\varphi}\, dr = \psi^{-1}$. Since ψ is not known, the use of (4.17) is too difficult. Accompanying measurements of the so-called scattering ratio $\eta_\pi = (\beta_\pi^{(R)} + \beta_\pi^{(a)})/\beta_\pi^{(R)}$ are assumed to be available in parallel with the bistatic sensing. The use of a pulsed monostatic lidar allows the profile $\eta_\pi(h)$ to be measured with high spatial resolution (at the wavelength λ_0), especially in the upper troposphere and lower stratosphere. The large amount of information on such measurements can be found elsewhere in the literature. Let us introduce the quantity $\xi_\pi = (\eta_\pi - 1)^{-1} = \beta_\pi^{(R)}/\beta_\pi^{(a)}$. It can easily be shown that

$$\psi = \beta_{sc}^{(R)}/S = K_{sc}(\lambda_0)\beta_{sc}^{(R)}/\beta_{sc}^{(a)} = 2\bar{K}_\pi(\lambda_0)\beta_\pi^{(R)}/3\beta_\pi^{(a)} \qquad (4.18)$$

where $\bar{K}_\pi(\lambda_0) = \int_R K_\pi(r,\lambda_0)\varphi(r)dr$ and $\bar{K}_{sc}(\lambda_0) = \int_R K_{sc}(r,\lambda_0)\varphi(r)dr$. Since $\bar{K}_\pi(\lambda_0) = \psi \int_R K_\pi(r,\lambda_0)\tilde{\varphi}(r)dr$, then, according to (4.18), one obtains the condition

$$\int_R K_\pi(r,\lambda_0)\tilde{\varphi}(r)dr = 2/3\xi_\pi \quad , \tag{4.19}$$

which can be used in (4.17) instead of $\int \tilde{\varphi}\, dr = \psi^{-1}$. Thus, one can formulate the following inverse problem

$$\bar{Q}_i\tilde{\varphi} = q_i \quad ; \quad (3\xi_\pi/2) \int_R K_\pi(r,\lambda_0)\tilde{\varphi}(r)dr = 1$$

$$\tilde{\varphi} \leq 0 \ , \quad r \in R \ , \quad i = 2,3,4 \quad . \tag{4.20}$$

In addition to the ratio $\beta_\pi^{(R)}/\beta_\pi^{(a)}$, one can employ information on the ratio $\beta_{sc}^{(R)}/\beta_{sc}^{(a)}$ in (4.17). Denoting this ratio as ξ_{sc}, one constructs a new scheme that is analytically equivalent to (4.20), viz.

$$\bar{Q}_i\tilde{\varphi} = q_i \quad , \quad \xi_{sc} \int_R K_{sc}(r,\lambda_0)\tilde{\varphi}(r)dr = 1$$

$$\tilde{\varphi} \geq 0 \ , \quad r \in R \ , \quad i = 2,3,4 \quad . \tag{4.21}$$

The inverse problems of bistatic lidar sensing (4.20) and (4.21) give equivalent information, and differ only in the forms of the a priori optical information used. The measurements of ξ_π and ξ_{sc} made at λ_0 complement, to a certain degree, the information on the optical properties of the scattering atmosphere obtained using a bistatic scheme. One can achieve some reduction of the inversion errors in (4.20) and (4.21) by varying λ and λ_0 in the above optical schemes, all other conditions being equal. Of course, such a choice of λ and λ_0 is possible provided the nature of noise in the measuring channels as well as the properties of the scattering medium are taken into account. Omitting a discussion of these questions, consider the problem on the determination of the absolute "scale" of the function s(r), since neither the full geometrical cross-section S nor the number density of particles can be estimated using the function $\varphi(r)$ (or $\tilde{\varphi}(r)$).

As has already been mentioned, $\xi_\pi(h)$ can be determined using a pulsed monostatic lidar. The values $T^{(a)}$ and $T^{(R)}$ can be neglected when only the upper troposphere and lower stratosphere are sensed, and therefore the lidar equation takes the simple form $P(h,\lambda_0)h^2/P_0B = \beta_\pi^{(R)}(h,\lambda_0) + \beta_\pi^{(a)}(h,\lambda_0)$. The profile $\beta_\pi^{(R)}(h,\lambda_0)$ can be calculated with errors of about 5 to 10% [4.12] and then $\xi_\pi(h,\lambda_0)$ can easily be determined. In some cases, altitude regions along the sensing path can be found for which $\beta_\pi^{(R)} \gg \beta_\pi^{(a)}$ which makes the estimation of ξ_π simple. There are no serious troubles with the application of (4.20) or with the estimation of $\tilde{\varphi}(r)$ using the lidar facilities.

Now, one can proceed to an estimation of the absolute values of the size-distribution functions s(r). The use of (4.20) enables one to determine the function $\tilde{\varphi}(r)$ which is related to s(r) according to $\tilde{\varphi}(r) = s(r)(S\psi)^{-1}$.

The constant ψ, from the inverse problems (4.20) and (4.21), can be determined from the condition $\int_R \tilde{\varphi}\, dr = \psi^{-1}$. The determination of the profile S(h) requires information about some of the optical characteristics. Let us consider methods for determining S(h) in more detail.

Let the wavelength used in a bistatic lidar be denoted as λ_{bist}. Since ψ is a function of λ, we introduce the value ψ_{bist}, assuming that $\psi_{bist} = \psi(\lambda_{bist})$ (this remark applies also to the value of ξ_π). The following relationship is extremely useful in practice

$$\bar{K}(\lambda) = \psi_{bist} \int_R K(r,\lambda)\tilde{\varphi}(r)dr \quad . \qquad (4.22)$$

This expression is valid for any of the factors from the Mie theory. If the polydispersed factors are known, then, in principle, the lidar equation for a monostatic scheme can be solved with respect to S(h) for any λ. The optical characteristics of Rayleigh scattering, which enter into this equation, can also be expressed in terms of S(h) according to the obvious relationship $\beta_{sc}^{(R)}(\lambda) = S(h)\psi(h,\lambda) = S(h)\psi_{bist}(h) \cdot (\lambda/\lambda_{bist})^4$. The determination of the profile S(h) completes the solution. This inverse problem of lidar sensing, comprises the determination of the altitude behavior of the aerosol microstructure, i.e., s(r,h), and the separation of the components of light scattering, i.e., $\beta_\pi^{(R)}(h,\lambda)$ and $\beta_\pi^{(a)}(h,\lambda)$, or the pair $\beta_{sc}^{(R)}(h,\lambda)$ and $\beta_{sc}^{(a)}(h,\lambda)$. We have thereby illustrated that lidar sensing, as a method for investigating the scattering media,allows one to construct consistent optical experiments which provide full measuring information for solving the complex inverse problems of atmospheric optics. An important applied problem is the separation of contributions of the aerosol and Rayleigh light scattering and we now consider the simplest variant of its solution, using lidar techniques. For this,let us recur to the reference scheme (4.17) and write the integral equation $\tilde{Q}\tilde{s} = q_i$ for the function $\tilde{s} = s/\beta_{sc}^{(R)}$. Determining the function $\tilde{s}$ from this equation, one can estimate such ratios of optical characteristics as $\beta_\pi^{(a)}/\beta_{ext}^{(a)}$; $\beta_{ext}^{(a)}(\lambda_1)/\beta_{ext}^{(a)}(\lambda_2)$ and so on. Then, taking into account the relation $\int_R K_{ext}(r,\lambda)\tilde{s}(r)dr = \beta_{ext}^{(a)}/\beta_{sc}^{(R)}$, and using a monostatic lidar, one can find from the corresponding lidar equation the profile $\beta_{sc}^{(R)}(h)$. This enables one to make an absolute calibration of the aerosol size-distributions s(r), on the one hand, and to infer the profiles of the aerosol optical characteristics, on the other. This method is advisable in those cases in which the main task of the study is to separate the components of light scattering rather than to investigate the microstructures of atmospheric aerosols. As has already been noted, the scheme (4.16) can provide reliable information on the size spectrum of aerosol particles. It is expedient in the practical use of (4.16) to apply simple inversion methods which provide an estimation of individual parameters of the microstructure. The method of model assessments is particularly useful (Sect.2.6).

Completing the presentation of the relation of the theory of bistatic lidar sensing to the "ill-posed" inverse problems of light scattering by polydispersed ensembles of particles, consider the accuracy of the determination of such functionals as $(K_\pi\tilde{\varphi})(\lambda)$ and $(K_{ext}\tilde{\varphi})(\lambda)$ defined by the set of regularized inverse solutions. The ratios of these functionals are used for solving the lidar equations for monostatic sensing schemes. It is important, for this reason, to know the errors which are inserted into these equations from the data of bistatic schemes. A numerical analysis has been carried out based on the results of the inversion of the component $c_2(\theta)$ of the vector $\mathbf{c}^{(s)}$ using (4.17). If the values $\bar{m}$ and $\varkappa$ are known exactly, an error $\varepsilon_q \leq 0.1$ yields errors in $\bar{K}_\pi(\lambda)$ and $\bar{K}_{ext}(\lambda)$ which are less than 0.1 of their maximum values. The presence of variations $\Delta m = \pm 0.05$ ($\bar{m}_0 = 1.55$) gives rise to errors in the above polydispersed factors of 16 and 10%, respectively. Variations of the imaginary part $\varkappa$ within the limits ±0.005 lead to a change in $\bar{K}_\pi(\lambda)$ that does not exceed 0.1 of its maximum value while the factor $\bar{K}_{ext}(\lambda)$ is practically unaffected. The above estimates correspond to variations of α within the limits of one order of the value of α_q. Figure 4.16 illustrates the dependence of the errors in the functions $\bar{K}_{ext}(\lambda)$ and $\bar{K}_\pi(\lambda)$ reconstructed, using (4.17). In addition to the error values, their signs are also given. Curve 1 represents the errors in $\bar{K}_{ext}(\lambda)$ and Curve 2 represents those in $\bar{K}_\pi(\lambda)$. It is seen that the second curve has a larger slope, which is caused by the fact that the backscattering efficiency factor is more sensitive to variations in the optical constants of the aerosol substance. Such initial uncertainties can be considered acceptable for the solutions of lidar equations in the case of monostatic schemes, if the errors which the lidar systems can provide are not lower than 10%. Recall that the above assessments of the efficiency factors $\bar{K}_\pi(\lambda)$ and $\bar{K}_{ext}(\lambda)$ are meant to be obtained from the data of bistatic lidar sensing. At the same time, the calculations show that the accuracy, with which the ratio of these parameters $b = \bar{K}_\pi/\bar{K}_{ext}$ (it is usually called the lidar ratio) can be determined, is, as a rule, higher. Thus, for example, the scheme (4.16) provides an estimation of b which is 10 to 20% accurate, re-

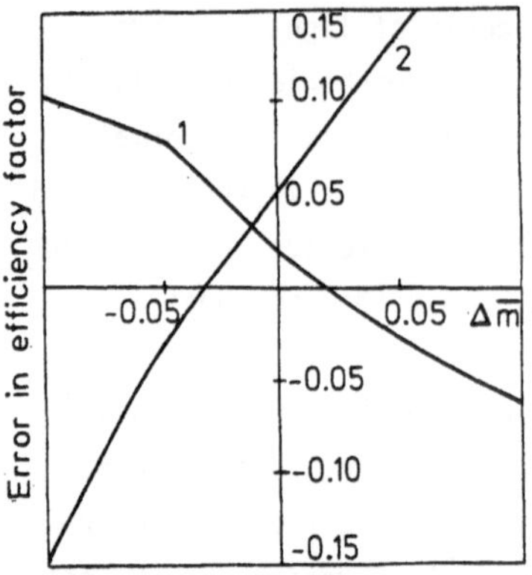

Fig. 4.16. Errors in the determination of polydispersed efficiency factors $\bar{K}_{ext}$ (Curve 1) and $\bar{K}_{\pi}$ (Curve 2) from polarization lidar measurements as functions of uncertainties in the refractive index, $\Delta\bar{m}$

gardless of the fact that the errors in $\bar{K}_{\pi}(\lambda)$ and $\bar{K}_{ext}(\lambda)$ are about 30 to 60%. The explanation of this fact is clearly due to a relatively weak dependence of b on the variations of size distribution s(r). We have just characterized, in general, the information potentialities of the analytical schemes discussed. The use of any of these schemes for the arrangement of optical experiments in the atmosphere should be determined by the parameters of measuring devices and by the total amount of optical information obtained in the course of a given experiment.

The above investigations can be regarded as a theoretical basis for sensing the scattering atmosphere using the bistatic scheme as well as of the combined scheme based on simultaneous use of monostatic and bistatic schemes. At present, these methods are not widely employed because of technical difficulties. Up to now, only the work in one example is known [4.12]. At the same time it is expected that improving the state-of-the art of lidar facilities will permit the combined scheme to take its due place among the methods for optical studies of the atmosphere.

4.6 Qualitative Methods for Interpreting the Angular Characteristics of Light Scattering

4.6.1 Use of the Logarithmic Derivative Method to Interpret the Aureole Part of the Scattering Phase Function

In addition to the above rigorous methods for inverting the angular characteristics of light scattering, some qualitative approaches can also be of certain interest in the interpretation of these angular characteristics. These methods are based on certain properties of these characteristics. First of all, the region of small scattering angles is very interesting from this point of view, since the properties of scattering phase functions for $\theta \to 0$ are well studied. As has been shown, the logarithmic-derivative method, suggested earlier for interpreting the spectral measurements, can be generalized for estimating the parameters of the aerosol microstructure from the measurements of the aureole part of the scattering phase function. Let us consider briefly such methods for the interpretation of measurements in the aureole region.

It is known that the kernel of the scattering phase matrix (4.8) for spherical particles has, in the region of small θ, the simple analytical form (Airy formula)

$$Q(r,\theta) = (x^2/4\pi)[2\, J_1(U)/U]^2 \quad ,$$

where $U = \theta x$, and $J_1(u)$ is the Bessel function of order 1. Formally this relation is valid for $\theta \to 0$ and $\theta \to \infty$. Since $Q(r,\theta)$ decreases rapidly with decreasing r when $\theta \to 0$, light scattering by polydispersed ensembles of particles is due mainly to scattering by large particles whose sizes are near the upper limit R_2 of the size spectrum. If $x_2 = 2\pi R_2/\lambda$ is large enough (e.g., $x_2 \geq 30$ [4.32]) then the above approximation of the kernel $Q(r,\theta)$ for small values of θ is quite satisfactory. In the following presentation, the region of r and θ values, at which the above approximation of the scattering phase matrix is acceptable, at least within the errors of experimental measurements, will be denoted as Ω. The region Ω is characterized by small values of θ and large values of x. In order to make use of this approximation in practice (i.e., to satisfy the second condition that x to be large enough) one should make a proper choice of the wavelength used for optical sensing.

If the size interval R of a polydispersed system under investigation belongs to Ω, then it can be shown that the following relation is valid

$$\frac{\partial Q(r,\theta)}{\partial\theta} = -2Q(r,\theta)f(u)/\theta \quad , \tag{4.23}$$

where $f(u) = (u/2)^2 - (u/2)^4/3 - 5(u/2)^6/12$ for $u \to 0$. The derivation of (4.23) is based on the analytical properties of the power series expansion of the Bessel function for small values of its argument. This formula can be used as a basis for studying the angular behavior of the characteristics of light scattering by polydispersed systems in the region of small scattering angles. In particular, the following three asymptotic relations can be obtained using (4.23)

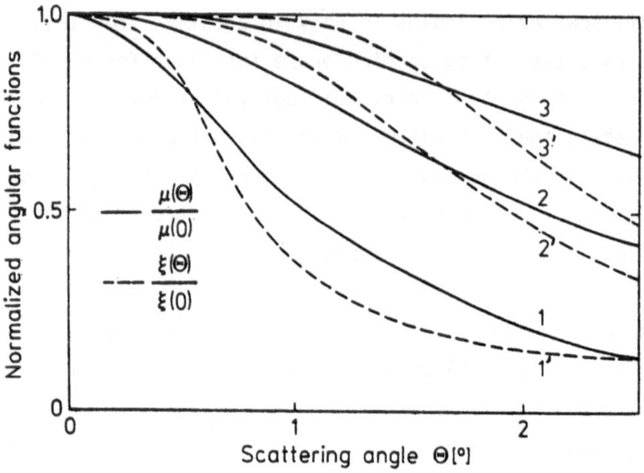

Fig.4.17. Polydispersed scattering phase functions (solid curves) and their logarithmic derivatives (dashed curves) in the range of small scattering angles for three wavelengths $\lambda = 0.345$; 0.69 and 1.06μm (Curves 1,2,3, respectively)

$$\lim_{\theta\to 0} \mu(\theta) = S(2\pi/\lambda)^2\overline{r^2}/4\pi \qquad (4.24)$$

$$\lim_{\theta\to 0} \mu'(\theta) = 0 \qquad (4.25)$$

$$\lim_{\theta\to 0} (-d\ \ln\mu/\theta\ d\theta) = (2\pi/\lambda)^2\overline{r^4}/2\overline{r^2} \qquad (4.26)$$

where $\overline{r^2}$ and $\overline{r^4}$ are the second and the fourth moments of the normalized size-distribution function $\varphi(r) = S^{-1}s(r)$. No assumptions of the size distribution $s(r)$ were used in the derivation of these relationships except the summability of the polydispersed integrals for $R_2 \to \infty$. The right-hand sides of (4.24,26) are determined by the parameters of the size-distribution functions $s(r)$. Note that the ratio $\overline{r^4}/\overline{r^2}$ can be regarded as a measure of "elongation" of the size-distribution function in the region of large particles. This value is of practical interest for estimating the upper boundary R_2 of the size spectrum. Taking this into account, let us consider techniques used for the interpretation of aureole scattering phase functions.

Let us introduce the function $\xi(\theta) = \mu'/\theta\mu$. It can easily be shown that for $\theta \to 0$, $\xi(\theta)$ approaches $\xi(0)$ uniformly from below and, as a consequence, for any finite but sufficiently small θ the function $\xi(\theta)$ gives a lower bound for $\overline{r^4}/\overline{r^2}$. The curves $\mu(\theta)$ and $\xi(\theta)$ for the wavelengths 0.345; 0.69 and 1.06μm are shown in Fig.4.17 (corresponding curves are numbered in the same order). The values of the scattering phase function $\mu(\Theta)$ have been calculated for the experimentally measured size-distribution function of aerosol particles of the ground layer. The curves show that the efficiency of the evaluation of $\overline{r^4}/\overline{r^2}$ depends on λ and it is greater for longer wavelengths. It is important to emphasize that (4.26) allows the evaluation of the parameter $\overline{r^4}/\overline{r^2}$ of the size distribution sought using s(r) independent of the index of refraction of the aerosol substance. This makes optical measurements in the aureole region advantageous.

4.6.2 The Use of Model Size-Distributions in the Method of Logarithmic Derivative

In practice, the parameter $\overline{r^4}/\overline{r^2}$ can be written in terms of other characteristics of the size-distributions sought and, in particular, in terms of those that are generally used in the optics of dispersed media. Thus, for the models (2.75,76) one has

$$\overline{r^4}/\overline{r^2} = \overline{r^2}\Gamma\left(\frac{\alpha + 7}{\gamma}\right)\Gamma^2\left(\frac{\alpha + 3}{\gamma}\right)\left[\Gamma\left(\frac{\alpha + 5}{\gamma}\right)\Gamma^2\left(\frac{\alpha + 4}{\gamma}\right)\right]^{-1} \qquad (4.27)$$

$$\overline{r^4}/\overline{r^2} = r_s^2 \exp(4/\delta^2) \quad , \qquad (4.28)$$

respectively. Substituting (4.28) into (4.26) one obtains the following expression for estimating the parameter δ^2 of the log-normal size-distribution of particles

$$\delta^2 = 4[\ln 2\xi(0) - 2 \ln x_s]^{-1} \quad , \qquad (4.29)$$

where $x_s = 2\pi r_s\lambda^{-1}$. The estimation of the mean radius $\bar{r}$ of the gamma size distribution can be obtained analogously once the values of α and γ are fixed. However, the estimation of R_2 from aureole scattering phase functions is helpful. This value is necessary, on the one hand, for solving the integral equations of optical sensing of polydispersed media and, on the other hand, its value is of independent interest, since any transformation of the microstructure of real atmospheric sols changes the value of R_2, especially when localized volumes of the atmosphere are studied. The changes in the particle size spectrum can be caused by the humidity of the air, or by the transfer of air masses. It is helpful to estimate R_2 using models of the size spectrum which are determined by the parameter R_2. The simplest example of such distributions are the power-law size-distribution functions $n(r) = ar^{-\nu}$ $(R_1 \le r \le R_2)$. Assuming that $R_1 \ll R_2$ and $\nu < 5$, one can easily show that

$$\overline{r^4}/\overline{r^2} \simeq R_2^2(5 - \nu)/(7 - \nu) \quad . \qquad (4.30)$$

Equation (4.30) allows one to estimate R_2 using $\xi(\theta)$. This can be done using the relation

$$x_2 = [2\xi(0)(7 - \nu)/(5 - \nu)]^{1/2} \quad , \qquad (4.31)$$

where $x_2 = 2\pi R_2 \lambda^{-1}$. The parameter ν of the size distribution should be set a priori or evaluated from optical measurements (Sect.3.1). These calculational methods provide an approximate determination of the microstructure parameters of aerosols using the aureole part of the scattering phase functions. For more details the reader is referred to [4.33-35].

4.6.3 Inversion of Experimentally Measured Scattering Phase Functions

Let us consider some examples of interpretation of experimental data in order to illustrate the above. These examples are based on data on the aureole part of the scattering phase functions of atmospheric haze, measured in the angular range from 40' to 3° at the wavelength $\lambda = 0.6328\mu m$ [4.36]. These data are shown in Fig.4.18. The values of the scattering phase functions are normalized to the value $\mu(5^{\circ})$. The values of $\xi(40')$, R_2, $\bar{r}$ and the meteorological parameters which occurred during the experimental studies are given in Table 4.4.

If one assumes that the variation of the scattering phase functions are caused by the influence of air humidity on the microstructures of the atmospheric haze, then the value $\xi(40')$ can be regarded as a quantitative measure of this influence. It is seen from Table 4.4 that this effect is strongest for the relative humidity larger than 50%. This fact is known quite well from purely qualitative assessments. In this example,a quantitative assessment of the influence of humidity on large aerosol particles is given.

Estimates of the upper bound R_2 of the aerosol size spectrum and $\bar{r}$ are also given in Table 4.4. These estimates have been obtained based on the use of models (the other parameters of the model are chosen according to [4.1]). The dependence of these parameters on air humidity is shown in Fig.4.19. This figure exhibits that the variations of the microstructure parameters are not, on the whole, very significant when the humidity varies within the given interval. It is evident for these reasons that an attempt made in the course of this experiment to estimate the effect of air humidity on the spectral behavior of the optical characteristics of aerosol was not successful.

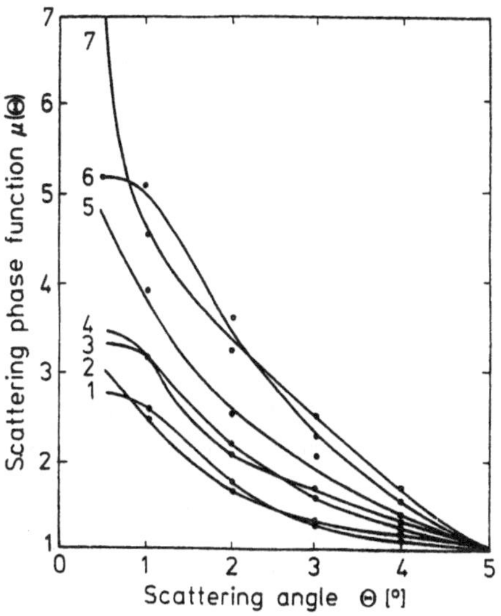

Fig. 4.18. Scattering phase functions of atmospheric hazes for scattering angles 40' to 5° [4.36] (see also Table 4.4)

Table 4.4. Results of the interpretation of the aureole scattering phase functions shown in Fig.4.18

No of curve in Fig.4.18	Temperature	Relative humidity	$\xi(40')$	R_2 [m] (4.20)	$\bar{r}^*$ (4.31)
I	17	29	1260	7.7	1.8
2	18	34	1230	7.6	1.8
3	18	35	1300	7.8	1.9
4	18	42	1340	7.9	1.9
5	17	51	1380	8.0	1.9
6	20	61	1510	8.4	2.0
7	20	86	1860	9.3	2.2

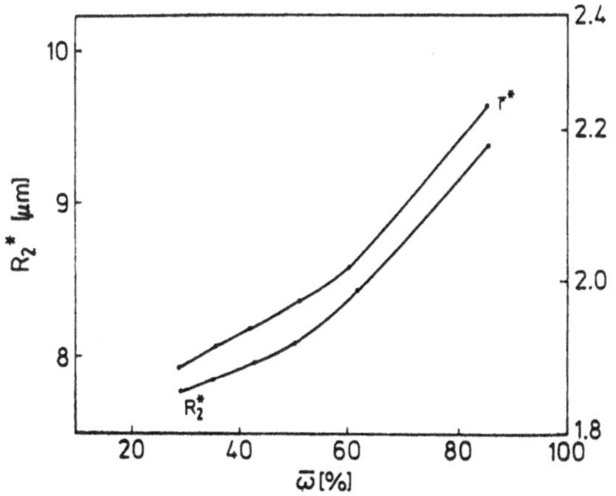

Fig. 4.19. Estimations of the mean $\bar{r}^*$ and the maximum R_2^* radii using data given in Fig.4.18

The example presented here shows that the interpretation of data on the aureole part of the scattering phase functions is of interest, primarily when transformations of the aerosol size spectra are to be investigated. Recall that the estimations of $\bar{r}$ presented in Table 4.4 concern the fraction of large particles in the atmospheric haze and are in no way related to the size region $\bar{r} \lesssim 1$ m due to the screening effect [4.37].

Some remarks concerning the calculations of $\xi(\theta)$ using experimental data should be made to complete the discussion of the methods for interpreting the function $\mu(\theta)$ for $\theta \to 0$. The determination of $\xi(\theta)$ requires the calculation of the function $\mu'(\theta)$ based on the measured values of $\mu(\theta)$. As is known, this problem is "ill-posed" and therefore it requires the development of algorithms stable with respect to measurement errors. Using the results of investigations on the numerical differentiation of the empirical functions, regularizing algorithms (packages of computer programs) for the operative processing of the nephelometric measurements by the method of logarithmic derivative have been developed.

A second peculiarity of the interpretation method suggested here is in the fact that the value $\mu(0)$ cannot be measured experimentally. Therefore, $\mu'(\theta)$ as well as $\mu(0)$ can be estimated only approximately. For this reason, the estimation $(\overline{r^4}/\overline{r^2}) > \xi(\theta) \cdot 2(2\pi\lambda^{-1})^{-2}$ is usually used in calculations with the value of θ being the smallest scattering angle available. It can easily be shown that the estimates $\bar{r}^*$ and $\bar{R}_2^*$ obtained in such a way, obey the conditions $\bar{r}^* < \bar{r}$ and $\bar{R}_2^* < R_2$, i.e., they underestimate the aerosol parameters.

Another method for estimating $(\overline{r^4}/\overline{r^2})$ equivalently $\bar{r}$ and R_2 can be based on the estimation of $\xi(0)$ by extrapolating the curve $\xi(\theta)$ from $\theta = \theta_{min}$ to $\theta = 0$. A quadratic formula can be used for this purpose, since

$$\xi(\theta) = \xi(0) - (2\pi/\lambda)^4\theta^2\overline{r^6}/2\pi\overline{r^2} \quad . \qquad (\theta \to 0)$$

Such a technique allows one to increase the efficiency of the estimations especially in cases for which θ_{min} is not small enough for a given aureole scattering phase function, i.e., when the measurements of $\mu(\theta)$ are terminated before the interval of sharp increase (Curve 7 in Fig.4.18).

One should emphasize that the extrapolation of the scattering phase function and its derivative into the region $\theta \cong 0$ is also of interest for studying the optical characteristics of light scattering by aerosols. It is important, in some optical problems, to estimate the coefficient of "pure" scattering of light fluxes by dispersed media using data from photometric measurements [4.38]. The method of logarithmic derivative enables one to solve such problems.

4.6.4 Estimation of the Boundaries of the Aerosol Size-Distributions Using the Logarithmic-Derivative Method

In addition to the interpretation of the aureole part of the scattering phase functions considered above, some other analytically more sophisticated methods can be used. These methods are based not only on the asymptotic property (4.26) but also utilize the analytical behavior of $\mu(\theta)$ when $\theta \to 0$. If $r, \theta \in \Omega$, then

$$(d\ \ln\mu/d\ \ln\theta) = -3 - \mu^{-1} \int_R Q(r,\lambda)s(r)(d\ \ln s/d\ \ln r)dr \qquad (4.32)$$

for $s(R_1) = s(R_2) = 0$. By analogy with (3.23), (4.32) can be regarded as a formula for differentiation of polydispersed integrals like (4.8) in the aureole region. Equation (4.32) can be derived in the same way as (3.23) for the case of spectral optical characteristics. It is based on the fact that the function $F(U) = (2J_1(U)/U)^2$, which is used to approximate the kernel $Q(r,\theta)$ in the region Ω, depends on the product of the variables r and $\theta(U = x\theta = 2\pi r\lambda^{-1}\theta)$. As a result, the derivatives F'_θ and F'_x are related by a definite functional relationship, viz., $xF'_x = \theta F'_\theta$. The further derivation of (4.32) is analogous to that of (3.23). The use of (4.32) for interpreting the aureole parts of the scattering phase functions requires, as in the case of (3.23), prior estimations of the logarithmic derivatives of the optical characteristics. It is possible, in this connection, to discuss the use of the logarithmic-derivative method developed in Chap.3 to interpret the angular characteristics of light scattering. Of course, the applicability of (4.32) is much more restricted than is the use for (3.23) and is limited to the vicinity of $\theta \simeq 0$.

Equation (4.32) gives an interesting inequality for the logarithmic derivative in the aureole region:

$$2\pi\lambda^{-1}(\overline{r^2}/2)^{1/2}\theta < |d\ \ln\mu/d\ \ln\theta|^{1/2} < 2\pi R_2\theta/\lambda\sqrt{6} \quad . \qquad (4.33)$$

The validity of (4.33) is proved using the Chebyshev inequalities for the definite integrals (Sect.3.1). The last inequality which we present here concerns the determination of the conditions under which θ can be regarded as being in the region Ω. It can be shown that for any $\theta \in \Omega$

$$|d\ \ln\mu/d\ \ln\theta| \leq 5/6 \quad . \qquad (4.34)$$

This condition is sufficient but not necessary. Note that the maximum angle defined by (4.34) depends on $x_2 = 2\pi R_2\lambda^{-1}$ and its value increases as x_2 decreases (Fig.4.17). Inequality (4.34) makes it possible to estimate the boundary of the angular region in the vicinity of $\theta = 0$ in which the Airy formula is applicable to the scattering phase functions under study. This region is characterized by sufficiently small values of the derivative μ'. Curves 1,3,4 and 6 in Fig.4.18 obey the condition (4.34). The estimation of $\overline{r^4}/\overline{r^2}$ from the Curves 2,5 and 7 is very approximate in this respect. It is evident that these methods are applicable to the interpretation of any elements of the polydispersed scattering phase matrix in the region of small scattering angles, provided the particles are nearly spherical. These methods can be used in connection with nephelometric measurements in the atmosphere. At the same time, the use of bistatic lidars requires the development of analogous methods for analyzing the angular behavior of light scattering characteristics in the region of Θ close to 180°. Unfortunately, nothing definite can be said on this question even for spherical particles. It is obvious, however, that the methods for interpreting the scattering characteristics in this region should take into account the possible deviations of the shapes of the aerosol particles from a sphere because it is for the region of $\Theta = 180^{\circ}$ that the particle shapes affect the angular behavior of the elements of polydispersed scattering phase matrices most strongly.

5. Light Scattering by Aerosols and Lidar Sensing of the Atmosphere

Preceding chapters have dealt with light scattering by aerosols for remote sensing of atmospheric aerosol formations. The main subject of the following discussion is remote determination of atmospheric parameters using lidar techniques that exploit light scattering by aerosols. Aside from aerosol characteristics, these parameters are humidity, temperature, vector of wind velocity, and atmospheric turbulence, which play an important role in radiation transfer and weather processes. As previously, we shall concern ourselves mainly with the numerical processing of lidar returns for extracting information on the respective parameters.

The lidar returns in schemes for monitoring the gaseous composition of the atmosphere also contain information on the aerosol involved. The inversion techniques for solving the system of lidar equations which are discussed below take this circumstance into account. Such an approach allows one to obtain more information from lidar measurements.

5.1 Lidar Sensing of Water Vapor and Other Atmospheric Gases Using Differential Absorption and Scattering (DAS)

5.1.1 General Theory of the Method

The development of methods and lidar facilities for the remote determination of the gaseous composition of the atmosphere is of great practical importance. This is due, on the one hand, to the needs of applied meteorology and, on the other hand, to the necessity for operative monitoring of air pollution. For the sake of simplicity, the discussion here will deal mainly with the determination of air humidity using lidars. Of course, the theory presented for the interpretation of lidar signals is applicable to other atmospheric gases, too.

The basic functional relations in the DAS method are

$$P(z,\lambda_{off}) = P_0 B z^{-2}\gamma(z)\beta_\pi(z,\lambda_{off})\exp\left[-2\int_{z_1}^{z_2}\beta_{ext}(z',\lambda_{off})dZ'\right] \tag{5.1}$$

$$P(z,\lambda_{on}) = P_0 B z^{-2}\gamma(z)\beta_\pi(z,\lambda_{on})\exp\left\{-2\int_{z_1}^{z_2}[\beta_{ext}(z',\lambda_{on})+k(z',\lambda_{on})\rho(z')]\right\}dz' \tag{5.2}$$

where $\gamma(z)$ is the lidar geometrical factor, $k(z,\lambda_{on})$ is the water-vapor absorption coefficient at the center of an absorption line and $\rho(z)$ is the profile of the water vapor density along the sensing path.

Equation (5.1) describes the lidar return at the wavelength λ_{off} located in a spectral region off the absorption line.

Since the absorption lines of atmospheric gases are very narrow,the wavelengths λ_{off} and λ_{on} can be chosen close to each other. As a consequence, the volume backscattering coefficients $\beta_\pi(z,\lambda_{on})$ and $\beta_\pi(z,\lambda_{off})$ as well as the extinction coefficients $\beta_{ext}(z,\lambda_{on})$ and $\beta_{ext}(z,\lambda_{off})$ are practically equal to each other. In this case,the simplest approach to the determination of $\rho(z)$ from the system (5.1,2) is to divide the amplitude of one signal by that of the other and then to differentiate this ratio numerically with respect to the spatial coordinate. Such a procedure gives a well-known (in the DAS method) expression for the profile $\rho(z)$

$$\rho(z) = [2k(z,\lambda_{on})]^{-1} d \ln[P(z,\lambda_{on})/P(z,\lambda_{off})]/dz \quad . \tag{5.3}$$

Since this equation determines the calculational scheme for handling the lidar data and is, to a great extent, responsible for the reliability of the results obtained,it should be investigated in more detail taking into account the peculiarities of optical experiments in the atmosphere.

First of all, it should be noted that (5.3) is valid only if the signals $P(z,\lambda_{on})$ and $P(z,\lambda_{off})$ are measured simultaneously along the sensing path. A time delay between measurements of these profiles which can occur, for example, due to the availability of a one-channel signal recorder can result in large systematic inaccuracies.

It is known that atmospheric aerosols are characterized by strong variations in time; this is revealed, for example, in the temporal behavior of $\beta_\pi(z,\lambda) = [\beta_\pi^{(a)}(z,\lambda) + \beta_\pi^{(R)}(z,\lambda)]$.

Fluctuations of local aerosol characteristics in time are due mainly to aerosol transfer in the atmosphere and are random in nature. Taking this into account,one can state that the ratio of two signals $P(z,\lambda_{on})$ and $P(z,\lambda_{off})$ in (5.3) includes statistical fluctuations if these signals are not recorded simultaneously. It is evident that the amplitude of such fluctuations does not depend on the experimental errors and is entirely determined by the optical stability of the atmosphere. Further differentiation of this ratio according to (5.3) is not justified due to the existence of such fluctuations. This statement follows from the fact that realizations of a random process, even if it is a stationary one, are not differentiable functions. Secondly, the irregular behavior of the ratio is due to an essentially nonuniform distribution of aerosols along any direction in the atmosphere (aerosol stratification). This is confirmed by numerous lidar observations of the atmosphere. Therefore, the profiles $\beta_\pi^{(a)}(z,\lambda)$ and, hence, the lidar returns $P(z,\lambda)$ are, in principle, nondifferentiable functions or, in the most favorable case, they are differentiable but not at every point of the sensing path.

Thus, the direct interpretation of DAS-lidar data on gaseous atmospheric components according to (5.3) is not valid due to the temporal variations and the stratified spatial distribution of aerosols in the atmosphere. Neglecting this fact results in an unjustified reduction of the sensing range and, as a consequence, inefficient use of the lidar facility. This makes clear the necessity for constructing inversion schemes which take into account the existence of irregular (nondifferentiable) components in the reference lidar returns. Since, physically, it is clear that the profile $\rho(z)$ in (5.2) is a continuous function which behaves fairly smoothly as a function of altitude, it is natural to use regularizing algorithms for the numerical solution of the functional equations in the above inverse problem. Such algorithms allow one to obtain sufficiently smooth inverse solutions suppressing the oscillating components regardless of their nature. A great variety of ways to construct such schemes exists. In the following, the technique of stable differentiation of empirical functions will be considered especially as applied to the ratio $P(z,\lambda_{on})/P(z,\lambda_{off})$ in (5.3).

A suitable regularizing algorithm based on the smoothing functional may be constructed by reducing the differentiation problem to the solution of a Fredholm integral equation of the first kind [5.1]. This equation is based on the integral relation between a function $f(z)$ and its derivative $\varphi(z)$

$$f(z) = f(z_0) + \int_{z_0}^{z} \varphi(y)dy \quad . \tag{5.4}$$

Using (5.4) one can easily write the following integral equation for $\varphi(z)$:

$$\int_{z_0}^{z_n} K(z,y)\varphi(z)dz - F(y) \quad , \quad z,y \in [z_0,z_n] \quad . \tag{5.5}$$

The kernel of this equation is

$$K(z,y) = \begin{cases} z_n - z & \text{if} \quad z_0 \le y \le z_1 \\ z_n - y & \text{if} \quad z \le y \le z_n \end{cases}$$

and $F(y) = \int_y^{z_n} f(z)dz - f(z_0)(z_n - y)$. Equation (5.5) can be solved by any of the numerical methods presented in Chap.2. In numerical simulations of sensing the humidity field, the scheme (5.6) will be used. We shall write it in the form

$$\varphi_\alpha(z) = (K^*K + \alpha D_2)^{-1}K^*F(y)$$

$$\varphi_\alpha(z) = \{d[\ln \tilde{P}(z,\lambda_{on})/\tilde{P}(z,\lambda_{off})]/dz\}_\alpha \tag{5.6}$$

$$F(y) = \int_y^{z_n} [\ln(\tilde{P}(z,\lambda_{on})/\tilde{P}(z,\lambda_{off})]dz - \ln[\tilde{P}(z,\lambda_{on})/\tilde{P}(z,\lambda_{off})]_{z=z_0}(z_n - y) \quad ,$$

where $\tilde{P}(\tilde{z},\lambda_{on})$ and $\tilde{P}(z,\lambda_{off})$ are the experimentally measured lidar returns.

5.1.2 Numerical Simulation and Estimations of the Method's Precision

We now consider some numerical experiments to illustrate the efficiency of the above regularizing algorithm. It was assumed that one sensing wavelength λ_{on} coincided with the water vapor absorption line λ = 694.38 nm, with the other being located in a transmission window at $\Delta\lambda$ = 0.05 ± 0.07 nm away. Figure 5.1 shows the absorption spectrum of H_2O molecules in this spectral region. Experimental studies were carried out using a tunable ruby laser and a spectrophone [5.2]. The halfwidth of the laser line was about 0.01 nm.

In order to find the profile $\rho(z)$ it is necessary, according to (5.3), to know the mass absorption coefficient k(z) as a function of altitude. Our subsequent calculations are based on the results given in [5.3]. Figure 5.2 shows water vapor absorption spectra for different heights. The calculations were made for a model of the summer atmosphere according to the Voigt formula (the designation k_V underlines this fact, see Fig.5.2). Curves 1 represent the "true" contours $k_V(\nu)$, while Curves 2 are the averaged ones. The interval of averaging was $\Delta\nu$ = 0.097 cm^{-1} (the assumed halfwidth of the Gaussian laser line). The value k(z = 0) is given in Fig.5.2 for the mean water vapor number density $\rho(z = 0)$ = 100 gm^{-3} and is about 0.54 ÷ 0.56 cm^2 g^{-1}. The effect of self-broadening was neglected in this model.

Figure 5.3 is an example of a water vapor profile reconstructed up to a height of 2 km. The reference profiles $P(z,\lambda_{on})$ and $P(z,\lambda_{off})$ used in these calculations have been constructed using the models of aerosol and molecular components of the atmosphere given in [5.4]. The profile $\rho_0(z)$ shown in Fig. 5.3a is taken as a reference since it is very close to a typical water vapor profile in the atmosphere. The profiles of expected errors $\varepsilon(z)$ in the signals measured by the device described below are shown in Fig.5.3a. At the beginning of the sensing path $\varepsilon(z)$ = 6 ÷ 8%, decreasing to 3 ÷ 5% and then increasing to 10 ÷ 12% at heights from 1.5 to 2.0 km. Considering the profile $\varepsilon(z)$ as the profile of random components of $\tilde{P}(z,\lambda_{on})$ and $\tilde{P}(z,\lambda_{off})$, one can compare the solutions $\rho(z)$ obtained with some mean profile $\tilde{\rho}_\alpha(z)$. For this purpose, a random number generator has been used in numerical simulations of the water vapor sensing. The variances (or more correctly,

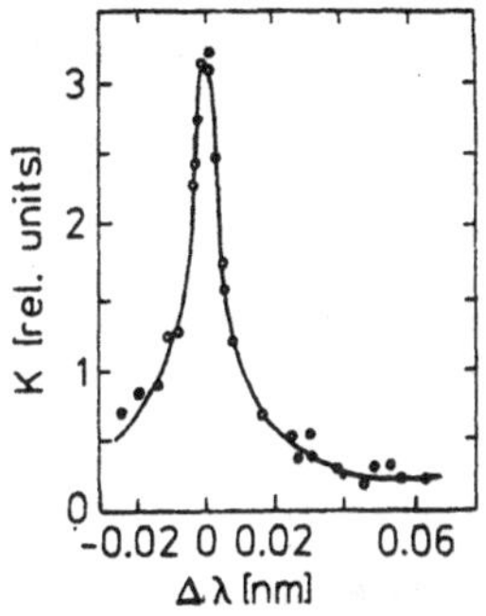

Fig.5.1. Spectral absorption coefficient of water vapor molecules in the vicinity of the 694.38 nm line [5.2]

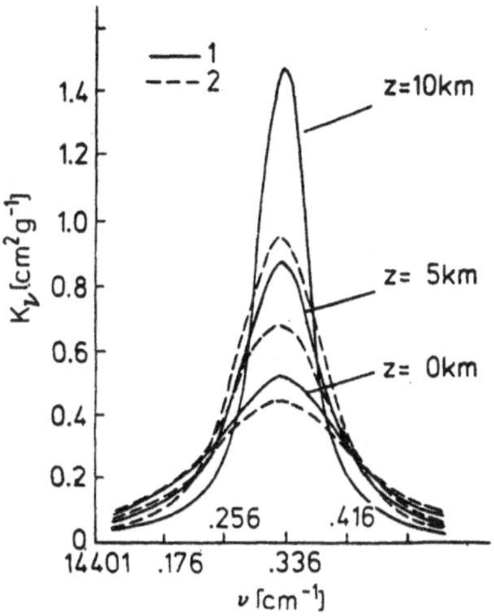

Fig.5.2 Spectral behavior of the water vapor mass absorption coefficient in the vicinity of 694.38 nm line for altitudes of [5.3]. The dashed curves represent the same data but averaged over spectral interval $\Delta\nu = 0.097\ cm^{-1}$

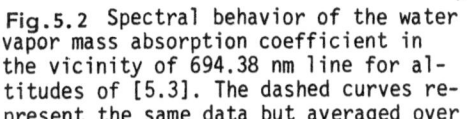

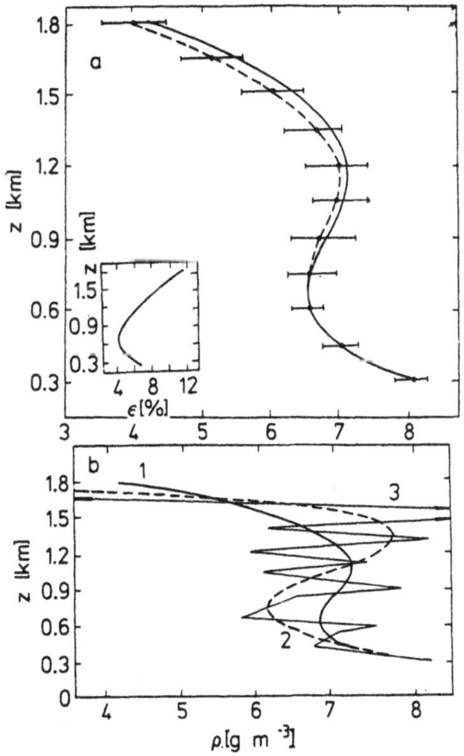

Fig.5.3 Results of numerical simulations on the restoration of the water vapor density ρ profiles. (a) The solid line is the reference profile; the dashed one represents the profile restored according to (5.6) (the altitude behavior of the error ε(z) is shown in the insert). (b) The same data, obtained using the least-squares method (Curve 2) and the method of finite differences (Curve 3); the solid line is the reference profile.

estimates thereof) of the solutions $\rho_\alpha(z)$ near $\tilde{\rho}_\alpha(z)$ are shown in Fig.5.3a. The maximum value of the variance does not exceed 20%. This corresponds to altitudes above 1.5 km, where the errors in the reference data are about 10 ÷ 12%. Preliminary numerical analysis has shown that the choice of the regularizing parameter α in the scheme (5.6) can be made according to the condition $\min_\alpha \| d\varphi_\alpha / d \ln \alpha \|$, i.e., the quasi-optimal value of α can be used. As is seen from Fig.5.3a, the deviations of the initial profile $\rho_0(z)$ from the mean one $\bar{\rho}_\alpha(z)$ do not exceed the corresponding variances along the sensing path. Therefore, the solution $\bar{\rho}_\alpha(z)$ may be considered an acceptable approximation to the solution $\rho_0(z)$ sought, at least if the errors in the measurements are taken into account.

Figure 5.3b shows examples of the interpretation of data using ordinary techniques for numerical differentiation, in order to illustrate the efficiency of (5.6) in processing the lidar returns that have random (nondifferentiable) components. The dashed curve in this figure shows the same profile $\rho_0(z)$ reconstructed using the least-squares method of differentiation. Corresponding values of the profile $\rho_0(z)$ have been obtained by the prior smoothing of the neighbouring readouts of P(z) using the quadratic approximation [5.5]. It is characteristic for this differentiation method that the largest errors in the reconstructed profile $\bar{\rho}_\alpha(z)$ correspond to the intervals in which $\rho'(z)$ reaches its maximum values. It is obvious that in just these z-regions the profile $\ln[P(z,\lambda_{on})/P(z,\lambda_{off})]$, as a function of z, deviates most strongly from the parabolic form. In this respect, (5.6) is more efficient.

The use of the method of finite differences to reconstruct the profile $\rho(z)$ from the data of this numerical experiment appears to be inconsistent. The error in the solution due to differentiating the irregular components of the data is so large that it becomes impossible to determine the profile $\rho(z)$. This is demonstrated by the broken line in the last figure. The use of the integral operator K introduced according to (5.5) is equivalent, in this respect, to the use of low frequency filters to suppress noise in the initial data.

Lidar sensing of the atmospheric water vapor using the DAS method at the ruby wavelength of 694,38 nm can be performed up to much greater heights than those mentioned in the previous example. However, lidar returns from such heights are, as a rule, very weak and can be detected using only photon counting techniques. The information obtained in such experiments is therefore inherently discrete. The length of a range increment cannot be chosen infinitesimal. Generally it is chosen a priori taking into account the

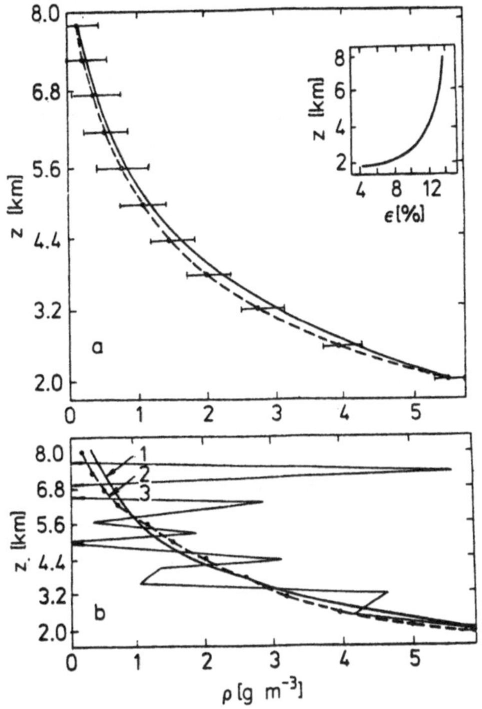

Fig.5.4 The results of numerical simulations analogous to those in Fig. 5.3, but made for altitudes up to 8 km and for the photon counting technique of the lidar return recording

characteristics of a recording electronics and the admissible time for signal storage. The discrete character of the experimental information requires, naturally, a different approach to the concept of and technique for numerical differentiation, if, as previously, the interpretation of the data is based on the scheme (5.3). These questions will be discussed below in more detail. Now we shall discuss the results of numerical experiments on the reconstruction of $\rho(z)$ according to (5.6) from the data obtained using photon counting techniques up to heights of about 8 km. The data are shown in Fig.5.4a. The altitude profile $\rho(z)$ of the relative errors in the optical measurements reflects the precision characteristics of a real lidar. The dashed curve in Fig. 5.4a is the mean profile $\bar{\rho}_\alpha(z)$. The deviations of $\bar{\rho}_\alpha(z)$ from $\rho_0(z)$ (solid curve) are in absolute value within the limits of admissible variance. Nevertheless, a significant increase of the relative error of the reconstruction of $\rho(z)$ with increasing altitude as compared with that in $\varepsilon(z)$ should be pointed out. The explanation of this fact is quite obvious. The density $\rho(z)$ decreases as the height increases, and hence the value $\exp\left[-2\int_{z_k}^{z_{k+1}}\rho(z')k(z')dz'\right]$ becomes very close to unity. This, in turn, results in poorer conditioning of the system of reference functional equations (5.1,2) and, as a consequence, the uncertainty in the inverse results increases. This is clearly illustrated by Fig.5.4a. Two contradictory requirements for the experimental device should be mentioned here. On the one hand, the possibly weak H_2O absorption lines must be used in lidar in order to achieve higher sensing altitudes, while, on the other hand, as is seen from the above, the initial functional equations are less well conditioned and, as a consequence, the data interpretation becomes more difficult. All this should be taken into account when constructing lidars for sensing the gaseous components of the upper atmosphere using DAS-lidar techniques.

The data presented in Fig.5.4b are equivalent to those in Fig.5.3b. The least-squares method allows one to reconstruct the profile $\rho(z)$ quite satisfactorily. This is caused by the fact that in the case considered here, the humidity profile is a very smooth function of altitude. Evidently the humidity profiles of the upper atmosphere can be inverted from optical information using simpler calculational schemes than (5.6). However, it is preferable to use more general and rigorous methods for signal processing in studies of the ground atmospheric layer since its physical parameters are inherently inhomogeneous time-dependent.

5.2 Experimental Studies on Sensing Atmospheric Humidity Profiles Using Differential Absorption

5.2.1 Lidar Parameters and Measuring Technique

At the Institute of Atmospheric Optics, SB USSR Academy of Sciences, Tomsk, a DAS lidar for sensing the humidity profiles of the atmosphere has been constructed. The lidar operates at wavelengths in the vicinity of the water vapor absorption line λ_{H_2O} = 694.38 nm. A detailed description of the lidar functional block-diagram as well as of its individual blocks has been presented in [5.2].

The basic parameters of this lidar are the following:

Range of wavelength tuning:	694.2 ÷ 694.5 nm
Energy per pulse:	0.1 J
Halfwidth of the radiation line	10^{-3} nm
Pulse duration	20 ns
Beam width	0.2 mrad
Diameter of the receiving mirror	0.3 m
Halfwidth of the interference filter	1 nm
Reproducibility of the laser radiation wavelength	better than $5\cdot10^{-4}$nm

The light source used in the lidar is a tunable ruby laser with a resonance reflector and a double-pass amplifier, constructed at the Institute of Atmospheric Optics [5.2]. The wavelength tuning was performed by changing the temperature of the resonance reflector. The temperature dependence of the wavelength shift is shown in Fig.5.5 and is nearly linear. The mean slope of the curve is about $6.0 \cdot 10^{-3}$ nm/deg.

It has already been mentioned that the profiles $P(z,\lambda_{on})$ and $P(z,\lambda_{off})$ should be measured within as short a time interval as possible. Otherwise (5.3) should not be used to interpret the sensing data. According to [5.6] the atmosphere can be regarded as "frozen", to a first approximation, for a time interval of about 1 ms.

Fluctuations of the backscattering coefficient at $\lambda = 0.69\mu$m having a frequency about 1 kHz and magnitude about 20% have been observed in [5.7]. Such fluctuations of $\beta_\pi(z)$ are usually observed in the atmosphere up to a height of 5 km [5.8]. Thus, the average time between two laser shots at λ_{on} and λ_{off} should not be longer than 1 ms. This requirement can be met with a variety of techniques, the simplest of which is the use of two synchronized lasers delivering pulses at different wavelengths.

The laser used in the lidar described generates two giant pulses in a 100 µs time interval. The radiation frequency of one of these pulses can be fixed at any point of the tuning region (about 7 cm^{-1} wide) with an accuracy of about $5 \cdot 10^{-2}$ cm^{-1}. The frequency shift of the other pulse can be continuously tuned with respect to that of the first one in the region ± 0.5 cm^{-1} by changing the sweeping speed of the single-mode generator. A detailed description of such a laser constructed for use in lidar sensing of the atmospheric humidity is given in [5.9,10].

The spectral adjustment of the laser line has been controlled with a spectrophone. The amplitudes of lidar returns have been measured using a 20-channel digital data acquisition system. The spatial resolution of the system was about 60 m for the analog signals and about 240 m in the photon counting scheme.

5.2.2 Some Experimental Results

The experimental data used below to illustrate the efficiency of the inversion methods suggested were obtained using the above lidar facility during May-July 1978 at the test field of the Institute of Atmospheric Optics, Tomsk, USSR.

The experiments were carried out, as a rule, at night under windless, fair weather conditions. The meteorological visual range during these experiments was generally $S_M \geq 20$ km. The measurements of analog lidar signals took about 10 ÷ 15 min for signals up to 2 km and about 25 ÷ 40 min for signals from altitudes up to 8 km using the photon counting scheme. In the case of sensing along horizontal paths, it was interesting to compare the lidar data with the data from standard humidity meters.

Figure 5.5a shows the results of a comparison of lidar data with those obtained using an optical hygrometer and calculated from in situ measurements with a wet-and-dry bulb thermometer. All the data were obtained along the same horizontal sensing path of length 980 m. The dots show the lidar data as compared with those of the hygrometer, while the crosses represent the comparison of lidar data and those of the wet-and-dry bulb thermometer. The standard deviation in the first case is about 0.24 gm^{-3} and in the second case about 0.55 gm^{-3}. A larger value of standard deviation in the second case is evidently caused by an inhomogeneous distribution of water vapor along the sensing path.

As is seen from Fig.5.5a, the standard deviations of lidar data from the corresponding data of a hygrometer and a wet-and-dry bulb thermometer are 1.5%-5% and 3.2%-7.7%, respectively (the absolute humidity range was from 5 to 11 gm^{-3}). These data clearly demonstrate the utility of the DAS technique for measuring atmospheric humidity.

Two humidity profiles along a horizontal path obtained in a 40 min time interval are shown in Fig.5.5b. The sensing path was chosen above a broken landscape and crossed a river and an island. The measurements of the humidity profiles $\rho(z)$ were carried out during the night of May 28, 1978 in a 40 min time interval (Curves 1 and 2, respectively). The spatial resolution was about 60 m. The air humidity near the lidar was determined from dry-and-wet bulb temperature measurements and was $\rho^{(1)}_{H_2O} = 15.9$ and $\rho^{(2)}_{H_2O} = 15.6$ gm^{-3}, respectively. A characteristic increase of the humidity above the water surface is readily seen from this figure.

The next example illustrates the lidar sensing of humidity profiles along a vertical path. The measurements shown in Fig.5.6 were made during the night of July 24-25, 1978 in a time interval of about 1 hour. The altitude profiles clearly show the presence of the inversion water vapor layer above 500 m. It is also seen that the inversion layer moved upwards. According to the sensing data the speed of this upward movement was about 110 m per hour before midnight and later about 40 m per hour. The maximum value of the water vapor density in the inversion layer is higher than (Curves 1,2) or comparable

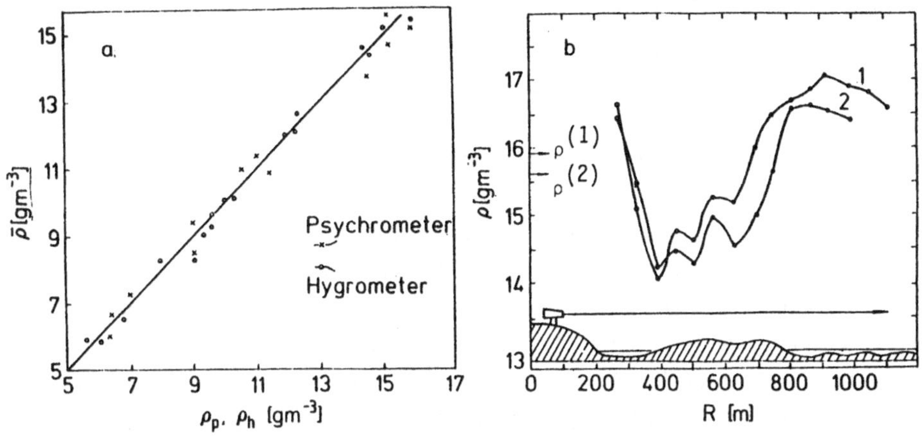

Fig.5.5a,b Examples of sounding the water vapor along the horizontal path. (a) Typical discrepancies between lidar data integrated over the horizontal path and data obtained with a psychrometer ρ_P and hygrometer ρ_H; b) Curves 1 and 2 represent the data obtained in a 40 min interval. $\rho^{(1)}$ and $\rho^{(2)}$ show corresponding data of in situ measurements using wet and dry bulb thermometers

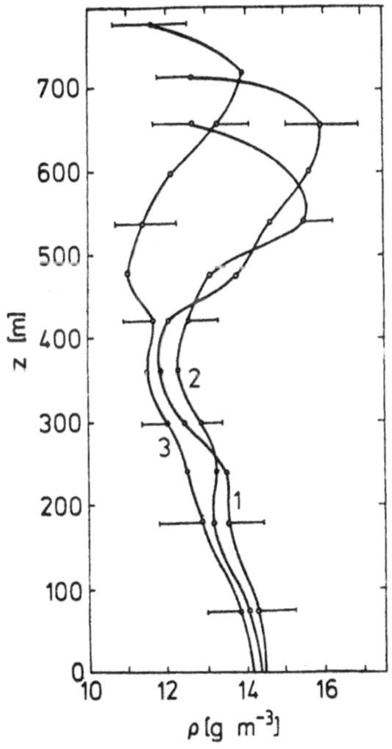

Fig.5.6. Water vapor profiles along the vertical path measured using DIAL lidar, demonstrating the capabilities of the lidar to reveal the inversion layers. Measurements represented by Curves 1,2 and 3 were carried out in a 1.5 hour time interval

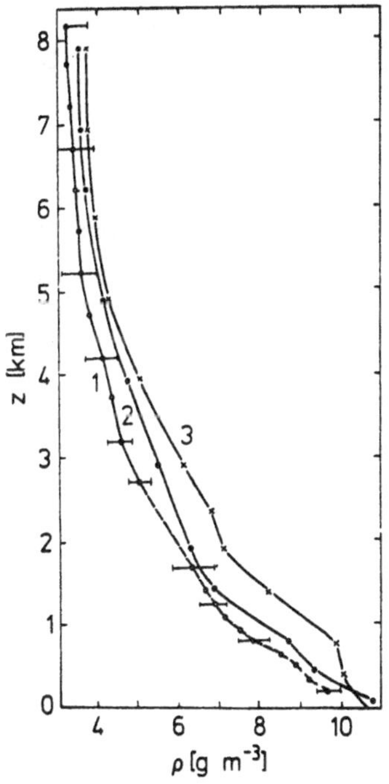

Fig.5.7 An example of restoring the water vapor profile up to 8 km altitude (Curve 1) from DIAL data obtained using a photon counting system. For comparison the model water vapor profile given in [5.4] (Curve 2) and the average profile for the region (Curve 3 [5.11]) are also presented.

to that at the earth's surface. The humidity decrease of about 0.5 gm^{-3}/100 m is seen explicitly up to an altitude of about 400 m.

This example demonstrates the capability of the lidar technique as applied to studies of the spatial-temporal dynamics of atmospheric processes to the extent that they are related to the humidity field.

Figure 5.7 shows results obtained by inverting the lidar data for the altitude region up to 8 km. The sensings were made during the night of July 27-28, 1978. Lidar returns up to 1.6 km were recorded with the analog recorder and in the altitude region from 3 to 8 km signals were measured using a photon counting device (Curve 1). The bars in the figure show the expected variances in the profile $\rho(z)$ due to noise in the electronics of the recorder. Curve 2 represents the humidity profile according to the summer model of McCLATCHEY [5.4] while Curve 3 characterizes the humidity profile for July in Novosibirsk statistically averaged over the period from 1961 to 1969 [5.11].

The data on humidity profiles along horizontal, vertical and slant paths obtained using the DIAL technique developed at the Institute of Atmospheric Optics clearly demonstrates its applicability in atmospheric studies. Specific features of this method are the use of a specially designed ruby laser with a narrow line and a highly sensitive spectrophone. An effective algorithm for data inversion was also specially constructed.

In a number of cases,the atmospheric humidity profiles have been obtained up to heights of 15-17 km. It should be noted that the lidar used in this investigation is not very sophisticated and could be significantly improved. An obvious improvement can be achieved by using a receiving mirror 1 m in diameter instead of the one used which had a diameter 0.3 m.

5.3 Lidar Sensing of Atmospheric Ozone

In recent years,increasing attention has been paid to studies of atmospheric ozone. The main reasons for this are the very important role of the atmospheric ozone layer in radiation transfer processes and the probable unthropogenic factors affecting this layer. In addition, atmospheric ozone is one of the most active components of industrial smogs.

The network of ozonometric stations can provide systematic data on the column density of atmospheric ozone as well as on its variations with geographical location and season. The ozone probes in current use provide information on the atmospheric ozone density profiles.

Table 5.1 Parameters of the lidars used for sensing the stratospheric ozone profiles in different studies

Parameters		Reference			
		[5.13]	[5.14]	[5.15]	[5.16]
Wavelength	[nm]	290-315	297/308	303.7/308	308
Halfwidth of a laser line	[nm]	0.5	0.2	0.15	0.7
Pulse energy	[mJ]	10-30	0.2	20-40	50
Pulse duration	[s]	$30 \cdot 10^{-9}$	$5 \cdot 10^{-6}$	$4 \cdot 10^{-6}$	$20 \cdot 10^{-9}$
Beam divergence	[mrad]	6	0.2×2	0.8	2×5
Pulse repetition frequency	[Hz]	0.1	1	0.25	0.07
Receiving area	[m^2]	0.19	0.6	0.5	0.19
Field of view		3-7	2	5	10
Suppression of signals from short ranges		transmitter--receiver spacing 25 m	transmitter--receiver spacing 4 m	mechanical shutter	transmitter--receiver spacing

However, certain important characteristic of the ozone layer behavior cannot be studied with the standard techniques. These include, for example, short-term variations in the ozone profiles, horizontal inhomogeneities in its concentration and, of course, local peculiarities of the ozone content which may be caused by anthropogenic factors. These features of the atmospheric ozone distribution can be studied using differential absorption lidar technique.

Only a limited number of papers on lidar studies of atmospheric ozone exist in the literature. A short summary of these works can be found in [5.12]. Table 5.1 presents the parameters of the lidars used in these investigations, with differential absorption lidar techniques [5.13-15], and with only one wavelength [5.16].

Figure 5.8 presents vertical profiles of atmospheric ozone reconstructed from the data of [5.13,15,16]. Horizontal bars in the figure denote the reconstruction errors. Other details of the results can be found in the figure captions.

As is seen from Fig.5.8 the lidars used in [5.13-16] do not contain enough information to permit useful information on the ozone layer of the stratosphere to be obtained. Nevertheless the results of these papers demonstrate that lidar methods are, in principle, capable of sensing the stratospheric ozone layer.

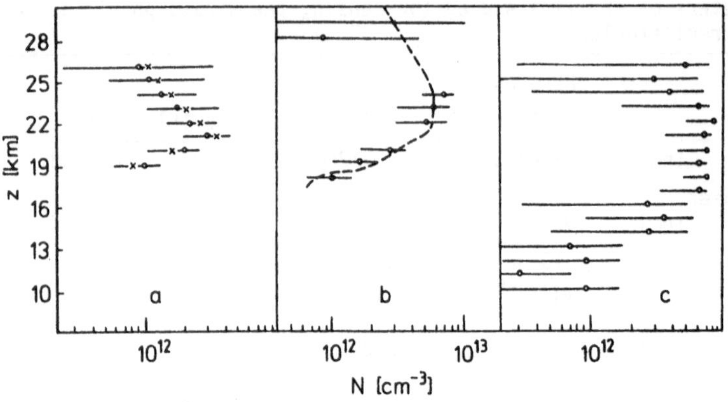

Fig.5.8. a) Vertical profiles of the ozone concentration [5.15] (dots represent the data obtained during the night of July 19 to 20 and crosses represent the data obtained during the night of July 20 to 21, 1977). b) Vertical profiles of the ozone concentration [5.16] obtained on June 12, 1978; dashed curve presents the ozonosonde data obtained on May 11, 1978. c) Vertical profile of the ozone concentration from [5.13] (October 16, 1979)

Further improvements in the methods for sensing the atmospheric ozone layer are expected with the development of new more efficient lasers. Thus, for example, the use of dye lasers with pumping from metal vapor lasers promise a significant improvement in the potentials of such lidar facilities. For example, copper-vapor lasers developed and manufactured at the Institute of Atmospheric Optics provided reliable operation with an output power of about 10 W at a 10^3 - 10^4 pulse repetition frequency. Also, the efficiency of laser sources as applied to lidar studies can be improved by reducing the line widths of the generated radiation to a value of about 10^{-2} cm^{-1}. The use of a spectrophone for fine adjustment of a laser line onto the centers of ozone absorption lines can also be very useful, as seen from the results on the water vapor sensing. Finally, the efficiency of the inversion algorithm used for reconstructing the ozone concentrations is also important. It should be noted that all the difficulties described in the previous section in connection with lidar studies of atmospheric water vapor, are also characteristic for the ozone measurements. Therefore, the use of better algorithms should undoubtly improve the reliability of the reconstructed data on the stratospheric ozone layer.

The use of all these possibilities in combination should evidently make the DAL technique the most sensitive and useful among the presently known methods for measuring atmospheric ozone profiles. Moreover, there are reasons to expect that such improvements can be achieved in the very near future.

5.4 The Method of Logarithmic Derivatives in Lidar Data Interpretation

5.4.1 Determination of Humidity Profiles and Profiles of Aerosol Optical Characteristics

It was assumed when deriving (5.3) that the profiles $P(z,\lambda_{on})$ and $P(z,\lambda_{off})$ are continuous and differentiable functions, i.e., the discontinuous character of an aerosol distribution along the sensing path was not taken into account. This fact suggests the utility of a preliminary processing of the lidar returns by constructing continuous and differentiable analogs $P(z,\lambda_{on})$ and $P(z,\lambda_{off})$ using regularizing algorithms. This idea can be realized in the logarithmic-derivative method, which is based on the functional transformation of (5.1) and (5.2).

The characteristic feature of the system of lidar equations are the exponential terms which appear in these equations. Often, it appears to be very helpful, for such functional equations, to use not the functions themselves, but their logarithmic derivatives. In this case the system of equations (5.1,2) will take the form

$$\begin{aligned}
[\ln \bar{S}(z,\lambda_{off})]' &= [\ln \beta_\pi(z,\lambda_{off})]' - 2\beta_{ext}(z,\lambda_{off}) \\
[\ln \bar{S}(z,\lambda_{on})]' &= [\ln \beta_\pi(z,\lambda_{on})]' - 2\beta_{ext}(z,\lambda_{on}) \\
&\quad -2k(z)\rho(z) \quad .
\end{aligned} \tag{5.7}$$

As before, the function $\bar{S}(z,\lambda)$ denotes in (5.7) the ratio $P(z,\lambda)z^2/BP_0\gamma(z)$. The system (5.7) is obviously defined only for differentiable functions. Therefore, this system should be used with the regularized analogs $\bar{S}_\alpha(z,\lambda)$ rather than the profiles $\bar{S}(z,\lambda)$. Taking this into account and subtracting the second equation from the first one, one can obtain the following equation

$$\rho(z) = [2k(z)]^{-1}\{[\ln \bar{S}_\alpha(z,\lambda_{off})]' - [\ln \bar{S}_\alpha(z,\lambda_{on})]'\} \quad . \tag{5.8}$$

Formally, (5.8) is identical to (5.3). However, the sequence of the data processing is different. It is very important that (5.8) does not involve the division of one signal by the other. Such an operation is quite undesirable if signals have random components, since it strongly affects the efficiency of inversion.

The use of $\bar{S}_\alpha(z,\lambda)$ instead of $\bar{S}(z,\lambda)$ can be regarded as an attempt to extract the regular component from the signal $P(z,\lambda)$. When doing this, one suppresses the oscillating components regardless of their nature. As a result, (5.8) is not so sensitive to the time delay between two successive measurements of $P(z,\lambda_{on})$ and $P(z,\lambda_{off})$. This circumstance is very important for the technical realization of the DAS lidar technique. Therefore, the method of logarithmic derivatives is preferable for use in lidar sensing as compared with the other methods for lidar data processing. Numerical methods for finding the logarithmic derivatives will be considered below.

Besides the humidity profile $\rho(z)$, one can obtain from (5.7) information on the aerosol optical characteristics. The first equation of this system can be used for this purpose. It is very important to note that, in some cases, a knowledge of the humidity profiles $\rho(z)$ can make the solution of this equation with respect to optical characteristics more informative. Such a possibility is provided by the correlation between the relative air humidity and optical characteristics of aerosols. Let us consider these questions in more detail.

Introducing the lidar ratio $b = \beta_\pi/\beta_{ext}$ and neglecting the contribution due to Rayleigh scattering at λ_{off} one can write the first equation in (5.7) as follows

$$[\ln \bar{S}_\alpha(z,\lambda_{off})]' = [\ln b(z,\lambda_{off})]' + [\ln \beta_{ext}^{(a)}(z,\lambda_{off})]' - 2\beta_{ext}^{(a)}(z,\lambda_{off}) \quad . \tag{5.9}$$

Let the analytical model of $b(\bar{\omega})$ relating the lidar ratio b and relative air humidity $\bar{\omega}$ be chosen a priori according to the particular optical situation in the atmosphere. Since $[\ln b(\bar{\omega}(z))]' = [\ln b(\bar{\omega})]' \cdot \bar{\omega}'(z)$, then omitting λ_{off} one can write (5.9) as follows

$$[\ln \bar{S}_\alpha(z)]' = [\ln b(\bar{\omega})]' \cdot \bar{\omega}'(z) + [\ln \beta_{ext}^{(a)}(z)]' - 2\beta_{ext}^{(a)}(z) \quad . \tag{5.10}$$

It is not difficult to find $\bar{\omega}(z)$ provided the profile $\rho(z)$ is known and, hence, (5.10) can be solved for the function $\beta_{ext}^{(a)}(z)$. This is done using the following iteration scheme

$$\ln \beta_{ext}^{(q)}(z) = \beta_{ext}(z_0) + \int_{z_0}^{z} \{[\ln \bar{S}_\alpha(z')]' - [\ln b(\bar{\omega})]' \, \bar{\omega}'(z) + 2\beta_{ext}^{(q-1)}(z')\}dz' \quad , \tag{5.11}$$

where q is the number of the iteration.

As earlier (Chap.3), the iteration scheme allows one not only to find the solution but also to estimate, to a certain degree, the acceptability of the models used in the data interpretation. In the problem under consideration, this concerns primarily the dependence $b(\bar{\omega})$. As investigations show, this dependence is strongest for $\bar{\omega} > 70\%$ [5.17]. The function $b(\bar{\omega}) = c_0 + c\bar{\omega}^{-1}$ can be used as a first approximation in this case. For the ground layer the values $c_0 = 0.0039$ and $c_1 = 1.6$ have been obtained in [5.17]. Under conditions of low humidity, the lidar ratio is practically independent of $\bar{\omega}$ and (5.11) leads to the iteration scheme for solving the ordinary lidar equation using the method of logarithmic derivatives. A simultaneous determination of the humidity profiles and profiles of the aerosol extinction coefficient gives new possibilities for studying the physical processes in the ground atmospheric layer.

The analytical advantages of the method of logarithmic derivatives as applied to lidar data interpretation can also be used for other problems of lidar sensing of the atmosphere. One of the more important of these is the theory of lidar sensing of atmospheric parameters employing the Raman effect.

5.4.2 Lidar Sensing of the Atmospheric Temperature and Aerosol Optical Characteristics Using a Pure Rotational Raman Spectrum

The use of Raman scattering in the lidar sensing of atmospheric gases has been analysed [5.18, 19]. Since we are concerned with inverse problems of lidar sensing, we shall give a brief explanation of the lidar equation

$$P(z,\lambda_J) = \sigma_J(z)\rho(z)B\gamma(z)P_0 z^{-2}T^{1/2}(z,\lambda_J)T^{1/2}(z,\lambda) \quad . \qquad (5.12)$$

Here $\sigma_J(z)$ is the differential Raman cross-section of the Jth rotational line of the gas sounded; $\rho(z)$ is the molecular number density; λ and λ_J are the wavelengths of the laser radiation and Raman line, respectively. Assuming λ and λ_J to be close enough, one can write (5.12) in terms of logarithmic derivatives as follows

$$[\ln \bar{S}(z,\lambda_J)]' = [\ln \rho(z)]' + [\ln \sigma(z,\lambda_J)]' + [\ln T(z,\lambda_J)]' \quad . \qquad (5.13)$$

Since $\sigma(z,\lambda_J)$ depends strongly on temperature, (5.13) can be used for the temperature ($\hat{T}$) sensing. In fact, [5.14]

$$\sigma(z,\lambda_J) = A(\lambda_J)D_J k^{-1}\hat{T}^{-1}(z)\exp[-BhcJ(J+1)/k\hat{T}(z)] \quad , \qquad (5.14)$$

where $A(\lambda_J)$ is proportional to λ_J^{-4}, $D_J = (J+1)(J+2)/(2J+3)$ for the Stokes rotational branch, and $J(J-1)/(2J-1)$ for the anti-Stokes branch; k is Boltzmann's constant; h is Planck's constant; B is a rotational constant and c is the velocity of light [5.20]. It is not difficult to prove that

$$[\ln \sigma(z,\lambda_J)]' = -[\ln \hat{T}(z)]'[1 + c(J)\hat{T}^{-1}(z)] \quad , \qquad (5.15)$$

in which $c(J)$ is $Bhck^{-1}(J+1)$.

For two pure rotational Raman lines J_1 and J_2 one can write the system of two lidar equations

$$\begin{aligned}[\ln \bar{S}_1(z)]' &= [\ln \rho(z)]' + \{-[\ln \hat{T}(z)]'[1 + c_1\hat{T}^{-1}(z)]\} - 2\beta_{ext}(z) \\ [\ln \bar{S}_2(z)]' &= [\ln \rho(z)]' + \{-[\ln \hat{T}(z)]'[1 + c_2\hat{T}^{-1}(z)]\} - 2\beta_{ext}(z) \quad .\end{aligned} \tag{5.16}$$

Since $\sigma(z,\lambda_J)$ of the principal atmospheric gases is uniquely defined by the temperature profile, (5.16) is, in principle, defined relative to $\hat{T}(z)$ and $\beta_{ext}(z)$. Thus, the use of the pure rotational Raman spectra of the main atmospheric gases provides an opportunity to determine simultaneously the atmospheric temperature profile as well as the profile of the aerosol optical characteristic $\beta_{ext}(z)$. Such an opportunity for certain applied researches is certainly important. Unfortunately, there are a lot of technical difficulties in the realization of this possibility in lidar sensing.

Consider briefly some of the most important questions relevant to the numerical solution of (5.16).

By analogy with the preceeding case [see (5.7)], one can obtain an equation for T(z), by subtracting one of the equations (5.16) from the other

$$[\ln \hat{T}(z)]' = -\hat{T}(z)(c_1 - c_2)^{-1}\{[\ln \bar{S}_1(z)]' - [\ln \bar{S}_2(z)]'\} \quad . \tag{5.17}$$

It should be mentioned here that the regularized analogs of the lidar returns $\bar{S}_{1\alpha}(z)$ and $\bar{S}_{2\alpha}(z)$ must enter into (5.17) in the case of experimental data inversion. Equation (5.17) is valid only in this case and it then defines a continuous temperature profile as one would expect from a physical point of view. Keeping this in mind,we shall use below the designation $\Delta_\alpha(z)$ for the expression early in brackets in (5.17). A variety of numerical methods for solving (5.17) can be used. By analogy with (5.11) one can write the simplest iteration scheme

$$\ln \hat{T}^{(q)}(z) = \ln \hat{T}(z_0) - (c_1 - c_2)^{-1} \int_{z_0}^{z} \Delta_\alpha(z')\hat{T}^{(q-1)}(z')dz' \quad , \tag{5.18}$$

where q is the number of iteration. It can easily be shown that (5.18) converges faster if $(c_1 - c_2)$ is large. On the other hand, if c_1 and c_2 are closer to each other, the initial system (5.16) is worse conditioned with respect to the temperature profile T(z). Therefore, the choice of proper lines (J_1 and J_2) is very important in the method considered, since the conditioning of system (5.16) is directly connected with the temperature sensitivity. A detailed analysis of these questions taking into account the physics can be found in [5.19,21].

Note that by analogy with (5.3) in the DAS-lidar method one can also use in this inverse problem the ratio of two signals $P_1(z)$ and $P_2(z)$ measured at two spectral lines corresponding to the rotational quantum numbers J_1 and J_2. The logarithm of this ratio is inversely proportional to the temperature, which makes it possible to construct a very simple method for interpreting the lidar signals. In contrast to previous inverse problems, the spatial and temporal fluctuations of aerosol formations along the sensing path do not affect the profile of this ratio. The irregular components of the signals are due only to noise. However, such simplicity also has drawbacks. The problem is that the use of the ratio $P_1(z)/P_2(z)$ instead of the signals themselves [equivalently $\bar{S}_1(z,\lambda_1)$ and $\bar{S}_2(z,\lambda_2)$] results in disappearance of the term $\hat{T}^{-1}(z)$, see (5.14). This means that the ratio of signals is less sensitive to temperature variations than are the signals themselves. At the same time, as is seen from (5.17), the method of logarithmic derivatives is free of this disadvantage.

The profiles $\hat{T}(z)$ determined from (5.18) allow one to find the density $\rho(z)$ of a gaseous component of the atmosphere from the following relationship

$$[\ln \rho(z)]' = -\mu g/R\hat{T}(z) + [\ln \hat{T}(z)]' \quad , \tag{5.19}$$

where μ is the molecular weight and R is the universal gas constant.

A knowledge of functions $[\ln \rho(z)]'$, $[\ln \hat{T}(z)]'$ and $\hat{T}(z)$ allows, in turn, the profile $\beta_{ext}(z)$ to be found from (5.16). The aerosol extinction component $\beta_{ext}^{(a)}(z)$ can then be separated out. In principle, by recording Raman signals at different rotational lines, spaced sufficiently far apart in the spectrum, one can obtain information on the spectral behavior of the aerosol extinction coefficient, which can be used for solving the inverse problems of aerosol light scattering. Thus, Raman lidar techniques present another version of the multifrequency method for sensing aerosol polydispersed systems. In practice,it is preferable to use two spectral intervals. As is shown in [5.22], one of these spectral intervals should include the nitrogen and oxygen pure rotational Raman lines whose intensity increases as the temperature increases (lines with $J > 10$), while the other interval should contain lines whose intensity decreases as the temperature increases (lines with $J < 10$). This fact is obviously connected with better conditioning of the system (5.16).

The logarithmic-derivative method can easily be generalized for the case of spectral intervals. The overall Raman scattering cross-section for the spectral interval can be written as follows

$$\sigma_1(z) = \sum_J [k\hat{T}(z)]^{-1} A^{(N)}(\lambda_J) D_J^{(N)} \exp[-c_J^{(N)} \hat{T}^{-1}(z)]$$

$$+ (N^{(0)}/N^{(N)}) \sum_{J'} [k\hat{T}(z)]^{-1} A^{(0)}(\lambda_{J'}) D_{J'} \exp[-c_{J'}^{(0)} \hat{T}^{-1}(z)] \quad (5.20)$$

where the indices "N" and "O" denote nitrogen and oxygen, respectively, $N^{(N)}$ and $N^{(0)}$ are the number densities of nitrogen and oxygen. It is assumed, in first approximation, that the ratio $N^{(N)}/N^{(0)}$ is constant along the sensing path. The logarithmic derivative of (5.20) is written as an expression analogous to (5.15) but instead of constants c(J) the profiles

$$c_1(z) = \sum_J A^{(N)}(\lambda_J) D_J^{(N)} c_J^{(N)} \exp[-c_J^{(N)} \hat{T}^{-1}(z)]$$

$$+ (N^{(0)}/N^{(N)}) \sum_{J'} A^{(0)}(\lambda_{J'}) D_{J'}^{(0)} c_{J'}^{(0)} \exp[-c_{J'}^{(0)} \hat{T}^{-1}(z)]$$

$$/\Big\{\sum_J A^{(N)}(\lambda_J) D_J^{(N)} \exp[-c_J^{(N)} \hat{T}^{-1}(z)] + (N^{(0)}/N^{(N)})$$

$$\cdot \sum_{J'} A^{(0)}(\lambda_{J'}) D_{J'}^{(0)} \exp[-c_{J'}^{(0)} \hat{T}^{-1}(z)]\Big\} \quad .$$

Using these expressions for two spectral intervals denoted as $c_1(z)$ and $c_2(z)$ one can rewrite the iteration scheme (5.18) in the form

$$\ln \hat{T}^{(q)}(z) = \ln \hat{T}(z) - \int_{z_0}^{z} \{\Delta_\alpha(z') \hat{T}^{(q-1)}(z')$$

$$/ [c_1(\hat{T}^{(q-1}(z')) - c_2(\hat{T}^{(q-1)}(z'))]\} dz' \quad . \quad (5.21)$$

Unfortunately, no results on the application of this method to the reconstruction of temperature profiles from lidar sensing data are available due to the absence of experimental data. It should be noted, however, that a temperature lidar now under development will soon provide a possibility for such an application of this method.

Thus, for example, tests of a pure rotational Raman lidar described in [5.23] have recently demonstrated the possibility of determining the atmospheric temperature remotely. The parameters of this lidar are as follows:

Transmitter:

Copper vapor laser with mean power	10 W
Wavelength	510.6 nm
Beam divergence	0.3 mrad

Receiver

Diameter of lens	0.3 m
Double monochromator	1 nm mm^{-1}
Data digital acquisition system:	
Photon counting device with a count rate	up to 40 MHz

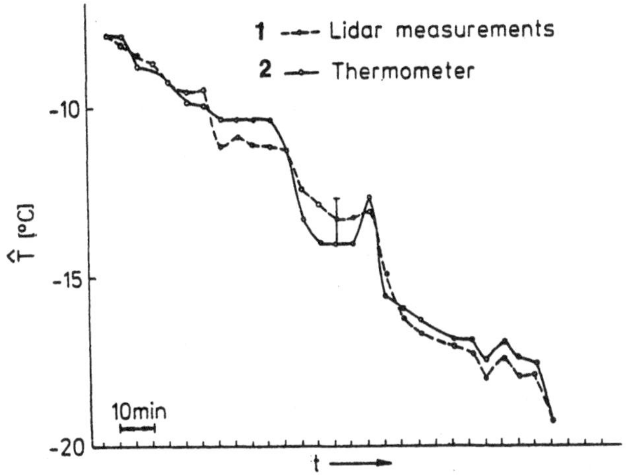

Fig.5.9. Temporal variation of the atmospheric temperature measured using a pure rotational Raman lidar (Curve 1) and the data of a thermoresistor thermometer (Curve 2)

Some results on the remote determination of the atmospheric temperature in a volume ~60 m long located about 150 m from the lidar are shown in Fig. 5.9. The atmospheric temperatures presented in this figure by the dashed curve were obtained from lidar measurements of the intensity ratio for two spectral intervals in a pure rotational Raman spectrum of air molecules. The solid curve shows the temperatures measured with a calibrated thermoresistor. The accuracy of the thermoresistor measurements was about 0.1 K, while the standard deviations of lidar measurements are shown in the figure by vertical bars. Both curves in Fig.5.9 represent the temperature change during the night of March 12, 1980 near Tomsk. The figure indicates that the largest discrepancy between the two curves only slightly exceeds 1 K.

5.4.3 Regularizing Algorithm for Processing the Sensing Data. Photon Counting Scheme

The use of the logarithmic-derivative method to interpret lidar sensing data requires efficient algorithms for numerical differentiation. An algorithm for differentiating the logarithm of the ratio $P(z,\lambda_{on})/P(z,\lambda_{off})$ using the Fredholm operator was described above. As numerical investigations have shown, it is quite efficient for handling continuous lidar signals. The data discretization in this case can be made with a very small increment $\Delta(z)$. Lidar data obtained with a photon counting technique are discretized according to the sensitivity of the measuring device and the conditions of the measurements. This occurs in problems of lidar sensing of such atmospheric parameters as temperature, humidity, pressure and molecular constituents, with the possible exception of aerosol characteristics. In this connection, it is more correct not to talk about differentiating the signals but rather about the construction of some continuous and differentiable profiles for the atmospheric parameters sought, as well as for the lidar signals themselves, from the discrete data. It is assumed that such profiles are close to the actual ones within errors comparable to the measurement errors. The function $\varphi_\alpha(z)$ determined using (5.6) in the DAS-lidar method can serve as an example of such a profile. The problem thus formulated can be solved using the regularizing algorithm suggested below.

Let us, as earlier, use the reference relationship (5.4). Taking into account the discrete character of lidar data, one can rewrite it as follows

$$f(z_i) = f(z_0) + \int_{z_0}^{z_i} \varphi(y)dy \quad , \qquad y \in [y_0,y_n] \quad i = 1,2,\ldots,n \tag{5.22}$$

where $f(z_i)$ are the experimental readouts. For seeking $\varphi(y)$, which is initially defined as an approximation to the derivative $f'(z)$ of a continuous profile $f_0(z)$, one has the system of equations

$$\int_{z_0}^{z_i} \varphi(y)dy = \tilde{f}(z_i) \quad , \qquad i = 1,2,\ldots,n \quad , \tag{5.23}$$

where $\tilde{f}(z_i) = f(z_i) - f(z_0)$. The system (5.23) has an infinite number of solutions; therefore, we shall use the smoothing functional method and choose a function $\varphi_\alpha(z)$ as a solution of (5.3) which minimizes the expression

$$T_\alpha(\varphi) = \sum_{i=1}^{n} \left[\int_{z_0}^{z_i} \varphi(y)dy - \tilde{f}(z_i)\right]^2 + \alpha p_0 \int_{z_0}^{z_n} \varphi^2(z)dz \quad . \tag{5.24}$$

The Euler equation for this variational problem has the form

$$\sum_{i=1}^{n} K(z,z_i) \left[\int_{z_0}^{z_i} \varphi(y)dy - \tilde{f}(z_i)\right] + \alpha p_0 \varphi(z) = 0 \quad , \tag{5.25}$$

where the kernel

$$K(z,z_i) = \begin{cases} 1 & \text{for} \quad z_0 \le z \le z_i \\ 0 & \text{for} \quad z_i \le z \le z_n \end{cases} \quad .$$

Assuming that

$$\tilde{f}(z_i) - \int_{z_0}^{z_i} \varphi(y)dy = \alpha p_0 c_i \quad , \tag{5.26}$$

where the c_i $(i = 1,2,\ldots,n)$ are constants, one can obtain from (5.25) that

$$\varphi_\alpha(z) = \sum_{i=1}^{n} c_i K(z,z_i) \quad . \tag{5.27}$$

Expression (5.27) represents the solution $\varphi_\alpha(z)$ sought in the form of linear combination of step functions $K(z,z_i)$. As a consequence, the function $\varphi_\alpha(z)$ is a piecewise constant (or steplike) function with discontinuities at the points z_i $(i = 1,2,\ldots,n)$. The primitive of this function is a broken line, i.e., a continuous function. This function is defined as follows

$$f_\alpha(z) = f(z_0) + \int_{z_0}^{z} \varphi_\alpha(y)dy = \begin{cases} f(z_0)+c_1(y-y_0) \quad \text{for} \quad y_0 \le y \le y_1 \\ \cdots\cdots\cdots\cdots\cdots\cdots \\ f(z_0) + \sum_{j=1}^{i=1} c_j(y_{j-1}-y_0) + c_i(y-y_0) \\ \qquad\qquad \text{for} \quad y_{i-1} \le y \le y_i \\ \cdots\cdots\cdots\cdots\cdots\cdots \\ f(z_0) + \sum_{j=1}^{n-1} c_j(y_{n-1}-y_0) + c_n(y-y_n) \\ \qquad\qquad \text{for} \quad y_{n-1} \le y \le y_n \end{cases} \quad . \tag{5.28}$$

Obviously, the function $f_\alpha(z)$ is nothing but the simplest analytic model of the lidar returns registered with a photon counting device. Nevertheless, it appears to be quite sufficient for solving many applied problems relevant to sensing using Raman lidars.

If necessary, the restriction can be imposed on $\varphi'(z)$ by adding the term $\alpha p_1 \int_{z_0}^{z_n} [\varphi'(z)]^2 dz$ in (5.24). This allows the construction of continuous profile $f'_\alpha(z)$ whose first derivative is also continuous everywhere between the points z_0 and z_n. It should be noted here that questions concerning the construction of differentiable functions are discussed in detail in papers on splines [5.24]. According to the spline terminology, we have constructed above the zero-order spline. Now, consider the determination of the coefficients c_i $(i = 1,2,\dots,n)$ in (5.27). Substituting (5.27) into (5.26) one obtains the system of equations

$$\sum_{j=1}^{n} A_{ij}c_j + \alpha p_0 c_i = f(z_i) \quad , \quad i = 1,2,\dots,n \quad , \tag{5.29}$$

where

$$A_{ij} = \begin{cases} x_j - x_0 \quad , & \text{if} \quad j \le i \\ x_i - x_0 \quad , & \text{if} \quad j > i \quad . \end{cases}$$

In the case of equidistant readouts (5.29) takes the simple form

$$\sum_{j=1}^{i} c_j j + (n\gamma + i)c_i + i \sum_{j=i+1} c_j = [\Delta(z)]^{-1}\tilde{f}(z_i) \quad , \quad i = 1,2,\dots,n \quad , \tag{5.30}$$

where $\gamma = \alpha p_0/\Delta(z)n$. The matrix of coefficients of this equation has the form

$$A_{ij} = \begin{Bmatrix} (1+\gamma n) & 1 & 1 \dots 1 \\ 1 & (2+\gamma n) & 2 \dots 2 \\ \dots\dots & \dots\dots & \dots\dots \\ 1 & 2\dots(i+\gamma n) & \dots\; i \\ \dots\dots & \dots\dots & \dots\dots \\ 1 & 2 & \dots\;(n+\gamma n) \end{Bmatrix} \quad .$$

The determinant of this matrix differs from zero and satisfies the inequality

$$|\det A| < n^{3n/2}(1+\gamma)^n \quad . \tag{5.31}$$

The determination of the c_i $(i = 1,2,\dots,n)$ completes the solution of the above problem. Thus, we have constructed a continuous profile $f_\alpha(z)$, its first derivative $\varphi_\alpha(z)$ and logarithmic derivative $\varphi_\alpha(z)/f_\alpha(z)$ based on the discrete set of reference data $\{f(z_i)\}$ $(i = 1,2,\dots,n)$. The ratio $\varphi_\alpha(z)/f_\alpha(z)$ is a piecewise continuous function with discontinuities of the first kind at the points z_i. For this reason, integrals in iteration schemes such as (5.18) are summed, providing continuous profiles of the atmospheric parameters sensed. The value of γ in (5.30) should be chosen taking into account the noise level of the measuring device.

It should be noted, in conclusion, that the logarithmic-derivative method for solving inverse problems can undoubtedly be much more widely used than the above examples. This concerns particularly the method of laser sensing based on the nonlinear effects of the interaction of laser radiation with the atmospheric constituents. The nonlinear character of the corresponding lidar equations requires their prior functional transformation and, at the same time, better regularity of the reference data. All this is possible with the logarithmic-derivative method.

These questions, although interesting, are beyond the scope of this book.

5.5 Laser Sensing of Wind Velocity and Atmospheric Turbulence Parameters

The methods of lidar sensing of wind velocity and the parameters of atmospheric turbulence are based on the assumption that aerosols are fully entrained in air mass motion caused by wind and atmospheric turbulence [5.25]. The aerosol particles themselves serve merely as tracers. Lidar signals backscattered from these aerosol particles bear information on their movements due to air mass motion.

5.5.1 Lidar Sensing of Wind Velocity

All lidar methods for measuring wind velocity can be divided into two classes: correlation and Doppler techniques.

The use of the Doppler effect for obtaining information on air mass movements has many advantages in the visual range as compared with the region of radio frequencies. The higher frequencies of optical waves provide relatively large Doppler shifts of the scattered radiation. At the same time, being shorter, optical waves provide higher spatial resolution.

The Doppler shift of the frequency of scattered radiation is described by

$$\Delta\nu_D = 2\lambda^{-1} V \sin(\theta/2) \cos\varphi \quad ,$$

where λ is the wavelength of sensing radiation, θ is the scattering angle, and φ is the angle between the wind velocity vector and the bisector of the scattering angle. In the single-ended lidar scheme,the Doppler shift $\Delta\nu_D$ is caused by the projection of the wind velocity onto the sensing beam direction.

Some particular methods for determining the Doppler shifts are considered in [5.26].

The idea of correlation methods for lidar determination of the wind velocity is that aerosol inhomogeneities moving through the illuminated volume of the atmosphere due to wind transfer cause fluctuations of the lidar-return amplitude. A correlation analysis of these fluctuations allows one to infer information about the wind velocity.

Let us consider briefly a correlation technique for measuring wind speed. Spatial and temporal inhomogeneities of atmospheric aerosols can be revealed and recorded from the fluctuations of backscattered radiation using lidars. Let us consider, for simplicity, the case of a multi-beam lidar sensing scheme and a plane field of aerosol inhomogeneities moving with the mean speed. The fluctuations of signals P backscattered from this layer may be characterized, in a Cartesian coordinate system (x,y) for the layer, by a spatial-temporal correlation function [5.27]

$$R(\bar{\xi},\bar{\eta},\tau) = \langle [P(\bar{x},\bar{y},t) - \langle P\rangle][P(\bar{x}+\bar{\xi},\bar{y}+\bar{\eta},t+\tau) - \langle P\rangle] \rangle$$
$$[P(\bar{x}_0, \bar{y}_0 t) - \langle P\rangle]^{-2} \quad , \tag{5.32}$$

where $\bar{\xi}$, $\bar{\eta}$ are the biases of the corresponding coordinates and τ is a time-lag between observations. In a small vicinity of the point ($\bar{\xi} = 0$, $\bar{\eta} = 0$, $\tau = 0$), the function $R(\bar{\xi}, \bar{\eta}, \bar{\tau})$ depends upon the value $(A_1\bar{\xi}^2 + A_2\bar{\eta}^2 + 2A_3\bar{\eta}\bar{\xi} + A_4\tau^2)$, where the constants A_i (i = 1,2,3,4) are determined by the sizes and shapes of the inhomogeneities [5.28]. In order to find the correlation function in the coordinate system (x,y) related to the point of observation (in the lidar location), one can use the coordinate transformation

$$x = \bar{x} + V_x t$$
$$y = \bar{y} + V_y t \quad ,$$

where V_x and V_y are the projections of the wind velocity vector on the corresponding coordinate axes. Taking this transformation into account one can write

$$R(\xi,\eta,\tau) = R(A_1(\xi - V_x\tau)^2 + A_2(\eta - V_y\tau)^2$$
$$+ 2A_3(\xi - V_x\tau)(\eta - V_y\tau) + A_4\tau^2) \quad . \tag{5.33}$$

By thus determining experimentally sections of the correlation function $R(\xi,\eta,\tau)$, one can easily determine the motion of the inhomogeneities. In a particular case when $\tau = \tau_0$ = const., one can find the components V_x and V_y by seeking the maximum of $R(\xi,\eta,\tau_0)$ with respect to variables ξ and η [5.29] as follows

$$V_x = \xi_m \tau_0^{-1} \quad , \quad V_y = \eta_m \tau_0^{-1} \quad , \tag{5.34}$$

where ξ_m and η_m are the coordinates of $R(\xi,\eta,\tau_0)$ maximum. Expression (5.34) can be used for determining the components V_x and V_y of the wind velocity vector from sensing data obtained with a single-ended lidar along the directions x and y in a time interval τ_0. The fraction $R(\xi_0,\eta_0,\tau)$ at fixed $\xi = \xi_0$ = const. and $\eta = \eta_0$ = const. has a maximum at a point τ_m on the time axis. From the condition $R'_\tau = 0$, it follows that

$$\tau_m = \xi_0(A_1V_x + A_3V_y)(A_1V_x^2 + A_2V_y^2 + 2A_3V_yV_x + A_4)^{-1} \quad . \tag{5.35}$$

This expression shows that the position of a maximum of the temporal correlation function $R(\xi_0,0,\tau)$ depends not only on the wind velocity but also on the coefficients A_i (i = 1,2,3,4), i.e., on the geometry of the inhomogeneities. If we assume the inhomogeneities to be isotropic (i.e., $A_1 = A_2$, $A_3 = 0$, $A_4 = 0$), then (5.35) takes the form [5.29]

$$\xi_m = \xi_0 V_x/(V_x^2 + V_y^2) \quad . \tag{5.36}$$

The product $\xi_0\tau_m^{-1}$ represents a component V'_x of the wind velocity that is related to the modulus of wind velocity by

$$V'_x = V/\cos\varphi \quad , \tag{5.37}$$

where φ is the angle between the x axis and the wind velocity. The value V'_x has a minimum at $\varphi = 0$, i.e., when the x axis coincides with the wind direction.

In summary, one can state that in the simplest forms the wind velocity can be determined using lidar techniques in two ways. The first is based on sensing the spatial structure of the field of aerosol inhomogeneities. The second requires observations of the temporal fluctuations of aerosol optical properties at two points. Naturally, these two variants do not cover all the possibilities of the correlation analysis in this particular application.

Some optical experiments on the verification of the above techniques have been carried out at the Institute of Atmospheric Optics, SB USSR Academy of Sciences [5.30].

Atmospheric conditions of stable windy weather with meteorological visual range $S_M \geq 30$ km were chosen for these experiments. The wind speed during these experiments did not exceed 1.5 to 8 ms^{-1}. According to the first variant [i.e., according to (5.35)] the component V_x was determined from the spatial shift of the first maximum of the correlation function $R(\xi,0,\tau_0)$

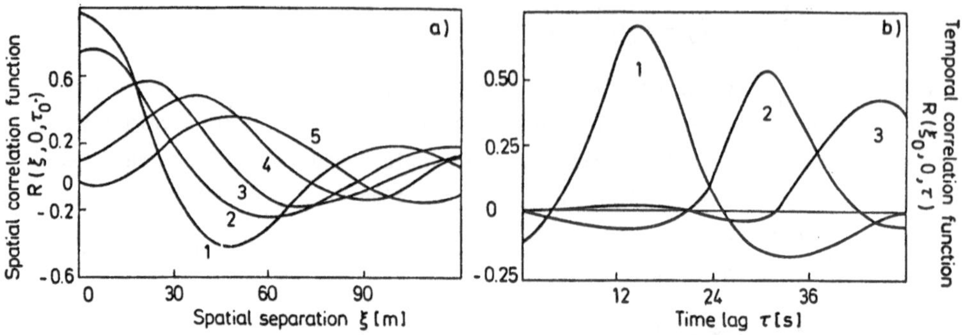

Fig.5.10 (a) Spatial correlation function $R(\xi,0,\tau_0)$ of the lidar returns for different values of τ_0. ($\tau_0 = 0$ for Curve 1 and $\tau_0 = 20$, 40, 60 and 80 s for Curves 2 to 5, respectively.) (b) Temporal correlation functions $R(\xi_0,0,\tau)$ of the lidar returns for $\eta = 0$ and different values of ξ_0 ($\xi_0 = 20$ m) for Curve 1, $\xi_0 = 40$ m and 60 m for Curves 2 and 3, respectively [5.30])

of backscattering coefficient fluctuations when sensing pulses are sent in a time interval τ_0. The laser used as the transmitter delivered light pulses at $\lambda = 0.53\mu m$. The pulse duration was about 10 ns and pulse energy was about 0.02 J. The laser beam divergence was 1 mrad. The pulse repetition frequency was up to 100 Hz. The receiving telescope-refractor had a diameter of 0.15 m and a field of view about 6 mrad. The post PMT electronics had a bandwidth of 50 MHz. The optical filter bandwidth was 4 nm.

At a range of about 800 m the signal-to-noise ratio was about 30 under the conditions of a slightly turbid atmosphere over a range interval of several meters. Simultaneous measurements of the wind velocity were carried out using a standard anemometer installed on a mast near the lidar. Figure 5.10a shows spatial correlation functions $R(\xi,0,\tau_0)$ for different time parameters τ_0.

It is seen from the figure that ξ_m increases with increasing time interval between sensings. At the same time, the sharpness of the maxima decreases, which leads to an increase of the errors in the determination of ξ_m and, as a consequence, in the measurements of wind speed. The mean value of $\xi_m\tau_0^{-1}$ was about 1.6 m s^{-1}. Standard anemometric measurements showed that the mean, over the time period of lidar measurements, wind speed was about 1.4 m s^{-1}. Such an agreement between two methods can be considered good enough. The errors in calculations of the correlation functions did not exceed 10%.

Consider now examples of the time correlation analysis of the lidar signals. As was shown above, the time lags of maxima of correlation function $R(\xi_0,0,\tau)$ can provide, if the aerosol inhomogeneities are isotropic, the determination of the component V'_x according to (5.35). Corresponding experimental studies have been carried out under conditions, such that no extraneous aerosol formations were observed within the sensing range.

Figure 5.10b shows the correlation functions $R(\xi_0,\tau)$ for different values of ξ_0. The sensings were performed along the wind direction, i.e., $\varphi = 0^0$. As in the above example, the modulus of wind speed was about 1.4 m s^{-1}. The observation time was about 2 min. It is seen from Fig.5.10b that τ_m increases as ξ_0 increases. Since the absolute correlation also decreases, the optimal value of ξ_0 may be regarded as being between 20 and 40 m. The sizes of aerosol inhomogeneities can be assumed to be within this interval. The mean value of V was estimated in terms of τ_m value using the ratio ξ_0, τ_m^{-1}. As numerical assessments have shown, the efficiency of this method is approximately the same as of the previous one.

The r.m.s. deviations between lidar and anemometric wind measurements were found to be near 0.5 m s^{-1} for wind velocities from 1 to 8 m s^{-1} [5.30].

It should be noted in conclusion that both correlation and Doppler techniques for measuring wind velocity are too complicated technically for use in conventional studies. This explains the small number of papers which have appeared on this subject.

5.5.2 Lidar Sensing of the Turbulent Parameters of the Atmosphere

Significant progress in studies of laser beam propagation through the turbulent atmosphere achieved in recent years has provided a basis for the successful solution of inverse problems of the optics of turbulent atmospheres. The fundamental parameter determining the intensity of micropulsations of the refractive index and finally the maximum number of statistical characteristics of a laser beam in the atmosphere is the structural parameter C_n^2 of the refractive index.

The idea of determining of C_n^2 using lidar suggested in [5.31] is the following. As is known, the field of the backscattered radiation is described in the single scattering approximation as follows

$$U^{(s)}(\mathbf{r}) = \sum_{j=1}^{N} U^{(0)}(\mathbf{r}_j)\ \exp[ik(\mathbf{r} - \mathbf{r}_j)]/|\mathbf{r} - \mathbf{r}_j| \quad , \tag{5.38}$$

where $\mathbf{r}$ and $\mathbf{r}_j$ are the radius-vectors of the observation point and of the jth particle in a scattering volume (N is the total number of particles). Using (5.38) one can easily determine the complex degree of coherence

$$\gamma_2 = |\Gamma_2(\mathbf{r}_1,\mathbf{r}_2)|/|\Gamma_2(\mathbf{r}_1,\mathbf{r}_1)\Gamma_2(\mathbf{r}_2,\mathbf{r}_2)|^{1/2} \quad , \quad \text{where}$$

$$\Gamma_2(\mathbf{r}_1,\mathbf{r}_2) = \langle U^{(s)}(\mathbf{r}_1) \cdot U^{*(s)}(\mathbf{r}_2)\rangle \quad .$$

Averaging the fluctuations over the sensing path and particle positions independently and taking into account the random phase difference in the incident and scattered waves one can find that

$$\gamma_2(r) = \exp[-D_s(r)/2 - (2zr/kd)^2] \quad , \tag{5.39}$$

where z is the path length, $r = |\mathbf{r}_1 - \mathbf{r}_2|$, k is the wave number, d is the mean size of a scattering volume, $D_s(r)$ is the structure function of a spherical wave phase propagating from the center of a scattering volume to the lidar receiver. It was assumed, when deriving (5.39), that the values of $\mathbf{r}_j$ are distributed uniformly, while the distribution of values of N follows the Poisson law. According to [5.31] one can write

$$D_s(r) = 2(r/r_c)^{5/3}$$

$$\bar{r}_c = \left[1.45\, k^2 \int_0^z c_n(\tilde{z})(\tilde{z}/z)^{5/3}\, d\tilde{z}\right]^{-3/5} \quad ,$$

where $\tilde{z}$ is measured beginning from a scattering volume. Determining the coherence radius r_{cd} of the reflected wave from the condition $\gamma_2(r_{cd}) = e^{-1}$ and using the approximation $D_s(r) \simeq (r/r_c)^2$ one can find that

$$r_{cd}^2 = (\bar{r}_c^{-1/2} + k^2d^2/4z^2)^{-1} \quad . \tag{5.40}$$

It is seen from (5.40) that $r_{cd} \to r_c$, if $d \to 0$, and then one can determine the integral value $\bar{c}_n^2$ for the path of length z by measuring the coherence radius in the backscattered wave. In such an approach,a limitation is imposed on the desired size of a scattering volume due to the turbulent diffusion of the sensing beam. However, the focusing saturation effect can be used to overcome this difficulty [5.32]. Let us take a single-mode focused sensing beam whose source has size $d \gg \bar{r}_c$. Then the mean size of the beam at a focus is $2z(k\,\bar{r}_c)^{-1}$. If the receiver field of view is larger than the divergence angle of the sensing beam, then the value d is determined by the mean size of the beam at a focus, i.e., $d = 2z(k\bar{r}_c)^{-1}$. Substituting this result into (5.40) one can find that $r_{cd} = \bar{r}_c/\sqrt{2}$. Thus, under the condition of saturated focusing ($d \gg \bar{r}_c$) the radius of a spatial coherent in the field of a wave scattered by aerosols is determined only by the atmospheric turbulence along the sensing path. This fact allows one to determine the integral value $\bar{c}_n^2$ by measuring r_{cd}. In principle, one can measure the profile $\bar{c}_n^2(z)$ by moving the focused region along the sensing path. The reference equation for this case can be written as follows

$$z^{-1} \int_0^z c_n^2(z')(1 - z'/z)^{5/3}\, dz' = \bar{c}_n^2(z) \quad . \tag{5.41}$$

Expression (5.41) is a Volterra equation of the first kind as is (5.23) and can be solved by the same numerical method (Sect.5.4.3). Omitting a discussion of these questions, we make some remarks on optical measurements of r_{cd}. The bistatic lidar scheme is the simplest for such an application. When the receiver field of view is centered at the focused region of a sensing beam, the receiving optics makes an image of this region in the form of a bright line. The width of this line is uniquely determined by the coherence scale r_{cd} and can be evaluated by image scanning with a narrow slit. Then one can find values c_n^2 according to

$$\bar{c}_n^2 = 0.165(y_{0.5}/F)^{5/3}(k^{1/3}z)^{-1} \tag{5.42}$$

where $y_{0.5}$ is the halfwidth of the light intensity distribution at the image plane of the receiving optics at a half-maximum level, F is the receiver focus.

In the case of atmospheric haze with meteorological visual range $S_M \geq 20$ km, the influence of aerosol light scattering on the value $y_{0.5}$ is negligible as compared with the turbulent effects and, hence, in the above method the aerosol is important only as a tracer or diffuse screen since it determines the intensity of scattered radiation.

An experimental verification of the method has been undertaken in [5.32] along horizontal ground paths of length 30 m and 130 m. Accompanying measurements of c_n^2 from the characteristics of the direct beam were also made. The measurements were made using a bistatic lidar with a cw He-Ne single-mode laser ($\lambda = 0.63 \mu$m) of 80 mW power. The sensing beam was focused with an optical system having output diameter 2d = 16 cm. A telescope AZT-7 with an optical unit for adjusting the imaging was used as a receiver. In the experiment, the reflection of a sensing beam from the diffuse scattering screen and from an artificial haze made by burning the wood sawdust near a sensing beam focus (z = 130 m) and from the natural atmospheric aerosols (z = 30 m) were investigated. A dissector and a slit were used for measuring the value of $y_{0.5}$ by scanning the image of a scattering volume. A comparison of data on c_n^2 obtained by the two independent methods revealed good agreement between them.

The use of lasers enables one to determine the analytical form of the turbulence spectra. As a matter of fact,the small wavelengths of laser radiation allows in principle the determination of fine structures of atmospheric turbulence which are impossible with other methods.

A more detailed description of the laser methods for sensing atmospheric turbulence, as well as the algorithms for reconstructing its characteristics together with the experimental results,are given in [5.26].

Note in conclusion that from simultaneous lidar measurements of wind velocity and turbulent characteristics of the atmosphere,one can obtain certain information on atmospheric aerosols as well. On the other hand, the above methods for solving the inverse problems of light scattering by aerosols can successfully be used for interpreting the data on air motion. It should be expected that further investigations will continue in just this manner.

5.5.3 Concluding Remarks

Let us give a short summary and discuss some perspectives of the further development of lidar methods for sensing atmospheric parameters based on light scattering by aerosols.

Note, first of all, the significant progress in solving the problem of lidar sensing of aerosols which has been achieved recently. Just during these years highly efficient algorithms for restoring the microphysical parameters of aerosols from the lidar sensing data have been developed. These algorithms were then successfully used for interpreting the data of multi-frequency lidar sensing of aerosols.

Thus the problem of reconstructing the size spectrum and number density of aerosol particles based on the single-scattering approximation may be regarded solved at least for spherical particles. Of course, these limitations restrict the potentialities of these algorithms as is discussed below.

As is shown theoretically and experimentally in [5.25,26], the applicability limits of the single-scattering approximation are uniquely determined by the optical properties of aerosols (volume scattering coefficient and scattering phase function) and by the geometrical parameters of a lidar scheme (transmitter beam divergence and receiver field of view).

The effects of multiple scattering for generally accepted geometrical parameters are shown in [5.26,33] to be negligible in the case of atmospheric haze regardless of the meteorological visual ranges. At the same time, these effects are essential when sensing clouds. However, even in this case they can be neglected if the lower layers of clouds are sensed, or narrow fields of view and narrow (with small divergence angles) sensing beams are used. These criteria allow one, at least, to estimate the optical depth of a cloud for which the single scattering approximation remains valid.

In a real atmosphere, nonspherical aerosol particles are frequently observed. However, the random orientations of such particles enable one to use the concept of equivalent spheres [5.33] and then to apply the apparatus developed for spherical particles to the reconstruction of microphysical parameters of ensembles of nonspherical particles.

One should note significant progress in differential absorption lidar techniques. The profiles of atmospheric water vapor have been measured using this method up to record heights with sufficient precision.

The results of sensing the profiles of atmospheric ozone are also very promising. The same can also be said of the lidar determination of atmospheric temperature profiles, profiles of wind velocity and turbulent characteristics of the atmosphere, although only preliminary successful attempts in these directions have been made.

The authors hope that the material presented in this monograph will be useful for obtaining new and more interesting results when solving certain problems of lidar sensing of the atmosphere with the use of aerosol light scattering.

The following problems are to be mentioned among the problems to be solved in the near future using the above algorithms for inverting the lidar sensing data:

1) Determination of both real and imaginary parts of the complex index of refraction of aerosol substance.

2) Determination of size spectra and concentration of aerosol particles from the bistatic lidar angular measurements of scattered radiation, as well as from the polarization measurements.

3) Reconstruction of profiles of atmospheric water vapor, ozone and other atmospheric gases using the differential absorption lidar technique. In the cases of ozone and water vapor, a further increase of the sensing range and experimental accuracy is required.

4) Establishing lidar monitoring of the atmosphere aimed at a detailed study of the processes of air pollution due to human industrial activities.

The methods presented in this monograph are expected to be especially useful and even to give qualitatively new results on various atmospheric processes if used simultaneously in proper combinations. Some typical and important examples are given below.

It is well known that atmospheric temperature inversions hinder the dissipation of air pollutants from the boundary layer by making the lower, contaminated atmospheric layer heavier. Moreover, as was observed during the Soviet-Bulgarian experiment near Razlog, Bulgaria, the heights of the inversion layers are lowered at night time, which led to additional contamination of the ground layer.

It is clear qualitatively that the process leading to temperature inversions are related to the humidity field, fields of wind velocity, turbulent diffusion and the aerosol microstructure. Evidently one can make an attempt at finding relationships among these fields by simultaneously sensing them using lidars. Consequently, recommendations of practical importance in air pollution control can be elaborated based on such relationships.

Another extremely interesting problem on the origin of noctilucent clouds, as well as on their dynamics and dissipation, can be solved by simultaneous sensing of different physical parameters of the atmosphere. The sodium vapor of the upper atmosphere can serve as a tracer in such investigations especially if the enormously large cross-section of resonance fluorescence for Na atoms is taken into account. On the other hand, the microstructure of noctilucent clouds can be determined using the above method of multifrequency sensing. Naturally, in order to obtain reliable and sufficiently precise data one should create a special receiving antenna.

Finally, simultaneous lidar measurements of different atmospheric parameters mentioned above, made in a local volume with high spatial and temporal resolution make it possible to study in detail the dynamics of various atmospheric processes, and their physical nature as well.

It goes without saying that these and other important studies of the atmosphere as a complicated dynamical medium using lidars will require corresponding improvements of the lidar facilities. First of all, lasers with the required spectral and energetic parameters should be developed. The development of the lidars themselves should evidently be undertaken in two directions: 1) stationary lidar systems and 2) mobile lidar complexes.

Stationary lidar systems may consist of special receiving antennas, a very efficient laser and other components. The use of such systems should be aimed at achieving the maximum possible measurement precision and the highest spatial and temporal resolution over as large a range as possible.

Mobile lidar systems should evidently provide high operativeness in solving applied problems of lidar sensing of the atmosphere as a primary aim. Both types of lidar systems should include sophisticated data acquisition systems allowing not only real time data processing but also computer control of the experiment.

Urgent needs to obtain reliable data on atmospheric parameters in such extremely important problems as weather forecasting and air quality control are a fundamental stimulus for further development of all the methods discussed here.

References

Chapter 1

1.1 K.W. Rothe, U. Brinkmann, H. Walther: Appl. Phys. *3*, 115-119 (1974)
1.2 E.D. Hinkley (ed.): *Laser Monitoring of the Atmosphere*, Topics in Applied Physics, Vol.14 (Springer, Berlin, Heidelberg, New York 1976)
1.3 W.H. Marlow (ed.): *Aerosol Microphysics I*, Topics in Current Physics, Vol.16 (Springer, Berlin, Heidelberg, New York 1980)
1.4 W.H. Marlow (ed.): *Aerosol Microphysics II*, Topics in Current Physics, Vol.29 (Springer, Berlin, Heidelberg, New York 1982)
1.5 J.W. Strohbehn (ed.): *Laser Beam Propagation in the Atmosphere*, Topics in Applied Physics, Vol.25 (Springer, Berlin, Heidelberg, New York 1978)
1.6 V.E. Zuev: *Laser Beam in the Atmosphere* (Plenum, New York 1981)
1.7 G.I. Marchuk, G.A. Mikhailov, M.A. Nazaraliev, R.A. Darbinjan, B.A. Kargin, B.S. Elepov: *The Monte Carlo Methods in Atmospheric Optics*, Springer Series in Optical Sciences, Vol.12 (Springer, Berlin, Heidelberg, New York 1980)

Chapter 2

2.1 H.P. Baltes (ed.): *Inverse Scattering Problems in Optics*, Topics in Current Physics, Vol.20 (Springer, Berlin, Heidelberg, New York 1980)
2.2 W.H. Marlow (ed.): *Aerosol Microphysics I*, Topics in Current Physics, Vol.16 (Springer, Berlin, Heidelberg, New York 1980) Chap.4
2.3 I.E. Naats: "'Ill-Posed' Inverse Problems of Lidar Sounding of the Atmosphere", in *Remote Methods for Atmospheric Research*, ed. by V.E. Zuev (Nauka, Novosibirsk 1980) (in Russian)
2.4 B. Grünbaum: *Measures of Symmetry for Convex Sets*, Symp. Pure Math. (Convexity), Providence, RI (1963) Vol.7
2.5 A.C. Holland, G. Gagne: Appl. Opt. *9*, 1113-1121 (1970); *10*, 1173-1174 (1971)
2.6 H.C. Van de Hulst: *Light Scattering by Small Particles* (Wiley, New York 1957)
2.7 M.E. Kerker: *The Scattering of Light and Other Electromagnetic Radiation* (Academic, New York 1969) p.606
2.8 I.E. Naats: The determination of a random chord's momenta in an ellipsoid. Izv. Tomsk. Politekh. Inst. *203*, 80-83 (1974)
2.9 I.E. Naats: On the determination of statistical properties of random chords in convex bodies. Izv. Tomsk. Politekh. Inst. *203*, 75-79 (1974)

2.10 I.E. Naats: *The Theory of Multifrequency Lidar Sounding of the Atmosphere* (Nauka, Novosibirsk 1980) (in Russian)
2.11 G.N. Plass, G.W. Kattawar: Appl. Opt. *10*, 1172-1173 (1971)
2.12 I.E. Naats: "On an Inverse Problem of Dispersed Media Optics", in *Atmospheric Aerosol Studies Using Lidar Methods*, ed. by V.E. Zuev (Nauka, Novosibirsk 1980) pp.55-70 (in Russian)
2.13 A. Deepak, O.H. Vaughan: Appl. Opt. *17*, 375-378 (1978)
2.14 I.E. Naats: "Questions on the Interpretation of the Spectral Behavior of the Aerosol Scattering Coefficient", in *Topics in Lidar Sounding of the Atmosphere*, ed. by V.E. Zuev (Nauka, Novosibirsk 1976) pp.74-83 (in Russian)
2.15 I.E. Naats: "On the Estimation of Size-Distribution Smoothness in Inverse Problems of Light Scattering by Aerosols", in *Topics in Lidar Sounding of the Atmosphere*, ed. by V.E. Zuev (Nauka, Novosibirsk 1976) pp.84-92 (in Russian)
2.16 I.E. Naats: "Qualitative Methods in the Theory of Light Scattering by Polydispersions and Their Use for Solving Inverse Problems of Lidar Sounding", in *Remote Sensing of the Atmosphere*, ed. by V.E. Zuev (Nauka, Novosibirsk 1978) pp.69-83 (in Russian)
2.17 E.G. Pinnick, D.E. Carroll, D.J. Hoffman: Appl. Opt. *15*, 384-393 (1976)
2.18 D. Deirmendjian: *Electromagnetic Wave Scattering on Spherical Polydispersions* (American Elsevier, New York 1969)
2.19 G.L. Knestrick, T.H. Cosden, J.A. Curcio: J. Opt. Soc. Am. *52*, 1010 (1962)
2.20 G.H. Hardy, W.W. Rogosinski: *Fourier Series* (University Press, Cambridge 1956)
2.21 I.E. Naats: "Questions of the Theory of Remote Determination of Atmospheric Aerosol Microstructures Using Lidar Methods", in *Laser Sounding of the Atmosphere*, ed. by V.E. Zuev (Nauka, Moscow 1976) pp.3-10 (in Russian)
2.22 A.N. Tikhonov: Dokl. Akad. Nauk SSSR *151*, 501-504 (1963); *153*, 49-59 (1963)
2.23 P. Chylek, R.G. Pinnick: Science *193*, 480-482 (1976)
2.24 H.C. van de Hulst: *Light Scattering by Small Particles* (Wiley, New York 1964)
2.25 C. Lanczos: *Applied Analysis* (Prentice Hall, Reading, MA 1956)
2.26 I.E. Naats: Opt. Spectrosc. *35*, 966-968 (1973) (in Russian)
2.27 I.E. Naats: "Estimation of the Efficiency of Optical Methods in Studies of Atmospheric Aerosol Microstructures" in *Propagation of Optical Waves Through the Atmosphere*, ed. by V.E. Zuev (Nauka, Novosibirsk 1975) pp.202-208 (in Russian)
2.28 I.E. Naats: "On the Interpretation of Optical Sounding of Atmospheric Aerosols in the IR", in *Topics in Remote Sensing of the Atmosphere* (Institute of Atmospheric Optics SB USSR Academy of Sciences, Tomsk 1975) pp.35-48 (in Russian)
2.29 V.V. Sobolev: *Light Scattering by Planets' Atmospheres* (Nauka, Moscow 1972) p.335 (in Russian)
2.30 V.A. Morozov: "Linear and Nonlinear 'Ill-Posed' Problems", in *Results of Science and Technology. Mathematical Analysis*, Vol.11 (VINITI, Moscow 1973) pp.129-173 (in Russian)
2.31 V.K. Ivanov, V.V. Vasin, V.P. Tanana: *The Theory of Linear "Ill-Posed" Problems and Its Applications* (Nauka, Moscow 1978) p.206 (in Russian)
2.32 E.K. Bigg: "On the Size Distribution and Nature of Stratospheric Aerosols", IAMAP/IAPSO Comb. 1st Special Assembly, Melbourne, Australia (1974); J. Atmos. Sci. *32*, 910-917 (1975)
2.33 B.S. Kostin, E.V. Makienko, I.E. Naats: "Investigations on the Information Content and Solution of Inverse Problems in Optical Sounding of Atmospheric Aerosols" in *Propagation of Optical Waves Through the Atmosphere*, ed. by V.E. Zuev (Nauka, Novosibirsk 1975) pp.208-213 (in Russian)

2.34 E.V. Makienko, I.E. Naats: "On an Algorithm for Inverting Spectral Optical Measurements", in *Topics in Lidar Sounding of the Atmosphere*, ed. by V.E. Zuev (Nauka, Novosibirsk 1976) pp.115-121 (in Russian)
2.35 B.S. Kostin, I.E. Naats: "Determination of Aerosol Size Spectrum from Optical Measurements Using Regularization Methods", in *Lidar Sounding of the Atmosphere*, ed. by V.E. Zuev (Nauka, Moscow 1976) pp.94-98 (in Russian)
2.36 I.Ya. Badinov, A.V. Poberovsky, L.V. Popova, A.P. Bobyleva: "Ground-Based Measurements of Atmospheric Optical Depth in the Region 0.3 to 2.5 μm", in Trudy GGO N276 (Gidrometizdat, Leningrad 1972) pp.89-91 (in Russian)
2.37 V.I. Dmokhovsky, A.S. Ivlev, A.Yu. Semenova: "Aerosol Measurements in the Ground Layer in Kara-Kum", in Trudy GGO N276 (Gidrometizdat, Leningrad 1972) pp.109-112 (in Russian)
2.38 R.J. Keyes (ed.): *Optical and Infrared Detectors*, Topics in Applied Physics, Vol.19, 2nd ed. (Springer, Berlin, Heidelberg, New York 1980) Chap.4
2.39 N.S. Laulsinen, A.I. Alkezweenny, I.M. Thorp: J. Appl. Meteorol. *17*, 615-626 (1978)
2.40 E.V. Makienko, I.E. Naats: "Questions on the Optimal Estimation of Aerosol Size-Distribution Parameters from Optical Measurements", in *Atmospheric Optics* (Nauka, Moscow 1974) pp.186-191 (in Russian)
2.41 E.V. Makienko, I.E. Naats: "Investigations of the Information Content of Lidar Sounding of Atmospheric Aerosols", in *Lidar Sounding of the Atmosphere*, ed. by V.E. Zuev (Nauka, Moscow 1976) pp.11-16 (in Russian)
2.42 M. Reed, B. Simon: *Methods of Modern Mathematical Physics*, Vol.I (Academic, New York 1974) p.357
2.43 V.E. Zuev, Yu.S. Balin, B.S. Kostin, I.E. Naats, I.V. Samokhvalov: "Optical Experiments and the Results of Data Inversion on Multifrequency Sensing of the Aerosol Microstructure of the Surface Layer", in *Problems of Remote Sensing of the Atmosphere* (IOA SO USSR Academy of Sciences, Tomsk 1976) pp.3-20 (in Russian)
2.44 V.E. Zuev, N.V. Kozlov, E.V. Makienko, I.E. Naats, I.V. Samokhvalov: Some results of sounding of stratospheric aerosol microstructure by a multifrequency lidar. Izv. Akad. Nauk SSSR. Fiz. Atmos. Okeana *13*, 548 (1977)
2.45 B.S. Kostin, E.V. Makienko, I.E. Naats: *"Determination of Aerosol Microstructure Parameters in the Upper Atmosphere Using Multifrequency Lidar Sounding"*, Preprint N21 (Institute of Atmospheric Optics SB USSR Academy of Sciences, Tomsk 1978) p.63 (in Russian)
2.46 V.E. Zuev, N.V. Kozlov, E.V. Makienko, I.E. Naats, I.V. Samokhvalov: "Some Results on Laser Sensing of the Stratospheric Aerosol Microstructure Using a Three-Frequency Lidar". Conf. Abstracts 4th All-Union Symp. Laser Sounding of the Atmosphere, Tomsk, 1976, p.142 (in Russian)
2.47 K.J. Kondratjev: *Atmospheric Aerosol and Its Influence on Radiation Transfer* (Gidrometeoizdat, Leningrad 1978) p.120 (in Russian)
2.48 V.V. Veretennikov, I.E. Naats: "On the Determination of Aerosol Characteristics by Inverting the Data of Polarization Measurements" in *Lidar Sounding of the Atmosphere*, ed. by V.E. Zuev (Nauka, Moscow 1976) pp.20-29 (in Russian)
2.49 R. Eiden: Appl. Opt. *10*, 749 (1971)
2.50 K. Bullrich, R. Eiden, G. Hänel: Scientific Rpt. N7, Contract F61052 (67 00461) 1969

Chapter 3

3.1 I.E. Naats: "Questions of the Theory of Remote Determination of Atmospheric Aerosol Microstructures Using Lidar Methods", in *Laser Sounding of the Atmosphere*, ed. by V.E. Zuev (Nauka, Moscow 1976) pp.3-10 (in Russian)
3.2 D. Deirmendjian: *Electromagnetic Wave Scattering on Spherical Polydispersions* (American Elsevier, New York 1969)
3.3 D.B. Rensch, R.K. Long: Appl. Opt. *10*, 1563-1573 (1970)
3.4 B.W. Fitch: J. Appl. Meteorol. *20*, 1119-1128 (1981)
3.5 K.J. Kondratjev: *Atmospheric Aerosol and Its Influence on Radiation Transfer* (Gidrometeoizdat, Leningrad 1978) p.120 (in Russian)
3.6 T.P. Toropova, A.B. Kasjanenko, K.M. Salomakhin, A.P. Ten, O.D. Tokarev: "Extinction of Light in the Ground Layer and Atmospheric Aerosol", in *Radiation Scattering Field in the Earth's Atmosphere* (Nauka, Alma-Ata 1974) pp.32-88 (in Russian)
3.7 V.V. Veretennikov, I.E. Naats, I.V. Samokhvalov, V.S. Shamanaev: "Determination of Profiles of Aerosol Optical Characteristics Using Lidar Sounding of the Atmosphere", in *Lidar Sounding of the Atmosphere*, ed. by V.E. Zuev (Nauka, Moscow 1976) pp.98-103 (in Russian)
3.8 R.T.H. Collis, M.C.H. Ligda: Nature *203*, 508 (1964)
3.9 R.T.H. Collis: Q. J. R. Meteorol. Soc. *92*, 220-230 (1966)
3.10 J.S. Reagan, J.O. Spinhirne, D.M. Byrne, D.W. Thomson, R.G. de Pena, Y. Mamane: Atmospheric particulate properties from lidar and solar radiometer observation compared with simultaneous in situ aircraft measurements, a case study. J. Appl. Meteorol. *16*, 911-928 (1977)
3.11 I.E. Naats: "Questions on the Interpretation of Spectral Behavior of the Aerosol Scattering Coefficient", in *Topics in Lidar Sounding of the Atmosphere*, ed. by V.E. Zuev (Nauka, Novosibirsk 1976) pp.74-83 (in Russian)
3.12 I.E. Naats: "On the Estimation of Size-Distribution Smoothness in Inverse Problems of Light Scattering by Aerosols", in *Topics in Lidar Sounding of the Atmosphere*, ed. by V.E. Zuev (Nauka, Novosibirsk 1976) pp.84-92 (in Russian)
3.13 I.E. Naats: "Qualitative Methods in the Theory of Light Scattering by Polydispersions and Their Use for Solving Inverse Problems of Lidar Sounding", in *Remote Sensing of the Atmosphere*, ed. by V.E. Zuev (Nauka, Novosibirsk 1978) pp.69-83 (in Russian)
3.14 I.E. Naats: "On the Interpretation of Optical Sounding of Atmospheric Aerosols in the IR", in *Topics in Remote Sensing of the Atmosphere* (Institute of Atmospheric Optics SB USSR Academy of Sciences, Tomsk 1975) pp.35-48 (in Russian)
3.15 G.L. Knestrick, T.H. Cosden, J.A. Curcio: J. Opt. Soc. Am. *52*, 1010 (1962)
3.16 W.M. Irvine, F.W. Peterson: J. Atmos. Sci. *27*, 62-69 (1970)
3.17 I.E. Naats: "Estimation of the Efficiency of Optical Methods in Studies of Atmospheric Aerosol Microstructures", in *Propagation of Optical Waves Through the Atmosphere*, ed. by V.E. Zuev (Nauka, Novosibirsk 1975) pp.202-208 (in Russian)
3.18 I.E. Naats: *The Theory of Multifrequency Lidar Sounding of the Atmosphere* (Nauka, Novosibirsk 1980) (in Russian)
3.19 E.V. Makienko, I.E. Naats: "An Investigation of the Information Content of Laser Sensing of Atmospheric Aerosol", in *Laser Sounding of the Atmosphere*, ed. by V.E. Zuev (Nauka, Moscow 1976) (in Russian)
3.20 I.E. Naats: "On the Selection of Wavelengths for Multifrequency Sounding of the Atmosphere", in *Methods and Devices for the Remote Sensing of*

Atmospheric Parameters, ed. by V.E. Zuev (Nauka, Novosibirsk 1980) pp.61-64 (in Russian)
3.21 S.T. Rasool: *Chemistry of the Lower Atmosphere* (Plenum, New York 1973)
3.22 T.E. Graedel, J.P. Franey: J. Geophys. Res. *79*, 5643-5645 (1974)
3.23 T.P. Toropova, A.P. Ten, G.N. Bushueva, O.D. Tokarev: "Optical Characteristics of the Ground Layer of the Atmosphere", in *Extinction of Light in the Earth's Atmosphere* (Nauka, Alma-Ata 1976) pp.33-112 (in Russian)
3.24 R. Reiter, W. Carnuth: Annual Report, Vol.1, Institut für Atmosphärische Umweltforschung (1976) p.42
3.25 W. Koechner: *Solid-State Laser Engineering*, Springer Series in Optical Sciences, Vol.1 (Springer, Berlin, Heidelberg, New York 1976)
3.26 A.N. Tikhonov, V.Ya. Arsenin: *Methods for Solving "Ill-Posed" Inverse Problems* (Nauka, Moscow 1974) p.203 (in Russian)
3.27 I.E. Naats: "Incorrect Inverse Problems of Laser Sounding of Atmospheric Aerosols", in *Remote Methods of Investigation of the Atmosphere*, ed. by V.E. Zuev (Nauka, Novosibirsk 1980) pp.41-89 (in Russian)
3.28 E.M. Patterson: Appl. Opt. *16*, 2414-2418 (1977)
3.29 J.P. Friend: Tellus *18*, 465-473 (1966)
3.30 I.E. Naats: Opt. Spectrosk. *35*, 966-968 (1973)
3.31 I.E. Naats: *Propagation of Laser Radiation Through the Atmosphere*, Conf. Abstracts 2nd All-Union Symp. (Tomsk, USSR 1973) pp.151-153 (in Russian)
3.32 E.V. Makienko, I.E. Naats: Opt. Lett. *5*, 135-137 (1980)
3.33 G.M. Krekov, M.M. Krekova, E.V. Makienko, I.E. Naats: Izv. Vyssh. Uchebn. Zaved. Radiofiz. *20*, 528-537 (1977) [English transl.: Sov. Radiophys.]
3.34 V.E. Zuev, G.M. Krekov, M.M. Krekova, I.E. Naats: "Theoretical Aspects of Lidar Sounding of Clouds", in *Topics in Lidar Sounding of the Atmosphere*, ed. by V.E. Zuev (Nauka, Novosibirsk 1976) pp.3-33 (in Russian)

Chapter 4

4.1 D. Deirmendjian: *Electromagnetic Waves Scattering on Spherical Polydispersions* (American Elsevier, New York 1969)
4.2 R.J. Perry, A.J. Hunt, D.R. Huffman: Appl. Opt. *17*, 2700-2710 (1978)
4.3 S. Asano, G. Yamamoto: Appl. Opt. *14*, 29-49 (1975)
4.4 V.V. Veretennikov, I.E. Naats: Izv. Akad. Nauk SSSR Fiz. Atmos. Okeana *15*, 560-563 (1979) [English transl.: Izv. Acad. Sci. USSR Atmos. Oceanic Phys.]
4.5 H.C. Van-de Hulst: *Light Scattering by Small Particles* (Wiley, New York 1957)
4.6 C. Lanczos: *Applied Analysis* (Prentice Hall, Reading, MA 1956)
4.7 I.E. Naats: *The Theory of Multifrequency Lidar Sounding of the Atmosphere* (Nauka, Novosibirsk 1980) (in Russian)
4.8 I.E. Naats: "'Ill-Posed' Inverse Problems of Lidar Sounding of the Atmosphere", in *Remote Methods for Atmospheric Research*, ed. by V.E. Zuev (Nauka, Novosibirsk 1980)
4.9 V.V. Veretennikov, I.E. Naats: "Inverse Problems of Lidar Sounding of the Atmosphere Taking into Account Polarization Measurements", in *Problems of Remote Sensing of the Atmosphere* (Institute of Atmospheric Optics SB USSR Academy of Sciences, Tomsk 1976) pp.86-97 (in Russian)
4.10 A.N. Tikhonov, V.Ya. Arsenin: *Methods for Solving "Ill-Posed" Inverse Problems* (Nauka, Moscow 1974) p.203 (in Russian)

4.11 I.E. Naats: "Estimation of the Efficiency of Optical Methods in Studies of Atmospheric Aerosol Microstructures", in *Propagation of Optical Waves Through the Atmosphere,* ed. by V.E. Zuev (Nauka, Novosibirsk 1975) pp.202-208 (in Russian)
4.12 J.S. Reagan, J.O. Spinhirne, D.M. Byrne, D.W. Thomson, R.G. de Pena, Y. Mamane: Atmospheric particulate properties from lidar and solar radiometer observation compared with simultaneous in situ aircraft measurements, a case study. J. Appl. Meteorol. *16*, 911-928 (1977)
4.13 T.P. Toropova, A.P. Ten, G.N. Bushueva, O.D. Tokarev:"Optical Characteristics of the Ground Layer of the Atmosphere", in *Extinction of Light in the Earth's Atmosphere* (Nauka, Alma-Ata 1976) pp.33-112 (in Russian)
4.14 L.S. Ivlev, S.I. Popova: Izv. Akad. Nauk SSSR, Fiz. Atmos. Okeana *9*, 1035-1043 (1973) [English transl.: Izv. Acad. Sci. USSR, Atmos. Oceanic Phys.]
4.15 G. Hänel: Tellus *20*, 371-379 (1968)
4.16 I.E. Naats: "Questions of the Theory of Remote Determination of Atmospheric Aerosol Microstructures Using Lidar Methods", in *Laser Sounding of the Atmosphere,* ed. by V.E. Zuev (Nauka, Moscow 1976) (in Russian)
4.17 A.L. Irisov, M.V. Panchenko, B.A. Saveljev, V.Ya. Fadejev: "Spectral Atmospheric Nephelometer", in *Propagation of Optical Waves Through the Atmosphere,* ed. by V.E. Zuev (Nauka, Novosibirsk 1975) pp.52-56 (in Russian)
4.18 S.T. Rasool: *Chemistry of the Lower Atmosphere* (Plenum, New York 1973)
4.19 I.E. Naats: "Lidar and Acoustic Sounding of the Atmosphere", 5th All-Union Symp. (Institute of Atmospheric Optics, Tomsk, USSR 1978) pp.3-11 (in Russian)
4.20 V.S. Kozlov, V.F. Panin, G.A. Rappoport, V.Ya. Fadejev: "Investigations of Microphysical and Optical Characteristics of Smoke Aerosols", in *Light Scattering and Refraction of Optical Waves in the Atmosphere* (Institute Atmospheric Optics SB USSR Academy of Sciences, Tomsk 1976) pp.78-95 (in Russian)
4.21 V.E. Zuev, B.V. Kaul, V.N. Kozlov, I.V. Samokhvalov: "Lidar Measurements of Light Scattering by Nonspherical Particles in the Upper Atmosphere", in Conf. Abstracts 6th All-Union Symp. Lidar and Acoustic Sounding of the Atmosphere (Tomsk, USSR 1980) pp.11-14 (in Russian)
4.22 A.S. Holland: Invited paper, 8th Int. Laser Radar Conference, Drexel University, Philadelphia, PA (1977)
4.23 B.H. McKellar: Q. J. R. Meteorol. Soc. *100*, 687-691 (1974)
4.24 A.L. Fymat: Appl. Opt. *18*, 126-130 (1979)
4.25 A.C. Holland, G. Gagne: Appl. Opt. *9*, 1113-1121 (1970); *10*, 1173-1174 (1971)
4.26 E.G. Pinnick, D.E. Carroll, D.J. Hoffman: Appl. Opt. *15*, 384-393 (1976)
4.27 G.N. Plass, G.W. Kattawar: Appl. Opt. *10*, 1172-1173 (1971)
4.28 I.E. Naats: "Lidar and Acoustic Sounding of the Atmosphere", 5th All-Union Symp. (Institute of Atmospheric Optics, Tomsk, USSR 1978) pp.3-11 (in Russian)
4.29 Yu.S. Balin, G.O. Zadde, G.G. Matvienko, I.V. Samokhvalov, V.S. Shamanaev: "Some Results of Sounding of Meteorological Formations with a Polarization Lidar", in *Propagation of Optical Waves in the Atmosphere* (Nauka, Novosibirsk 1975) pp.183-186 (in Russian)
4.30 R.H. Kohe, M.I. Flaherry, R.L. Portin: "Measured Backscatter and Attenuation Properties Including Polarization Effects of Various Atmospheric Dispersions at 0.9 m", Conf. Abstracts 8th Int. Laser Radar Conf., Drexel University, Philadelphia, PA (1977)
4.31 R.A. McClatchey, R.W. Fenn, J.E. Selby, F.E. Volz, J.S. Garing: "Optical Properties of the Atmosphere", AFCRL-71-0279, Environmental Research Papers No.354 (1971) p.85

4.32 K.S. Shifrin, V.A. Punina: Izv. Akad. Nauk SSSR Fiz. Atmos. Okeana *4*, 784-791 (1968) [English transl.: Izv. Acad. Sci. USSR Atmos. Oceanic Phys.]
4.33 I.E. Naats: Opt. Spektrosk. *39*, 595-596 (1975) [English transl.: Opt. Spectrosc.]
4.34 I.E. Naats: Izv. Akad. Nauk SSSR Fiz. Atmos. Okeana *12*, 329-332 (1976) [English transl.: Izv. Acad. Sci. USSR Atmos. Oceanic Phys.]
4.35 B.P. Ivanenko, J.E. Naats: Izv. Akad. Nauk SSSR Fiz. Atmos. Okeana *13*, 312-314 (1977)
4.36 T.P. Toropova, A.B. Kasjanenko, K.M. Salomakhin, A.P. Ten, O.D. Tokarev: "Extinction of Light in the Surface Layer and Atmospheric Aerosol", in *Radiation Scattering Field in the Earth's Atmosphere* (Nauka, Alma-Ata 1974) pp.32-88 (in Russian)
4.37 I.E. Naats: "Qualitative Methods in the Theory of Light Scattering by Polydisperions and Their Use for Solving Inverse Problems of Lidar Sounding", in *Remote Sensing of the Atmosphere*, ed. by V.E. Zuev (Nauka, Novosibirsk 1978) pp.69-83 (in Russian)
4.38 A. Deepak, M.A. Box: Appl. Opt. *17*, 3169-3176 (1978)

Chapter 5

5.1 A.N. Tikhonov, V.Ya. Arsenin: *Methods for Solving "Ill-Posed" Inverse Problems* (Nauka, Moscow 1974) p.203 (in Russian)
5.2 B.P. Ivanenko, V.N. Marichev: Izv. Vyssh. Uchebn. Zaved. Fiz. *3* (132-80), 39 (1980)
5.3 O.K. Vojitsekhovskaya, Yu.S. Makushkin, V.N. Marichev, A.A. Mitsel, A.V. Sosnin, I.V. Samokhvalov: On laser sounding of water vapor of the atmosphere by the resonance method. Izv. Vyssh. Uchebn. Zaved. Fiz. *1*, 62-70 (1977)
5.4 P.A. McClatchey, R.W. Fenn, J.E. Selby, F.E. Volz, J.S. Garing: "Optical Properties of the Atmosphere", AFCRL-71-0279, Environmental Research Papers No.354 (1971) p.85
5.5 C. Lanczos: *Applied Analysis* (Prentice Hall, Reading, MA 1956)
5.6 S.A. Achemed: Appl. Opt. *12*, 901-903 (1973)
5.7 R.M. Shotkong: J. Appl. Meteorol. *13*, 71-77 (1974)
5.8 A.I. Grishin, G.G. Matvienko: "Lidar and Acoustic Sounding of the Atmosphere", Conf. Abstracts 5th All-Union Symp. (Institute of Atmospheric Optics SB USSR Academy of Sciences, Tomsk 1978) pp.111-115 (in Russian)
5.9 V.G. Gladyshev, V.S. Gulev, A.S. Kasjanov, V.N. Marichev, N.V. Nedel'kin, V.D. Ugojaev, K.G. Folin: "Ruby Laser for Remote Sounding of the Humidity of the Atmosphere", in *Apparatus and Methods of Remote Sounding of Atmospheric Parameters*, ed. by V.E. Zuev (Nauka, Novosibirsk 1980) pp.64-75 (in Russian)
5.10 W. Koechner: *Solid-State Laser Engineering*, Springer Series in Optical Sciences, Vol.1 (Springer, Berlin, Heidelberg, New York 1976)
5.11 V.S. Komarov: "The Statistical Structure of the Air Humidity Field in a Free Atmosphere Above the USSR", in Proc. NIIAK No.70 (Gidrometeoizdat, Moscow 1971) (in Russian)
5.12 O.K. Kostko: "Lidar and Acoustic Sounding of the Atmosphere", Conf. Abstracts 6th All-Union Symp., Vol.1 (Institute of Atmospheric Optics SB USSR Academy of Sciences, Tomsk 1980) pp.119-204 (in Russian)
5.13 V.P. Gusarov, O.K. Kostko, A.P. Prokhorov, N.D. Smirnov: "Lidar and Acoustic Sounding of the Atmosphere", Conf. Abstracts 6th All-Union Symp., Vol.1 (Institute of Atmospheric Optics SB USSR Academy of Sciences, Tomsk 1980) pp.222-224 (in Russian)

5.14 A.J. Gibson, J. Thomas: Nature *256*, 561-563 (1975)
5.15 G. Megie, J.Y. Allain, Chanin, J.E. Blamont: Nature *270*, 349-351 (1977)
5.16 O. Ushino, M. Maeda, J. Kohno, T. Shibata, C. Nagosawa, M. Hirono: Appl. Phys. Lett. *33*, 807-809 (1978)
5.17 Yu.S. Balin, I.V. Samokhvalov: "Lidar Sounding of the Atmosphere", Conf. Abstracts 4th All-Union Symp. (Institute Atmospheric Optics SB USSR Academy of Sciences, Tomsk, 1976) pp.59-62 (in Russian)
5.18 J.A. Cooney: J. Appl. Meteorol. *11*, 108-112 (1972)
5.19 Yu.F. Arshinov, S.A. Danichkin: "Pure Rotational Raman Spectra of Nitrogen and Oxygen and Air Temperature Measurements", in *Propagation of Optical Waves Through the Atmosphere,* ed. by V.E. Zuev (Nauka, Novosibirsk 1975) pp.169-173 (in Russian)
5.20 A. Weber (ed.): *Raman Spectroscopy of Gases and Liquids,* Topics in Current Physics, Vol.11 (Springer, Berlin, Heidelberg, New York 1979)
5.21 Yu.F. Arshinov, S.M. Bobrovnikov, S.V. Sapozhnikov: Zh. Prikl. Spektrosk. *32*, 725-731 (1980) [English transl.: J. Appl. Spectrosc.]
5.22 J.A. Cooney: J. Appl. Meteorol. *11*, 108-112 (1972)
5.23 Yu.F. Arshinov, S.M. Bobrovnikov: "Lidar and Acoustic Soundinq of the Atmosphere", Conf. Abstracts 6th All-Union Symp.(Institute of Atmospheric Optics SB USSR Academy of Sciences, Tomsk 1980) pp.242-245 (in Russian)
5.24 P.Sh. Loran: *Approximation and Optimization* (Mir, Moscow 1975) p.496 (in Russian)
5.25 P. Bradshaw (ed.): *Turbulence*, Topics in Applied Physics, Vol.12, 2nd ed. (Springer, Berlin, Heidelberg, New York 1978)
5.26 V.E. Zuev: *Laser Beams in the Atmosphere* (Plenum, New York 1981)
5.27 B. Saleh: *Photoelectron Statistics,* Springer Series in Optical Sciences, Vol.6 (Springer, Berlin, Heidelberg, New York 1978)
5.28 B.N. Briggs, G.J. Phillips, B.H. Shinn: Proc. Phys. Soc. London B *63*, 106-121 (1973)
5.29 V.E. Zuev, G.G. Matvienko, I.V. Samokhvalov: Izv. Akad. Nauk SSSR, Fiz. Atmos. Okeana *12*, 1243-1250 (1976) [English transl.: Izv. Akad. Sci. USSR Atmos. Oceanic Phys.]
5.30 G.G. Matvienko, I.V. Samokhvalov: "Lidar Measurements of Wind Velocity Using the Correlation Technique", in *Remote Sensing of the Atmosphere,* ed. by V.E. Zuev (Nauka, Novosibirsk 1980) pp.113-124 (in Russian)
5.31 M.S. Belenky, V.L. Mironov: "Lidar and Acoustic Sounding of the Atmosphere", Conf. Abstracts 5th All-Union Symp., Vol.3 (Institute of Atmospheric Optics SB USSR Academy of Sciences, Tomsk 1978) pp.3-6 (in Russian)
5.32 M.S. Belenky, A.A. Makarov, V.L. Mironov: "Lidar and Acoustic Sounding of the Atmosphere", Conf. Abstracts 6th All-Union Symp., Vol.1 (Institute of Atmospheric Optics SB USSR Academy of Sciences, Tomsk 1980) pp.164-167 (in Russian)
5.33 I.E. Naats: *The Theory of Multifrequency Lidar Sounding of the Atmosphere* (Nauka, Novosibirsk 1980) (in Russian)

Subject Index

Approximation
of kernels of integral equations 45,46,192
of polydisperse scattering operators 22,25,46
piecewise-square 59,131
Coefficient of
absorption 215
backscattering 82,132
directed scattering 91,166
polydisperse extinction 82,132
scattering 82
transmission 94
Conditionality of functional equation 67
Density of distribution
bimodal 97,118,138
model 71,107
normalized 107,200
piecewise-continuous 59
uniform 32,107
Differentiation formula of polydisperse integrals 102,213
Diffraction parameter 35
Discrepancy of integral equation 53
generalized 126,136
Distribution component of optical characteristics
irregular 40
nondifferentiable 220
oscillating 100
Effective boundaries of size spectrum 106,136,144,213
Euler lidar equation 56,62,237
see also Lidar equation
Functional
smoothing 53
stabilizing 53
Incorrectness of inverse problem 42
Integral equation *see also* Volterra integral equation
Fredholm of the first kind 10
in the form of Riemann-Stieltjes integral 10,14
Interval of local smoothing 45,100
Lidar equation 50,160,196
see also Euler lidar equation
Lidar ratio 87,172
Matrix for polydispersed ensemble 165
Measure of particle symmetries of polydisperse system 15
Method of
differential absorption 215
generalized discrepancy 126
histogram 58
integral equation 10,22,217
inverse operator 52
iterational 232
logarithmic derivative 121,208, 229
model estimates 71

Method of (cont.)
  multifrequency sensing 81
  optical models 25
  qualitative interpretation 84, 85,92,96,206
  smoothing functional 53
  spectral transmission 96
Mie theory of scattering 50,165
Operator
  differential 62
  integral 21
  inverse 51
  of local smoothing 45
  optical 51,115,176
  regularizing 56
Optical depth 47
Optical equivalence of scattering particles 23
Optically active sensing interval 96
Polarization ratio 199
Polydisperse scattering phase function 167
Quasi-solution of inverse problem 135
Rayleigh theory of scattering 50
Regularizing parameter 55
  optimal 55,138
  quasi-optimal 171,220
Scattering efficiency factor
  monodisperse 8
  polydisperse 107,205
  smoothed 44,100
Set of distributions 21
  compact 53
  reference data 21
Smoothness of distributions 40,103
Variational principle 54,71,115,176
Volterra integral equation 28,31, 245 *see also* Integral equation

Laser Monitoring of the Atmosphere

Editor: **E.D.Hinkley**
With contributions by numerous experts
1976. 84 figures. XV, 380 pages. (Topics in Applied Physics, Volume 14). ISBN 3-540-07743-X

Contents: Introduction. - Remote Sensing for Air Quality Management. - Laser-Light Transmission Through the Atmosphere. - Lidar Measurement of Particles and Gases by Elastic Backscattering and Differential Absorption. - Detection of Atoms and Molecules by Raman Scattering and Resonance Fluorescence. - Techniques for Detection of Molecular Pollutants by Absorption of Laser Radiation. - Laser Heterodyne Detection Techniques.

Laser Speckle and Related Phenomena

Editor: **J.C.Dainty**
With contributions by numerous experts
1975. 133 figures. XIII, 286 pages. (Topics in Applied Physics, Volume 9). ISBN 3-540-07498-8

Contents: Introduction. - Statistical Properties of Laser Speckle Patterns. - Speckle Patterns in Partially Coherent Light. - Speckle Reduction. - Information Processing Using Speckle Patterns. - Speckle Interferometry. - Stellar Speckle Interferometry. - Additional References with Titles. - Subject Index.

The Monte Carlo Methods in Atmospheric Optics

By **G.I.Marchuk, G.A.Mikhailov, M.A.Nazaraliev, R.A.Darbinjan, B.A.Kargin, B.S.Elepov**

1980. 44 figures, 40 tables. VIII, 208 pages
(Springer Series in Optical Sciences, Volume 12)
ISBN 3-540-09402-4

Contents: Introduction. - Elements of Radiative-Transfer Theory Used in Monte Carlo Methods. - General Questions About the Monte Carlo Technique for Solving Integral Equations of Transfer. - Monte Carlo Methods for Solving Direct and Inverse Problems of the Theory of Radiative Transfer in a Spherical Atmosphere. - Monte Carlo Algorithms for Solving Nonstationary Problems of the Theory of Narrow-Beam Propagation in the Atmosphere and Ocean. - Monte Carlo Algorithms for Estimating the Correlation Function of Strong Light Fluctuations in a Turbulent Medium. - References. - Subject Index.

Ocean Acoustics

Editor: **J.A.DeSanto**

1979. 109 figures, 5 tables. XI, 285 pages
(Topics in Current Physics, Volume 8)
ISBN 3-540-09148-3

Contents: *J.A.DeSanto:* Introduction. - *J.A.DeSanto:* Theoretical Methods in Ocean Acoustics. - *F.R.DiNapoli, R.L.Davenport:* Numerical Models of Underwater Acoustic Propagation. - *J.G.Zornig:* Physical Modeling of Underwater Acoustics. - *J.P.Dugan:* Oceanography in Underwater Acoustics. - *N.Bleistein, J.K.Cohen:* Inverse Methods for Reflector Mapping and Sound Speed Profiling. - *R.P.Porter:* Acoustic Probing of Space-Time Scales in the Ocean. - Subject Index.

The Stratospheric Aerosol Layer

Editor: **R.C.Whitten**

1982. 62 figures. XI, 152 pages
(Topics in Current Physics, Volume 28)
ISBN 3-540-11229-4

Contents: *R.C.Whitten, P.Hamill:* Introduction. - *E.C.Y.Inn, M.H.Farlow, P.B.Russell, M.P.McCormick, W.P.Chu:* Observations. - *R.G.Keesee, A.W.Castleman, Jr.:* The Chemical Kinetics of Aerosol Formation. - *R.P.Turco:* Models of Stratospheric Aerosols and Dust. - *O.B.Toon, J.B.Pollack:* Stratospheric Aerosols and Climate.